HOUSE DAVION

The Federated Suns

HOUSE DAVION

Writing
Boy F. Petersen, Jr.
C. R. Green
J. Andrew Keith

Editorial Staff
Editor-In-Chief
L. Ross Babcock III
Senior Editor
Donna Ippolito
Editorial Assistant
C.R. Green

Production Staff
Production Manager
Jordan Weisman
Art Director
Dana Knutson
Production Coordinator
Tara Gallagher
Uniform Design and Illustration
David R. Dietrick
Color Illustration
Jeff Laubenstein
Todd F. Marsh
Jim Nelson
Cover Art
Dana Knutson
Illustration
Todd F. Marsh
Jim Nelson
Jeff Laubenstein
Dana Knutson
Tim Bradstreet
Rick Harris
Chris Palm
Bill Perry
David Zenz
Typesetting
Patrice A. Jones
Tara Gallagher
Layout
Tara Gallagher

Si Fecisti Nega

Published by
FASA Corporation
P.O. Box 6930
Chicago, IL 60680

INTRODUCTION

In every society, you will find the dualism of Man and his works. There is no good without evil, no civilization without barbarism, no life without death. The inevitable progression of Humanity from primitive culture to advanced techno-society is in fact but half the truth, for the forces that would tear down these bastions of progress are no less inevitable. In the end, History is not so much a chronicle of years as it is a story of cycles. Perhaps the ancients recognized this best with their stories of Armageddon, Ragnarok, and Nuclear Holocaust...but they reckoned without the turn of the wheel that will start the whole cycle over once more.

—Dr. Ernst Meyer, *Shadows of the Future*, ComStar Publications, Terra, 3024

The Federated Suns is one of the largest and most powerful of the Successor States. Along with the Draconis Combine, it stands as a dominant influence in the politics and military strategies of the 31st century. Indeed, few objective observers dispute the fact that these two interstellar states are most likely of all the Successor powers to achieve the elusive goal of reunification of the Inner Sphere, which has been at the root of the seemingly endless wars of the modern era.

The Federated Suns is the largest alliance in the Inner Sphere, comprising over 500 inhabited star systems spread out over some 20,000 square parsecs of space, and it is this size that gives House Davion preeminence in Human-occupied space. Beginning as a loose association of 20 worlds just over 700 years ago, it has been one of the most stable of the political bodies in explored space. Although occasionally plagued by internal disruptions, and often heavily involved in wars with its two neighboring states—the Draconis Combine and the Capellan Confederation—the Federated Suns has managed to grow and prosper through most of its long and eventful history. Yet, no other Successor State has been as badly hurt by the loss of technology during the Succession Wars. In some ways, its economic condition might be summed up as a few overdeveloped worlds in an underdeveloped realm. The Davion government has, therefore, placed a high priority on scientific research, which may spell trouble for ComStar in the future.

Of all the Successor States, the Federated Suns makes the most of "keeping alive the principles of democracy" and "promoting the cause of personal freedoms," and takes a highly pragmatic view of the matter. Prince Hanse Davion, the current House leader, believes that whether a government is "free" or "totalitarian," it must spend about the same amount of energy to maintain order, either by taking pains to protect its citizens' rights or by creating agencies to suppress those rights. Prince Davion reasons that there is more to be gained in loyalty and patriotism by protecting the peoples' rights then by oppressing them. However, the Draconis Combine is no more evil because it is generally perceived as a brutal, repressive police state than the Federated Suns's image automatically makes it "good." Like all Human endeavors, the Federated Suns has its strengths and weaknesses, and owes much of what it is now to the political and social evolution that is History in action.

Though outward appearances may suggest that the Federated Suns is a champion of democratic freedoms, the realities are not so simple. Like the other interstellar states that have survived the chaos of post-League war and the continuing devolution of social and economic order, the Federated Suns has been forced to adopt expediencies rather than cling to idealism. If democracy ever was the ideal of the Federated Suns—and there are those that would dispute this—it certainly has been forced to give way in the harsh light of necessity. Led by a warrior aristocracy, locked in constant wars over territory, this realm would not appear much different than other States to an unbiased observer. The fact that idealism once counted for so much, and still forms the mainstay of the government's propaganda effort, may be more damning than any other fact. In looking back at Federated Suns history, however, starting even before the Crucis Pact that marked the political beginnings of that realm, it is hard to see how it could have developed differently, given all the factors that shaped Man's patterns in the Third Millennium.

The Federated Suns grew from the vision of a single man, Lucien Davion, whose Crucis Pact was one of the earliest and most successful interstellar unions forged in the wake of the collapse of the Terran Alliance. The Crucis Pact, though, was the development of one statesman's hopes for a better future. A scion of the unique society of New Avalon, Lucien Davion brought to his vision the traditions of his family, his planet, and his life. And the Federated Suns that grew out of the Crucis Pact reflected the essence of these traditions, a mixture of independence and authoritarianism, democracy and aristocracy, soaring aspirations and practical, sometimes even paranoid fears. The history of the state is the story of all these elements writ large against the background of Human civilization, and like the author of that first Pact, the Federated Suns has passed through hope to ambition to the edge of collapse and back again.

What the future holds for the Federated Suns depends on its recent alliance with House Steiner's Lyran Commonwealth, whose economic power is a perfect complement to Davion's military might. Though some believe that this alliance may be the first step toward reunification of the Inner Sphere, or at least the establishment of a new Star League, it is far too early to tell. The biggest obstacle is that the other three powers of the Inner Sphere banded together in an alliance of their own almost before the ink was dry on the Steiner-Davion Pact.

We must remember, too, that the five Successor States, vast as they are, are only the largest powers of the modern galaxy. Those of us from the Periphery are more than a little skeptical of the recorded history of the Great Houses of the Inner Sphere. Some of it has been lost to the destruction of the Succession Wars that have dominated our race for almost three centuries. Some of it is purely self-serving. With ComStar's access to the best and most extensive source materials available, Our Blessed Order's researchers have unearthed the truth about House Davion if anyone could. Where contradictions appeared, I have attempted to resolve them as sensibly as possible. Where that was not possible, I have decided to include equally plausible versions of the same story so that the reader can judge for himself.

No one text can hope to cover the whole intricate tangle of military, political, economic, and social factors that have shaped the Federated Suns over the course of some 700 years. Only the most significant forces, the crucial causes and effects, can be analyzed here. In the history of the House of Davion, New Avalon, and the Federated Suns, we can see vividly the influences that have been at work in all of the Successor States through the centuries, and from them recognize the turn of the wheel as it comes around once again in our times—and beyond.

On a personal note, I would like to thank ComStar Director Julian Tiepolo for entrusting me with this project. Not only was it a great personal honor to serve ComStar—and truth—in this way, but it also gave me the opportunity to research the factual history of my purported ancestor, Stefan Amaris.

—Anastasia Marcus, Historical Director, Davion Research Project, ComStar Archives, Terra, 3028

HANSE DAVION, LANCE COMMANDER (BATTLE OF BANTON HILL 2998)

HISTORY

INTRODUCTION

It took Humankind some 200 years to evolve from a relatively primitive, planetbound race to a spacefaring empire of over a thousand colony worlds. The greatest irony of the Succession Wars is that, in a mere two centuries of warfare, the Human Sphere has taken enormous strides backward instead of forward.

The roots of the five Successor States and dozens of Periphery states go back to primitive rocketry experiments during World War II in the 1940s. In the following decades, two victors of that war, the United States of America and the Union of Soviet Socialist Republics, competed to see who could explore "outer space" more quickly. Though these early space probes only carried explorers about a hundred miles into Terran orbit, these tentative efforts would one day lead to full-scale colonization.

[EDITOR'S NOTE: It has also been remarked that the end of Terra's feasibility as home for the entire Human race began when the names of nations got longer and longer. At that time, there was even a country called the People's Democratic Republic of South Yemen, but ComStar researchers have been unable to find its location on any ancient map of Terra.]

In the 21st century, political alignments on the Terran homeworld began to shift drastically. After the Second Soviet Civil War (2011–2014), the Western Alliance (America and Europe) became the preeminent world power. The Western Alliance's highest priority was peaceful scientific research and development to overcome disease, hunger, overpopulation, and environmental damage.

Meanwhile, in 2018, one of the greatest scientific breakthroughs of the ages occurred when two theoretical physicists named Takayoshi Fuchida and Thomas Kearny found a slight anomaly in the work on prototype fusion reactors. Extrapolating from this, they postulated that it was possible for mass, in short bursts, to exceed the speed of light. Fellow scientists ridiculed Kearny and Fuchida's work, and drove them from the profession. As a result, their groundbreaking discovery would remain forgotten for almost a century.

Meanwhile, other scientific breakthroughs such as asteroidal mining, recombinant DNA organisms, and free-floating space factories (which permitted industrial procedures impossible in an atmosphere or under gravity) ensured a standard of living higher than ever before. Without enough food to feed people on the ground, however, famines became increasingly widespread among the world's population.

[EDITOR'S NOTE: Indeed, the most famous artwork of the 21st century may well be Alessandro Mekam's *Threnody,* a portrait of a child dying of hunger as she watches an early holovid. It is greatly to be regretted that so much technology, even of the 21st century, has been lost in the Succession Wars.]

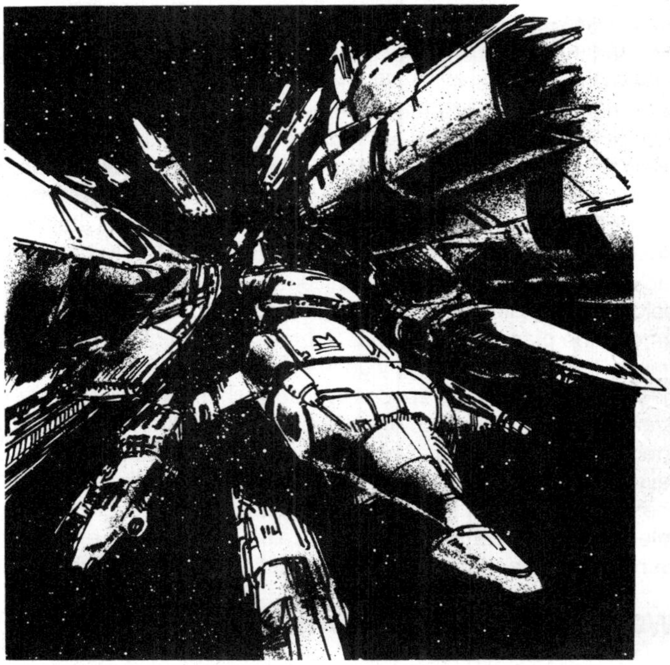

A GREAT LEAP

By 2050, Alliance scientists were using the first fusion drive, developed 24 years earlier, to spearhead the Magellan Program. Under the auspices of this project, automated probes were sent to dozens of neighboring star systems to seek out habitable worlds. The probes discovered three such planets in the Tau Ceti, Epsilon Eridani, and Epsilon Indi systems. In retrospect, this event dwarfs even the most important political event of that era, the replacement of the old Western Alliance with the Terran Alliance, in 2086.

In 2102, two separate teams of research physicists surprised the world by simultaneously announcing their findings that Kearny and Fuchida had been right. It was theoretically possible to transport mass instantaneously to a spot several light years away if a sufficient energy field could be generated to do so. The new research led to the Deimos Project, whose goal was to develop a working faster-than-light (FTL) hyperpulse drive. The Project was a success, and by 2108, the Terran Alliance had succeeded in launching the *TAS Pathfinder,* the first manned interstellar vessel using the Kearny-Fuchida FTL drive. The *Pathfinder*'s maiden voyage to Tau Ceti allowed scientists to survey that planet for future colonization.

A mere eight years later, the first Human colony in space was officially established on New Earth (Tau Ceti IV) in 2116.

SETTLING THE STARS

The great adventure had begun, as more and more FTL ships took colonists to new homes on new worlds. The Grand Survey of 2172 reported more than 100 Human colonies spread across a sphere 80 light years in diameter. By 2235, the fourth Survey showed the settlement of more than 600 worlds. In these years of expansion into the galaxy, the major political parties on Terra were known as the Expansionist and Liberal parties. Their debate centered on whether it best served humanity to tap the economic possibilities of space travel and colonization or whether the government should first solve the pressing problems of poverty and hunger on the homeworld. It was this conflict that would one day erupt into a tragic civil war and a total collapse of the Terran Alliance government.

[EDITOR'S NOTE: This intense expansion almost stopped dead after the collapse of the Terran Alliance in 2314. Perhaps the lack of a single home base nullified the need to found further colonies. Despite the changing political fortunes and loyalties among the planets of the Inner Sphere, the size of the explored galaxy has remained largely unchanged for nearly seven centuries. Perhaps there are epochs in history when humanity finds an area large enough to grow into—first a country, then a continent, then a world, then four thousand worlds—and remains at that level of expansion until it, too, has been outgrown. We may be on the verge of a new outreach within the next century, judging by the intensity of the demand by certain elements of the Federated Suns government for a deeper exploration into the Periphery.]

WORLD OF HOPE, WORLD OF SORROW

Trusting Divine Providence to be our guide, we, the Sovereign Citizens of New Avalon, do this day ordain and approve this Covenant, that all Peoples upon the face of this planet shall be forever equal under the Law, that Justice shall rule the Strong as it does the Weak, that Freedom shall be our Sword and Hope our strongest Shield.

—From *The Covenant of New Avalon,* 2239

FOUNDING OF NEW AVALON

New Avalon, a garden planet in the heart of the Crucis Reach, was surveyed in the first decade of the 23rd Century. As the world was lush and did not have to import water via Ice Ships, it did not take long for three successive waves of colonists to settle the region in the years from 2213 to 2221. Accompanying the first colonists were the usual trappings of Terran power—three regiments of Colonial Marines, a Governor-General with full administrative entourage, and a long list of Terran-formulated laws, regulations, and quotas. Colonists searching for the oft-advertised "new freedom of new worlds" quickly experienced the familiar restrictive control prevalent on Terra. Perhaps it was that the mother planet of man, with its dwindling resources, expanding population, and antiquated governments, really did need such restrictions to survive. Out in the unspoiled frontier of space, however, terrestrial ways were sadly inadequate to solve extraterrestrial problems.

The colonists of New Avalon had been selected by quota from volunteers in various Western European nations, with Scotland, Ireland, England, Germany, and Pan-Scandanavia contributing the largest percentage. In the first 15 years, eight cities were established on three continents, with hundreds of farming communities scattered throughout the vast, fertile farm lands.

By 2231, several Governor-Generals had come and gone on New Avalon. In that fateful year, Alliance Fleet Rear-Admiral Emil Varnay arrived to take charge of planetary affairs. Varnay was the alleged villain of the Smolensk Incident, in which Colonial Marines fired upon a crowd of protestors during a drought. He was not the monster that later generations would remember, however, but merely an unimaginative, inflexible military man trying to do his duty. Is he to blame for the unfeeling orders issued to him by masters too many light years away?

THE GRAIN REBELLION

New Avalon was now one of the thousand Human colonies that flourished across a sphere 500 light years in diameter. With each expansion of the Human Sphere, the time it took for messages to travel to and from Terra to its colonies increased. Indeed, it now took a full eight months for communications to reach the most distant Human outpost. It is not surprising that the resourceful, independent breed of Human colonists should eventually want independence from the distant rulers of Terra. In 2236, a coalition of colonies along the outer reaches of known space declared their independence. During the next 18 months, the Terran Alliance fought to hold onto her farflung colonies, but it was a losing battle.

VARNAY'S FOLLY

Few leaders in history have managed as well as Emil Varnay to take the wrong stand at the wrong time. Painted as the "Sadist of Smolensk" by some, Varnay will probably always be a figure of controversy. Could he really have been the beast that historians claim, or was he simply the victim of circumstance and political manipulation? Very likely, he was just a rigid disciplinarian too immured in military ways to realize that mere obedience to orders is not always enough. The Smolensk Incident was certainly a tragic waste of life, but few modern histories mention that Varnay, as Military Commander on Smolensk, was subject to the orders of Alliance Colonial Office Representative Heinicke at the time. We need only look at the verdict of the Court, however, to see that Varnay was reprimanded and put on half-pay for excessive use of force, though charges of willful murder were dropped. Heinicke committed suicide on Smolensk, and so his side of the story will never be known.

As for Varnay, his appointment as Governor-General on New Avalon must have been a last chance. Caught between an unreasonable Colonial Office, a mob of protestors, and the memory of past misdeeds, is it any wonder that he froze? Fast, conciliatory action might have headed off an ugly incident and allowed the old spacer to die among family and friends with his honor intact. But Varnay was not capable of such action, and the retreat of Terran marines from New Avalon would have happened with or without Emil Varnay. It must have been painful for him when his eldest son defected to the New Avalon rebels or when the Expansionists and Isolationists heaped scorn on him in his later years. The character that Chartiers and Garuda created for their trilogy on the end of the Colonial Era (*Three Faces of Empire,* 2658) was not really Emil Varnay. His failure was never the stuff of "might-have-been," only the basis for holoshows and vidtapes about a monster who never really lived.

—From *Tumultuous Times*, by Erid Wohlspahn, Remagen Press, 2983

In 2237, with Terran military forces already trying to put down uprisings in the colonies, the Alliance Parliament ordered its loyal worlds to increase their production quotas to cover losses due to the Colonial Rebellion. The teeming billions of Terra continued to need food and raw materials, and could only look to their colonies to supply them. The Terran leaders found it convenient to overlook the fact that most of the colonies were themselves already stretched to the limit.

On New Avalon, demonstrators protested against the new work quotas by tearing down the pronouncements from their posting places and burning them publicly. With winter coming on after a poor harvest season, the colony needed its food stores simply to survive. Though it was clearly impossible to meet the new quotas, Varnay chose to obey his orders rather than listen to his advisors. Armed marines rounded up the quota shipments and then escorted them to the spaceport, where they were to be loaded aboard Terra-bound freighters. Those freighters would never arrive home, however. The angry colonials stormed the port that night, taking heavy casualties before they seized the ships, unloaded the shipments, and then sabotaged the vessels beyond hope of repair in New Avalon's poor facilities. The Grain Rebellion is still commemorated on New Avalon each year.

Governor-General Varnay found himself in an untenable position, unable to meet his quota and unable to get assistance from the Alliance. Deserted by most of his marines (who sympathized with the colonial position) and even by members of his own family, Varnay and a handful of senior staff personnel fled New Avalon. The Colonial Vice-Governor surrendered the Residency peacefully to protestors when it became clear that the colonists were ready to starve out the staff.

Varnay reached Terra, but by the time he presented his case to the Alliance Colonial Office, political upheavals were already toppling the Expansionist Party. No relief expedition was dispatched to New Avalon. The leaders of the Isolationists chose to retire Emil Varnay from public life and to sweep the whole unfortunate incident under the rug of "colonial readjustments."

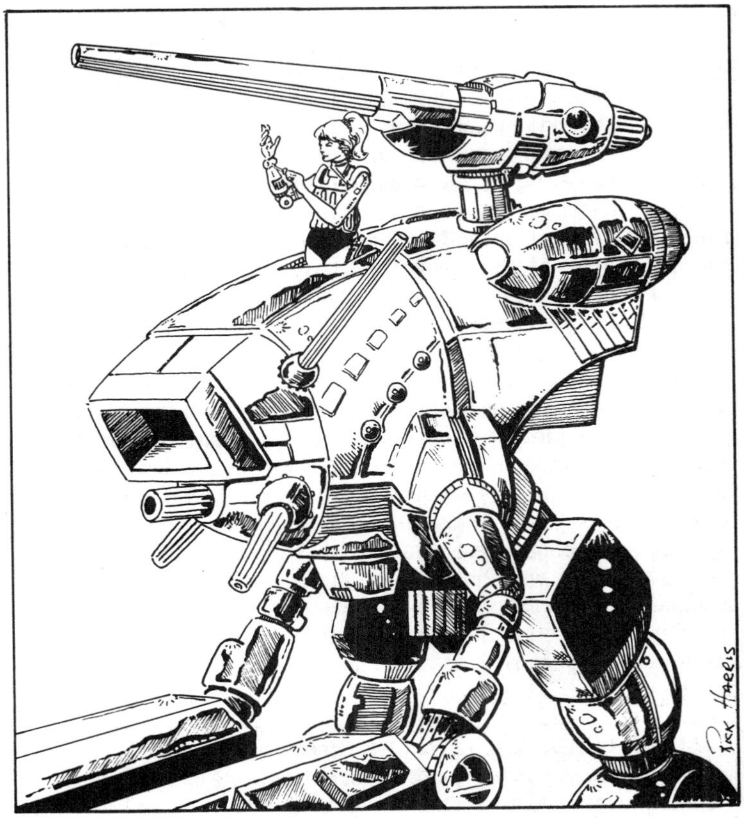

GRAIN REBELLION DAY

In 2740, I journeyed to New Avalon to observe the natives on their Grain Rebellion Day. Avalon City was aglow all day and all night, partly from the beautiful weather and roseate street lighting and partly from the spirit in the air. Beginning late in the morning, there is a four-hour reenactment of the Grain Rebellion itself. People don colorful period costumes and play the roles of their ancestors storming the spaceport, which has no scheduled traffic for twelve hours in deference to the ritual. Many of the adults taking part will fall to the ground as if they have been shot, only to jump up and continue the charge seconds later, like children playing Rangers 'n Bandits. In early afternoon, the spaceport is captured, grain is distributed from a dummy ship, and the second phase of the celebration begins.

The solemn portion of the holiday begins when the politicians and priests take over. There are speeches, followed by an ecumenical religious observance in honor of the souls of those who died in the Rebellion.

As night is coming on, the festivities begin again. People return to their homes, banquet halls, or churches to hold the biggest feast of the year in memory of how the Rebellion kept the people of New Avalon from starving. After that, there is a wild, planet-wide party reminiscent of the ancient Terran tradition of Mardi Gras. Dressed in bizarre costumes that range from the sinister to the sexy, the people dance in the streets all night.

All this reflects the Davion people's devotion to their history, but it also reveals how their decadent foolishness defiles the memory of their ancestors' profound sacrifices. The ritual does not celebrate the primeval mystery of the renewal of the seasons or the triumph over atavistic or occult dangers. The poor fools think that all they are doing is having a good time.

—From *Celebration of the Slave Goddess: An Anthropological Approach to Holidays,* by Whitecomb Sando, Capella University Press, 2759

BRAVE NEW GOVERNMENT

Is this why we fought the Governor—for the right to destroy ourselves?
—Colonel Jason Hasek, 2249

As it became obvious that the leaders of the Terran Alliance lacked the military resources and the popular support to crush the Colonial Rebellion, a new political crisis erupted. When the Expansionist Party toppled from power in favor of the more isolationist Liberal Party, the new leaders ordered that both troops and assistance be withdrawn from their former colony worlds. Whether they liked it or not, the colonists now had their precious freedom. Meanwhile, the Terran Alliance gradually withdrew its boundaries to a single 30-year FTL jump from home.

Like many other colony worlds, the inhabitants of New Avalon had been expecting a full-fledged War of Independence that never came. Instead, they found themselves cut off from both the power and the aid of the mother planet. Seeing that they must now govern themselves, the Avalonians established a Provincial Government under the leadership of Colonel Jason Hasek, a militia leader who had distinguished himself in the Grain Rebellion. Hasek called for a Constitutional Convention, with delegates elected by the colony. This ad-hoc Congress drafted the Covenant of New Avalon, which was ratified by a general vote in 2239.

Full of high ideals and a sweeping condemnation of Terran tyranny, the Covenant also became the Preamble to a Constitution that provided a democratic form of government for New Avalon. A Prime Minister, a Judiciary Committee, and a Chamber of Deputies were established as the main arms of the new government, with the populace at large deciding all issues through computer voting. It was a government that might have suited an advanced and totally self-sufficient world. For a comparatively young colony like New Avalon, the system was doomed to fail.

In 2239, Jason Hasek was voted the colony's first Prime Minister by an overwhelming majority. The colony functioned fairly well during his one-year term, mostly because the populace was willing to sacrifice for the common good. Nine successive Prime Ministers took office after Hasek, who retired to a peaceful farmer's life while retaining the honorary post of Colonel-General of the planetary militia.

By 2249, as the tenth annual planetary elections drew near, the flaws in the administrative apparatus were beginning to show. New Avalon might boast of its independence, but the colony's agricultural base was insufficient to the needs of the people. Further, the necessity of submitting all issues to a general democratic vote was a stumbling block to progress. Poorly informed or easily misled citizens usually outnumbered those capable of making rational judgements and of accepting short-term hardship for long-term achievements. Though rapid industrialization was desperately needed, the prohibition against a Prime Minister serving more than one successive term in office kept New Avalon's leaders from being able to implement a forward-looking policy in any area.

JASON HASEK: THE NEW CINCINNATUS

In *Shadows of the Future*, Ernst Meyer wrote that "every culture must have its folk heroes, whether those heroes are real or not." The career of Jason Hasek of New Avalon is a case in point.

It is unlikely that Hasek possessed any true statesmanship. A Colonel in the ad hoc militia formed during the rebellion against the Terran Alliance, he held his rank on the basis of popularity rather than talent. What he lacked in tactical skill, however, Hasek made up for with bravery, leading his men in a daring attack on the control tower at Avalon Landing. Luckily for Hasek, the defenders were unprepared. Otherwise, New Avalon's first hero might have received his honors posthumously. As it was, Jason Hasek, like Napoleon at the Bridge of Lodi, earned a reputation for courage and fighting skill that made him the darling of the rebel cause. Thus was he propelled into the limelight and into public life.

The record of Hasek's political career reveals that he was a disastrous choice for high office. It was Hasek's failure to understand just where the new constitution was leading that sowed the seeds for later civil war. Strong leadership was called for in those early days, but Hasek appeared willing to accept the Prime Ministry as a reward for faithful service rather than as a responsibility for the future. It was only in his last days, with New Avalon nearly crumbling around him, that Jason Hasek finally seemed to understand. By then, however, it was too late.

—From *Political Frauds*, by Andrew Hoexter, Miflin-Gowers Publishing, 2659

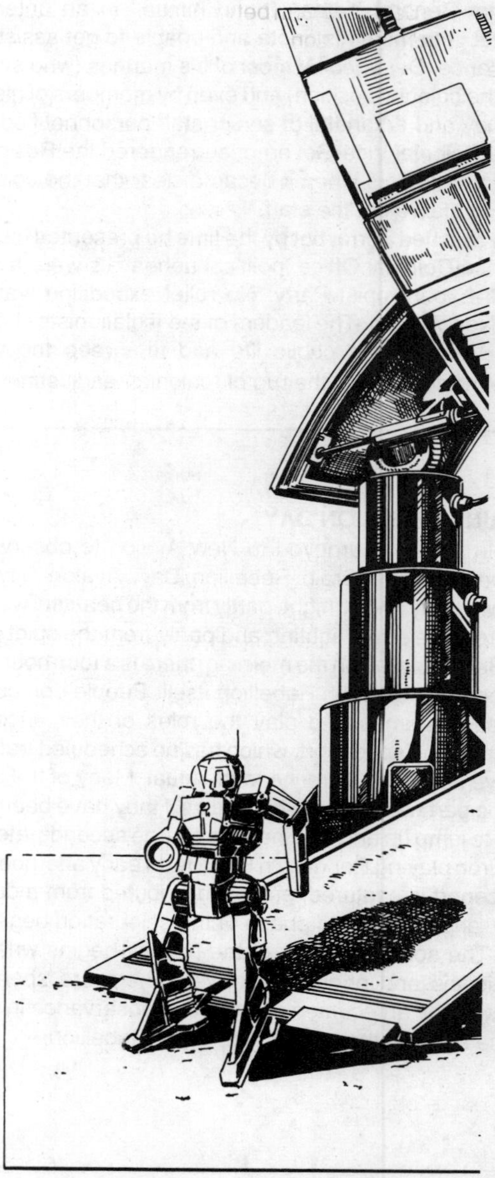

THE POWER OF ADVERTISING

In 2253, Ingram, the second-largest town on New Avalon, was holding municipal elections. The front-runners for mayor were two men named Boyce and Janovic. Also on the ballot were several other lesser-known candidates and three or four more write-in campaigns—all in a town of about five thousand people. There were no residency requirements to run for office because people tended to move back and forth between the handful of early settlements on New Avalon.

One of the newer residents of Ingram was an entrepreneur named Louis Smithson, who had just arrived to open a branch of his firm, Smithson Fertilizer. As the political campaign heated up, Smithson had a public relations brainstorm that would have planetary repercussions. Smithson hung a sign outside his office window that blended right in with the political banners in the main town square. The sign read, "For good city government, vote for whomever you like. For crop yields of 10 to 40 percent higher per acre, vote for Louis Smithson!"

Most of the people of New Avalon were second- and third-generation pioneers who had never seen a modern advertising campaign. Needless to say, Smithson won the election on write-in ballots. Janovic, Boyce, and a couple of other candidates disputed the election, petitioning that an undeclared candidate could not be allowed to take office. Smithson, who had no political ambitions, supported their suit, and he suggested that Boyce, the number two vote-getter, be declared the winner. The judge's decision was completely unexpected. He not only declared Smithson the winner of the election, but proclaimed that if he did not serve his term of office, he would be liable to prosecution for defrauding the public trust. Though the mayor's job was part-time, Smithson's fertilizer business suffered both from his neglect and the suspicion of the townspeople.

Ultimately, Smithson Fertilizer went bankrupt, Janovic retired from public life, and Boyce succeeded Smithson as mayor of Ingram. The town began to flourish under Boyce's guidance, but fell into decline once again when a competing settlement lured most of the population away to new jobs. As Avalon City continued to grow, it eventually annexed Ingram, which is now one of the oldest and most prestigious neighborhoods in the capital.

—From *Democratic Exigencies: Experiments in Government During the Exodus,* by Marty Fiore, Federated Press, 2589

RISE OF NEO-FEUDALISM

Compounding the problem was the rise of Neo-Feudalism across the planet. The few centers of industry that did exist on New Avalon generated vast personal fortunes for their owners. In one decade, a handful of industrialists had built huge power bases. Economic power translated into political power, for the common citizen tended to vote for candidates who had the power to put bread on the table and tools in their hands. By 2248, most of these "Industrial Aristocrats" had formed their own private armies to further their ambitions and as protection against rivals. As the tenth annual planetary elections drew near, sporadic violence and civil unrest broke out, but the planet's central government was too weak to put it down.

Three months prior to the scheduled Independence Day elections, Jason Hasek mustered the planetary militia to impose martial law and to bring the more aggressive power groups to heel. In a clash at the estate of the Jorgensson family, the Jorgensson "retainers" routed the militia and killed Jason Hasek. Almost on signal, there followed a complete breakdown of order, with each of the so-called "First Families" making a separate bid for power.

It was left to two surviving officers of the militia, Colonel Adam Davion and Colonel Nathan DuVall, to save New Avalon from chaos and seemingly inevitable collapse. Both of these officers were themselves from powerful First Families, but each knew that if the feudalistic rivalry continued, it would spell doom for their world. Backed by their own militias, Davion and DuVall set out to unite supporters from their families so that they could negotiate from a position of strength.

SCHEDRIN'S NEO-FEUDALISM

The idea behind Schedrin's theory is that previous economic systems have failed for one of two reasons, always related to population movement. A system in which vast numbers of the population were mobile could never become stable enough to develop its economic potential. With large population shifts, demographics become difficult, if not impossible, to determine. Jobs in one part of an economic area go begging while people in another area do the same, and neither wages nor taxes can be fairly apportioned.

At the opposite extreme are governments that require people to remain in the same place indefinitely. In ancient times, this principle was known as having "pass laws" (the practice still exists throughout the Draconis Combine and the Capellan Confederation), and it was an economic disaster. Though an overly mobile society decreased stability, no mobility at all imposed such rigidity that the economy ground to a halt.

Schedrin's solution was both simple and elegant. Through Neo-Feudalism, a government could assure its citizens' right to move around at will, while ensuring social stability though enormous tax incentives for families remaining in a particular locality. The system, in other words, combined the best aspects of capitalism and socialism. That is why Schedrin's theories will never die, and why they continue to offer us so much in the present age.

—From *Neo-Feudalism: Ripple from the Past, Wave of the Future,* by Lawrence Robert Head, Perspectives Press, 2832.

REFORMATION

No government in history has ever balanced freedom with stability. You may create order or you may encourage freedom, but you shall never do both.

—Prime Minister Nathan DuVall, in a letter written to his son, 2281

After five years of civil war and two years of uneasy negotiations, the crisis on New Avalon finally ended with the drafting of the Second Covenant of New Avalon. Though the document was an attempt to harken back to the spirit of the Rebellion, it was a hollow effort. The Second Covenant and the resulting new constitution poorly disguised the fact that New Avalon's democratic experiment had failed. In its place was an oligarchy, with control vested in the hands of the First Families.

The new constitution strengthened New Avalon's central government. Where a Prime Minister had previously been elected for a one-year term, he would now inhabit the office for life. Representatives of the First Families, who sat on the Chamber of Deputies on behalf of the people who owed them fealty, selected the Prime Minister from among their number. The Prime Minister was given extensive powers, but the Chamber of Deputies could overrule him. Popular voting was a thing of the past. The new governmental apparatus was intended to provide greater stability to the administrative machinery of the government while keeping the First Families happy. Life-terms of office for Prime Ministers suppressed active political rivalries among the Families, while permitting the Chief Executive to carry out long-term policies. It was not, perhaps, the ideal arrangement, but compared to the preceding seven years of fear and strife, the new constitution might have been made in heaven.

The First Families extended services and economic support in return for the fealty of the masses. Though "representatives" to the Chamber of Deputies were not elected and could rarely be eclipsed, the arrangement did leave the oligarchy open-ended. As economic conditions on New Avalon stabilized and the First Families lost their monopoly on essential functions, there would be room for fresh blood and liberal ideas. This aspect was an essential part of the plan formulated by Davion and DuVall, but it was kept under wraps to avoid setting off reactionary elements jealous of personal power or influence.

FIRST PRIME MINISTERS

The first Prime Minister elected under the new laws was Sandra Lockhart (2255–2261), an elderly compromise candidate selected more because she was not expected to live long than because of any particular qualifications. During her term of office, the government was concerned mostly with establishing itself as a viable, practical organ of society. Following her was Joseph Cartwright (2261–2280), who appointed Colonel DuVall as Deputy Prime Minister, and then retired for most of his term to the secluded island of Yves. The Cartwright-DuVall-Davion alliance was cemented during this era and quickly rose to dominate the Chamber of Deputies. They obtained their power by clever manipulation of new legislation aimed at sweeping improvements of the New Avalon economy. DuVall, a political genius unsurpassed in the use of patronage and legal incentive, also worked behind the scenes to stimulate industry and offworld contacts.

When Cartwright died in 2280, the 70-year-old DuVall was himself elected to the Prime Ministry, despite his failing health. He was in office less than two years, but those years would have resounding repercussions on the government of New Avalon. This was mainly due to DuVall's appointment of his son Martin to the same Deputy Prime Minister post that Colonel DuVall had held under Cartwright. Martin DuVall used this office to build up his own power base, which allowed him to step into his father's position when the elder DuVall died in 2282. This marked the first time that the Prime Ministry had passed from one member of a First Family to another of the same clan, and it set a precedent that would continue for over a century.

Martin DuVall's Prime Ministry was a period of rapid expansion, not just for New Avalon's burgeoning economy but also for the planet's space industry. Though New Avalon had no shipbuilding industry and few repair facilities at the time of the Grain Rebellion, the colony had never totally lost contact with the rest of the Human Sphere. The handful of courier vessels and small transports left behind by Varnay had managed to keep interstellar communications channels open, at least intermittently.

By the time of the Second Covenant, age had taken its toll on the tiny Avalonian fleet, but shipbuilding was expanded under the Cartwright/DuVall

AN OPPOSING VIEW

By all accounts, Edward Allis Schedrin was a good, modest man, who was kind to his family and never threw litter in the street. Unfortunately, he was a 19th-century man in a 24th-century galaxy. He never seems to have realized that his social ideas were out of date centuries before they were formulated.

Schedrin's simple, elegant bunkum had to do with tying down workers for large portions of their working lives in order to increase social stability. Even though this helped maximize profits for management and profit-sharing for labor, there are several factors that Schedrin did not take into consideration, and these make his beautiful theory collapse like a house of cards in a Chakachmna windstorm.

One is the problem of incentive. The workers in a Neo-Feudal situation have no more potential for lateral mobility than they have for geographic mobility. They cannot change jobs, their promotions are scheduled years in advance, and they know what they will be doing and where they will be doing it so far ahead of time that they are robbed of all adventure and all sense of economic possibility. They become not only lackluster producers, but also lackluster consumers.

Another problem Schedrin does not address is that of new products and services. With the stable populations he envisioned, the creation of new business ventures, particularly new fields or new products, was risky. Under his plan, entire populations of consumers get so set in their ways that the laws of supply and demand apply only to the demand for what is familiar. Thus, Neo-Feudalism prevents any real economic growth.

Schedrin's theories combine the worst of both Mercantilism and Communism. A facile approach to complex problems is very tempting, which explains why Neo-Feudalism keeps coming back every couple of generations or so. On the surface, the theory appears to solve the most basic problems of economic theory, but a deeper analysis shows why it must always fail.

—From *Fallacies and Foofooraw,* by Hyman Rockwell, Perspectives Press, 2832

DEBATE ON THE SECOND COVENANT

I tell you, gentlemen, I fear for our colony. We are told that there is no emotional bond so great as having labored and fought for one's native soil, but that is just Terran nonsense. There is no bond like having labored and fought for a land you have traveled dozens of light years just to approach. Those who, like myself, are immigrants to this world, know that I speak the truth. The provisions of the Covenant will bankrupt those of us with anything to lose, they will lay our most basic liberties before a newly created class of tyrants, and they will plunge us into a civil war that will make the recent Terran conflict look tame. We must not let this new class of tyrants hold sway on our world, for if they do, only bloodshed can come of it.

—Brendan Nagy, Alderman for the Fifth Municipal District of Avalon City, in a speech to his constituents, 2282

Critics of the proposed Covenant have warned that it will lead to loss of liberty or bloodshed, or both. With all due respect to well-meaning individuals, we think they haven't got a clue.

The hard fact is that we have already had bloodshed, and we have also had to deal with tyranny. The Covenant, whatever its flaws, is designed to stop a civil war while leaving our liberties intact. And, let's face it, we have had no liberties under the martial law imposed on New Avalon in recent years. The exigencies of military rule leave little room for human values. If we are to remain truly human, we must do away with martial law, and make sure that it never rears its ugly head again. The Second Covenant is not only our best chance to do so, it may be our last.

—From an editorial by Grant Sumner, *Avalon City Daily Dispatch,* 2282

administrations. In 2293, Martin DuVall dedicated the first home-built freighter, the *NAS Hopewell.* Within the next ten years, New Avalon's trading fleet grew quickly, reaching out to other worlds of the Crucis Reach. Some of these, being more self-sufficient than New Avalon, could supply superior goods and higher technologies, while the less fortunate worlds were ideal markets for export goods.

The rebirth of widespread trade in the Crucis region was the salvation of New Avalon. Never again would the planet be cut off, and never again would lack of resources lead to political crises at home. Few people on New Avalon understood how important was this development, or that it would forever change their society. Within two generations of the Second Covenant, it had become inevitable that the First Families would one day have to surrender their claim to power.

When Martin DuVall died in an untimely hunting accident in 2309, Lucien Davion, DuVall's nephew and the grandson of Colonel Adam Davion, was Deputy Prime Minister. Young Davion's accession to the post of Prime Minister might have been considered dynastic succession, except that the Deputies regarded him as the most logical choice at the time. Indeed, the Davion family was heir to the DuVall political legacy as well as quite powerful in their own right. If anything, Colonel Adam Davion was even more popular with the Militia than was Nathan DuVall.

It was in this same era of New Avalon's earliest Prime Ministers that a whole new wave of colonists had begun to flee Terra once more, seeking to escape the political chaos created by the ever-deepening rift between the Isolationists and the Expansionists. These newest voyagers to the stars were among Terra's best and brightest, and their departure ushered in the era now known as the Exodus.

A WORD TO THE WISE

There are those who call it going from the frying pan to the fire, and those who call it blowing hot and cold. Whether you call it being between a rock and a hard place or Scylla and Charybdis, Rome was not built in a day. Having to choose between our Prime Ministers serving a pathetic one-year term or having them lord it over us for a lifetime is not exactly cricket, even if the game is not over till the last man is out. Somewhere along the line, we have to look for the Golden Mean. To paraphrase an old Terran politician, extremism in defense of liberty is no vice, but extremism in setting terms of office is no virtue—especially when the politicians themselves talk in such insipid cliches.

It might have been better to elect the Prime Minister for a four-, five-, or six-year term. Indeed, whichever of these choices might have been best, there can be little question that any of them would have been better than what we got. Going from one extreme to the other is never desirable, especially when the common good is at stake.

—Quentin Dexter Rambuzie, editorial in the *New Avalon Daily Herald,* 2255

RISE OF HOUSE DAVION

How do you make sweeping statements about about a family with a thousand-year history? No set of catch-phrases can describe every individual in the Davion line accurately. Yet the Davions themselves have an old saying about their clan, one that echoes what was said of the Claudian emperors of ancient Rome: "The Davions breed two kinds of children, the fair and the foul, but it's not always clear which is which!"

—From *History of the Davions*, by Enrico Rosario, Nova Press Interstellar, 3015

ORIGINS

Only a family born to war could produce the soldier who has given us peace.

—From a speech given by King Charles III, 2014

The background of the Davion family before the settlement of New Avalon is obscure, but certain family traditions and fragmentary histories do suggest some of their origins. A 16th-century Frenchman, Jean-Baptiste d'Avion, was the first of the line to use a recognizable form of the family name. Family records claim that d'Avion was born in the village of Avion in Gascony, the bastard son of the Count de Foix. With a romanticism typical of the Davion family's long history, d'Avion supposedly turned his back on his father's noble birth, enlisted in the French army as a common soldier, and was selected to be among the bodyguard that accompanied Mary Stuart back to her homeland when she became Queen of Scots. Service in her defense was usually rewarded with lands in the border country of Scotland. Later Davions would claim to be of Scots rather than French blood.

Other family traditions are equally romantic. The Davion family on Terra must have had a flair for embracing lost causes and other idealistic movements. As loyal subjects of the Stuart kings, the family was granted a patent of nobility by Charles I. One "Baron Davion" is recorded as falling in battle at the head of a volunteer regiment led by Montrose in the battle of Preston. His grandson fought under Dundee at Killicrankie, and another Davion was at Sherrifmuir. The family titles passed to a cadet branch for their faithful allegiance to the Stuarts, and after the rising in 1719, the titular head of the line, Charles Francis Davion, was exiled to Europe. A Davion also turns up in the young Pretender's army at Culloden. Pardons issued in the latter part of the 18th century permitted the Davions to return home, but they did not regain their titles or lands.

Most family stories pass over the next several generations except to assert that Davions did their share of military service in the three World Wars. A strong tendency toward militarism seems to have been a part of Davion psychology. One family saying dating back to this period claims that, like Achilles, the Davions were ever fated to short but glorious lives.

The name of Brigadier Arthur Davion, hero of the Battle of Moscow in 2013, became especially important in the family annals. Created a Baron in that same year, the Brigadier was one of the first five officers selected to form the Alliance peace-keeping forces that were the direct ancestors of Terra's Colonial Marines. Lord Davion revived his old family coat of arms, that of a fox crouched above a blood-red shield bearing the motto "Audacity, Bravery, Destiny." House Davion retains this personal crest (as opposed to the official star-and-sunburst of the Federated Suns) to this day. It is as much for this as for his cunning that Hanse Davion has been nicknamed "The Fox" by friend and enemy alike.

DESTINY'S CHILD

We owe our freedom to the dedication of our patriots, to the determination of our citizens, and to the valor of our soldiers. But for the support and the bravery of all who have fought for our cause, none would have lived to see this day.

—From the inaugural speech of Prime Minister Jason Hasek, Provisional Government, New Avalon, 2237

The Davion family served the Alliance well for two centuries. Its members fought in almost all the brushfire wars and peace-keeping actions on Terra during this era, and there were Davions in several of the most important expeditions to new planets. Thomas, the fifth Lord Davion and second Fleet General of the Alliance Colonial Marines, was responsible for important reforms in the Alliance military structure during the 22nd century. Two members of the family were Admirals, and there were a dozen more who attained the ranks of regimental commander or ship-captain.

Captain Lord Robert Davion, of the Sixth Colonial Marines, was the great-grandson of Fleet General Davion. He was an angular man with a ruddy complexion, gray hair, and bushy eyebrows. As a younger son, he was not in line to inherit the privileges or the duties of the Barony, but his name helped advance his military career. He served as a staff officer on Terra until the death of his wife when he was 32.

With a seven-year-old son in tow, Robert volunteered for duty on the frontier. His motivations will probably never be known, but such impulsiveness is typical of many Davions. Having used all his influence to obtain a post in a colony world, Lord Robert gathered up his son, his sister, and a young cousin just out of the Alliance Military Academy and traveled to New Avalon in 2232. There, he and his family would one day play a role that could only have been forged by destiny.

A CHANGE OF HEART

Robert Davion quickly became attached to the beautiful world of New Avalon and to the bold, adventurous people who were settling her. Within a year of his arrival, he had met and fallen in love with Suzanne Marsin, heiress to the wealthy Marsins who controlled much of the iron and steel industry on the planet's north continent. As matters began to come to a head on New Avalon in 2237, the Captain found himself in sympathy with the colonists rather than with the administrators.

There are many tales of how Robert Davion came to side with the colonists, the most popular being the one about how he prevented some marines from abusing a farmer and his family while trying to confiscate their produce. The plain truth, however, is that Davion's decision was cold and calculating. Shortly before the outbreak of the Grain Rebellion in 2237, Davion became acquainted with the Sons of New Avalon, the militant underground group responsible for most of the anti-Terran agitation on the planet. As a member of the Colonial Liaison staff, he staffed the First New Avalon Colonial Militia with SNA malcontents to give them access to marine weapons and equipment on the night before the raid on Avalon Landing. He also supplied information that the rebels needed to attack Avalon Landing and to seize the transports berthed there. Robert Davion deserted his post at the Residency six hours before the outbreak of the Rebellion, and is believed to have acted as a military advisor to the rebels when they began their siege of the Residency.

DAVION TO THE RESCUE

Robert Davion's part in the Grain Rebellion is probably best seen in the incident that first brought him fame on his adopted planet. It was a deed that might never have been noticed were this not a time when our patriotic Sons of New Avalon felt little trust for offworlders.

Riding native Terran horses like the Hussars of old, Davion and a brace of other young officers on patrol came upon a farm where the Colonial tax collectors were harassing a poor farmer named Koliak. They had already confiscated Koliak's crops for export to Terra and were beating up the protesting Koliak when Davion and his men arrived.

The chief tax collector called out to the three officers to join the fun. "I see you have a whip there, Captain," the villain called out, pointing to Davion's riding crop. "In the name of the rightful Terran government, I command you to thrash this fool and teach him a lesson about resisting our power!"

Davion said not a word, but sat nearly motionless in the saddle. The only sound was the restless stamping of the horse's feet. The tax collector repeated his order.

Slowly, Davion dismounted, and the other two officers did the same. None of them bothered to tie their horses, which remained standing in perfect formation. Although Davion stood at least six inches shorter than the tax collector, he raised his crop and lashed him three times, once quickly across each cheek and a third time across the back of the head as the man ducked.

No one else moved for a moment, because Davion's action was as surprising as it was dramatic. Then the tax collector's minions took to their heels, running away and leaving their horses for the officers and Koliak. During the entire encounter, Robert Davion never spoke a word.

—From *Glory Days: Early Independence from Terra,* by Justin Frederick Pullea, Davion Historical Press, 2820

Much nonsense has been written about Robert Davion's "rebellion" during the real Rebellion of 2237. Though his courage during that time is undisputed, the truth about his motives is somewhat murkier.

Take, for example, his famous "rescue" of a New Avalonian grain farmer from the clutches of Terran Marine tax collectors. He was riding with two ranking members of the Sons of New Avalon, who were sure to report the incident to their superiors and thus establish his credibility as a sympathizer with the group. Moreover, the act of provoking the Colonial Government at that point was sure to spark new rebellion elsewhere, which could only benefit Robert's political ambitions.

More to the point, historians overlook the fact that Davion and his two confederates had ridden all day with inadequate provisions, trusting to the generosity of any farmers they might encounter on their patrol. Indeed, Mikhail Koliak treated them to a fine feast that day after the thrashing Robert Davion had given the tax collector.

—From *Truth or History: Lies Our Forefathers Told Us,* by Trudi Simms, NAIS Press, 3021

KATHERINE ANNE DAVION: AN APPRECIATION

Some have wondered how a person listed in the *Guinness Book of Galactic Records* as having one of the 20 highest IQs ever recorded could have been simply gathered up and transported off Terra "like a sack of Cartago potatoes," as one historian put it. Katherine Anne Davion not only allowed her brother to do this, but subsequently showed herself to be one of the most dynamic women of her generation.

There is some dispute as to whether Katherine was emotionally disturbed during her life on Terra. In her late teens, from 2226 to 2228, she was confined to a mental hospital, for two reasons. First, the continuous fighting between the Liberals and the Expansionists made family members of military men like Robert Davion frequent targets of terrorist assassination attempts. Yet Katherine also did and said things that set her apart from others, both children and adults.

When Katherine was eight, a neighbor discovered her talking to her dolls in the sort of gibberish that some children make up as languages of their own. It later came out that she had found some textbooks around the house, and was talking to the dolls in poorly pronounced, but grammatically correct, Portuguese, Pali, Polish, and Plattdeutsch. (A linguist friend of the family, fearful of unrest in his neighborhood, had begun transferring his library to the Davion home, and Katherine had happened upon a stack of books in the "P" languages.) The neighbor knew nothing of this, however, and scolded the little girl, saying that she must never make up something and then insist that it was real. Incidents like these eventually led to Katherine's stay in a mental hospital.

Moving to New Avalon was a chance to start a new life. There, no one would know of either her eccentricities or her medical record, and so Katherine jumped at the chance to leave with her brother. Once settled on the new world, Katherine found it child's play to complete her doctorate at the newly established Outworlds University (now New Avalon University), and soon thereafter married Dr. Morton Conyers. Together, they did groundbreaking work on the social dynamics of mental illness. No symptomatic behavior ever recurred in Katherine Anne Davion.

—From *Unsung Heroines: Women Who Made a Difference*, by Lesli Karen Gris, Pankhurst Press, 2871

A WOMAN OF INFLUENCE

In 2237, Katherine Anne Davion left the Outworlds University to enter politics. As an Avalon City alderwoman, she was responsible for drafting a unique city charter, which remained in effect up to the time of the Reunification War. It called for a peculiar form of profit-sharing among residents of the city ("If the citizens have to pay taxes for city services," she argued, "it is only fair that they should get something back if there is a surplus."). Avalon City grew by leaps and bounds and soon became the largest and most cosmopolitan city in the colony worlds. This is one of the reasons why New Avalon became the capital of the Federated Suns.

To exploit mineral resources in the hills outside Avalon City, Katherine Anne Davion negotiated contracts with the companies that owned the mineral rights, assuring the best possible deal for the industry workers that lived in the city. During an inspection tour of a new mine in the spring of 2242, she was killed in a suspicious cave-in. No evidence of foul play was ever found, but the owners of the mines were known to have opposed certain provisions of Katherine's contracts. She died at the age of 34. Though Katherine Davion had barely begun her work, she influenced the course of Davion history more than some other leaders who lived twice as long.

—From *Unsung Heroines: Women Who Made a Difference,* by Lesli Karen Gris, Pankhurst Press, 2871

THE DESDEMONA INCIDENT

One of the turning points of New Avalon's history occurred near the end of the civil war, when the local militia of a town called Desdemona invited a would-be dictator named Jorgensson to be military protector of their district. Flushed with his success in a few recent battles, Captain Adelbert Jorgensson was on the verge of declaring himself military dictator, which took most residents of Desdemona completely by surprise. Those who did speak out in protest were split as to the means they should take to oppose him.

The two factions were called the Gandhists and the Activists. The former practiced passive resistance, as prescribed by their hero, the Terran social theorist Mohandas Gandhi. The Activists insisted on immediately taking up arms against any potential dictator who came along.

At the time of Jorgensson's coup, Desdemona was already a city of 150,000 people, with a permanent, professional police force in addition to its volunteer militia. The Captain entered the city like a Roman emperor, trailing an entourage half a kilometer long, and the people of the city lined the sidewalks along his route. Jorgensson was standing up in a car, waving to the procession, while the police did their traditional duty of holding back the crowd and otherwise keeping order.

As Jorgensson's procession approached the town square, one of the Gandhists sneaked past the police line and moved to sit down with great dignity in the middle of the road before Jorgensson's lead car. Several vehicles back, Jorgensson nearly toppled face-first into the front seat of his car as the vehicles pulled up sharply. The young protestor announced in a loud, clear voice that he had no weapons to oppose the soldiers entering his town, but that they would only rule over his dead body.

When Jorgensson was informed of the announcement, he said, "Very well, then. Make him dead."

The soldiers opened fire and killed the young man, and the parade rolled on. Then the same thing happened again. When a young woman sat in the road and announced that the soldiers would rule over her dead body, she was killed, and the parade rolled on. After an old woman and another young man were killed, one of the Activists took more forceful action. Slipping past the police line, he drew a gun, and took a policeman hostage.

Dragging the officer to the center of the road, the Activist made the same announcement that the Gandhist martyrs had made, but he shielded his body with the policeman's. By this time, Jorgensson was exasperated and bellowed that anyone who got in his way, voluntarily or not, would be cut down.

No one expected such ruthless stupidity from the man who was supposed to put an end to the violence of the civil war. When the soldiers opened fire on the young activist and his hostage, both the police and the town militia began to return fire. The crowd on the sidewalk scattered for cover, as the local militia routed the better-armed and better-trained soldiers by the sheer fury of their response. Jorgensson was killed, along with half his force, and the soldiers retreated back into the countryside.

Jorgensson's aide-de-camp, who took over upon Jorgensson's death, was more intelligent. He sent apologies to the town of Desdemona and especially to the relatives of those who were killed. He also pledged that his soldiers would never again enter the town except as protectors of the peace. Desdemona had shown that a mere show of force was not enough to subjugate a people who love their freedom.

The Desdemona Incident, as it came to be known, was an inspiration for the planet's freedom fighters for centuries to come.

FOUNDATIONS OF A DYNASTY

Davion's role in the rebellion has been largely suppressed, partly by "patriots" who preferred not to give too much credit to a Terran malcontent, and later by the Davion family itself. Despite Davion's invaluable assistance in freeing New Avalon from Terran domination, some of his new compatriots never trusted him entirely. They believed that if he could betray one government, he might just as easily betray another. This did not worry Robert Davion, who married Suzanne Marsin in 2239. In one stroke, he became one of the richest and most powerful men on the planet. When no children were born to this union, Robert's first son Adam became the scion of a First Family, eventually to emerge as the head of the Marsins.

Few official Davion family histories deal with Robert, the man who laid the foundation for all the family's fortunes. As a Captain in the Alliance Colonial Marines, he did not easily fit the mold of a hero of the rebellion. His motivations were always too obscure, his intentions too uncertain, and his actions too blatantly self-serving to elevate him to the status of a Washington, a Lee, or a Hasek. Lord Robert proved to be a master opportunist, however. The plans he laid during New Avalon's struggle for independence would yield far more than he would ever know, for it took several generations for his seeds to reach fruition.

Davion's ambitions and goals seemed to have shifted and turned with each new development. There is no doubt that he sympathized with the colonists on New Avalon. He never behaved as anything but a friend to the locals and avoided the marine pose of aloof superiority. It was through such fraternizing that he met Suzanne Marsin, fell in love, and gained even more attachment to his new home and its people. It may be that he never considered what Suzanne's wealth and influential family connections would mean to him.

Robert Davion never attempted to enter public life, knowing that his adopted people would find it difficult to accept leadership from someone associated with their former oppressor, no matter how many services he had rendered them. Instead, he used the Marsin family's influence to groom his own son for power. By 2245, the handsome 20-year-old Adam Davion had already been elected titular Colonel of the Fourth Regiment of the New Avalon Militia. Though it was common practice for First Families to secure such appointments for their children, Robert had his own motives. Among his private papers are many expressions of contempt for the First Families and their power blocs. Their only real use to Davion were as steppingstones to advance the fortunes of his own kin.

Shortly after Adam's election to Militia Colonel, the political tensions on New Avalon exploded. Robert Davion responded to Jason Hasek's call-to-arms by becoming a military adviser at the battle of Lockhart Manor. It is said that Davion advised Hasek to attack before the opposition could bring up reinforcements, but Hasek ignored the advice and lost the battle. Captain Lord Robert Davion was killed in the fight to protect Hasek when the militia command post was overrun. That left Adam as heir to Robert's plans and aspirations, a young man of 21 with the authority, but not necessarily the experience, to become a leader of men.

FIRST FAMILIES WAR

Adam Davion was an enigma. Though nepotism shaped his early life, it seemed unlikely that he could turn his advantages into personal success. When the Civil War broke out between the First Families, young Davion had a wife and an infant son. His stepmother, the last of the blood Marsins, was still alive but could contribute little protection to the tiny family. Disaster loomed before young Adam, who responded by taking the only course to safety. He saved himself and his family, but at the price of being called a coward.

Although the Marsins were a strong force in the pre-war economic life of New Avalon, they were ill-prepared to fight. Robert Davion had not believed in private armies, and Adam had neither the time nor the talent to forge a fighting force out of the Marsin retainers when the crisis came. He and his stepmother chose instead to adopt an awkward neutrality. In exchange for steel and Marsin mills, Adam negotiated pledges of protection from the three strongest factions. Thus protected from the rivalries of the other First Families, Davion bided his time and remained aloof from the fighting that his militia commission required him to oppose. He spent his time courting a fourth faction, that of Colonel Nathan DuVall, which had considerable military strength but a poor industrial base.

DuVall, ten years senior to Davion, had been a popular militia colonel before the war. After the Battle of Lockhart Manor, he had rallied remnants of the Colonial Militia and joined them to his own family army. This gave DuVall a dubious claim to legitimacy as the senior surviving representative of the planetary government. By himself, however, he still lacked the power to take on any of the other factions.

Davion approached DuVall with an offer of support following the latter's defeat at Owen's Ford on Southcont. In a secret alliance, Davion/Marsin retainers began to swell the ranks of DuVall's "Planetary Militia." Davion also began to negotiate with other less powerful First Family factions (starting with the Bulows, his wife's relations) to forge an alliance. To the horror of his stepmother, Adam even brought in the Lockhart family, whom she blamed for Robert Davion's death. After several months of behind-the-scenes maneuvering, this alliance managed to coordinate their military activity enough to begin to turn the tide of war. DuVall maintained titular leadership of the coalition, but Adam Davion was never far from the seat of power. It was Adam's hand, not DuVall's, that signed the Truce of Loch Rivenval that ended the Civil War.

HIDDEN COUNSELOR

In the years that followed, Davion continued to take a back seat to Nathan DuVall, but his wishes played an important part in the writing of the Second Covenant, the new Constitution, and the reorganization of the New Avalon government. Davion held no post in the new government, and appeared to settle down to a quiet life on his stepmother's estate. Nonetheless, it was Adam Davion, not Nathan DuVall, who held together the coalition of First Families while DuVall was building his political power base.

DuVall never forgot Davion's importance to his success. Throughout the Lockhart and Cartwright administrations, Davion received the best contracts and the most government support for economic expansion. The alliance between the two families was completely sealed in 2268, when Adam's son Jerome married DuVall's daughter Jennifer. This placed Jerome in a position nearly equal to that of Nathan's own son Martin as heir to the DuVall political mantle.

Neither Adam nor Jerome Davion took leading roles in Avalonian politics thereafter, but each contributed in his own way. It was a Davion-owned plant that assembled the *NAS Hopewell* and three other ships in New Avalon's trading fleet. From 2297–2306, Jerome Davion served as Ambassador to El Dorado, a vital trading partner in New Avalon's growing commercial interstellar empire. He was killed when his ship, the jump-courier *Jason Hasek,* blew up during the journey home to New Avalon in 2306. Jerome left behind four sons, Lucien, Joseph, Louis, and Charles.

At the time of his father's death, Lucien Davion was already high in the government of Martin DuVall. In 2307, the next year, the childless Prime Minister appointed Lucien to the key post of Deputy Prime Minister. With this act, another major crisis nearly erupted among the First Families. As the son of Nathan DuVall's daughter Jennifer, Lucien was Martin DuVall's closest blood relative. Yet many DuVall cousins insisted that any "dynasty" of DuVall Prime Ministers should pass down through those bearing the family name. Most of the argument was secret, for no one had yet acknowledged the concept of an inherited Prime Ministry.

Lucien strengthened his power base in the Chamber of Deputies by marrying Marie Fabier, whose grandfather had been the strongest of the anti-coalition warlords in the Civil War. When his son was born, Lucien named him Reynard, after his maternal grandfather, a clear indication of where Lucien was counting on getting fresh support. In a single generation, the power of the Davions on New Avalon had eclipsed that of the DuValls.

BIRTH OF A FEDERATION

I fear the growth of a subtle dictatorship far more than the return of the Terrans. We have long watched the DuValls forging shackles for the rest of us to wear. I say that the choice of this Ambassador's son as Prime Minister is the final link in forging the chains that we shall be compelled to wear for all time to come.

—Jason Hasek III, in a letter to an unknown correspondent, 2307

Lucien Davion was elected Prime Minister of New Avalon within a week of Martin DuVall's death. It is testimony to Davion's reputation that not even his harshest critics credited rumors that he had something to do with the hunting accident that took DuVall's life. Though Lucien's talent for political maneuvering made him both hated and feared, he was also respected as a man of principle and honor who would never stoop to betrayal.

Lucien had little in common with other members of his family. Soft-spoken and intensely idealistic, he was a sincere proponent of reform and of expansion to the stars. He was also an orator of considerable skill, with a flair for statesmanship and compromise. Nevertheless, he owed much of his early political success to his younger brothers. Louis Davion, in particular, was a born schemer who had concocted the idea of marriage with the Fabier family to win Lucien the post of Deputy Prime Minister. United, the Davion brothers were a formidable political team.

The New Prime Minister devoted his first few years in office to developing the policies begun under the two DuValls. New Avalon was well on its way to extensive industrialization, and a growing trading fleet was plying the star systems within two jumps of the planet. Trade brought economic prosperity and an inflow of raw materials for the expanding industries. Even though the First Families were steadily losing their monopoly on economic power, Lucien's popularity increased among the oligarchs as their profits mounted. Fear of a stagnant, polarized feudal society taking root on New Avalon soon began to fade. In its place, however, came concern over the long-term effects of runaway inflation as interstellar wealth debased local currency and the upper classes became increasingly decadent.

MCKENNA'S HEGEMONY

In 2314, six years after Lucien became Prime Minister, a bloody civil war broke out on Terra. In the wake of an armed clash between the Expansionist and Liberal parties, the Terran Alliance collapsed. To restore order, Fleet Admiral James McKenna stepped in at the head of the Alliance military. The charismatic McKenna created a new order from the death of the old, naming it the Terran Hegemony. Dedicated to the restoration of Terra as the center of the Human Sphere, McKenna's new government stimulated interest in the former Terran colonies, which could only spell trouble for those worlds. Where the Terran Alliance had withered in influence, the Hegemony began to expand and soon had regained control of more than 100 worlds. Lucien Davion knew only too well that if a resurgent Terra decided to pick off the colonial worlds one by one, there would be little any single planet could do to stop her. Independent for less than a century, New Avalon was as vulnerable to takeover as any other world.

In a series of swift and decisive moves, he began to contact other worlds of the Crucis Reach, who had no reason to love Terra. Though none had gone through armed revolt, all had suffered when the Alliance had withdrawn from her colonies in 2237, leaving them to sink or swim. "*Homo stellaris,*" whom McKenna had extolled in speeches on Terra, had grown up without help from Terra. Now his offspring were hostile to the idea of surrendering their autonomy to the Terrans again, after making it through the hard years of what had already become known as the Exodus. It did not take much to manipulate those hostilities and fears, and Lucien Davion was the man who could do it best.

In a whirlwind interstellar tour begun in 2316, Davion visited each world of the Crucis Reach. His theme was constant: one world could not resist Terra, but many standing together could. He stressed the idea that a show of independence early in the crisis would keep the threat of interstellar conflict from materializing. The fledgling Hegemony was not yet strong enough to fight a major war at the end of a light-years-long supply line. McKenna's government would try instead to pick off weak planets through threats, economic sanctions, or small armed demonstrations. Lucien was persuasive, and it won him many supporters.

THE CRUCIS PACT

Our one world cannot stand alone any longer; we must reach out, forge new bonds, and seek friends among the stars, so that our children can stay strong and free.

—Lucien Davion in a speech in support of the Crucis Pact, 2317

It is important to understand that Davion took special pains to identify himself completely with the plan. On some worlds, he conducted his meetings with local authorities, not as Prime Minister of New Avalon, but as a private citizen concerned for the future of the Reach. By 2317, he was ready to convene a summit conference of leaders from 23 planets. The conference lasted for three long months, but when it finally broke up, all but three participating governments had agreed to sign the historic treaty that became known as the Crucis Pact. Creating a mutual defense and trade union known as the Federated Suns, the treaty ensured complete local autonomy for all member planets. It also offered the benefits of favorable trade terms, military assistance, and an assembly known as the High Council for the discussion of issues and the settlement of disputes between members. Lucien Davion was unanimously chosen as President of the new body, a tribute to his personal popularity and his crucial role in calling together the summit and drafting the terms of the Pact.

The Crucis Pact embodied a number of high ideals that Lucien Davion genuinely valued. At the same time, it was so loaded with amendments ensuring local jurisdiction that it was virtually useless. One 24th-century historian referred to it as "the only document to make the League of Nations charter look practical."

Despite its flaws, the Pact worked well, thanks to Davion's tireless efforts. He delegated most of his Presidential duties on New Avalon to his brothers, which freed him to travel from planet to planet in pursuit of a workable interstellar governmental system. His reputation for impartiality and his statesmanship helped Lucien Davion obtain the cooperation of even the most unruly member, while adding several new worlds to the federation. The Crucis Pact of 2317 sparked a number of similar mutual defense and trading pacts elsewhere in the Human Sphere, but few developed as smoothly or functioned as well in these early years as the Federated Suns.

In the 15 years between the signing of the Crucis Pact and Lucien Davion's death in 2332, the first President of the Federated Suns turned a loose association of planets into a union of closely allied worlds. In the process, the Presidency became synonymous with the Davion name. After Lucien's death, there still existed enough petty rivalries between member worlds to make it difficult for them to agree on anything without their President's mediation, but the name of Davion was magic. Lucien's youngest brother Charles, though hardly in a class with his respected elder, became a prime candidate for President of the Federated Suns. It was only *after* Charles had secured that post that he announced his intention to stand as New Avalon's Prime Minister.

Lucien had not appointed any of his family members as Deputy Prime Minister, for his brothers had acted as his personal representatives without benefit of specific office. Now the Chamber of Deputies was once more reluctant to allow a "dynasty" to control the Prime Ministry. For a time, they seemed bent on selecting one of the Hasek line to show their defiance of the Davion power base. It was the talented Louis who changed their minds in a speech full of veiled references to the growing importance of the Federated Suns. His threats were unspoken, but they came through loud and clear. No Avalonian wanted to see his world abandoned as the capital of the newly created Federated Suns, and many feared that the Davions would use their influence to stir up bad feelings between New Avalon and the other member-worlds of the Pact.

After making the last serious attempt to stand up to the Davion family, the Chamber of Deputies finally backed down. Though the First Families of New Avalon would continue to produce important leaders, their political power dwindled, leaving the Chamber of Deputies little more than a rubber stamp for Davion decisions and policies.

It is ironic that Lucien Davion, the idealist who used all his skill and charm to forge a workable union of worlds, was the one responsible for the death of individual liberty on his homeworld. Was he blind to where his actions might one day lead, or did he believe that his own family's dominion would be less damaging to New Avalon and the Federated Suns than would a Terran takeover?

In the years of the Davion family's consolidation of power on New Avalon, other influential leaders might have offered a challenge to their meteoric rise. That these others failed to block the Davion advance is testimony both to the skill of four successive generations of Davion politicians and proof of the fear of infighting in an oligarchic society. From start to finish, the Davion family's single most powerful weapon was the fact that their potential opposition was unwilling to unite against them.

Among the major First Families, several names stand out. The Bulows and the Fabiers were particularly influential, between them dominating New Avalon's communications and electronics industries. The Davions chose to make ties of marriage with each of these two families, marriages that strengthened the Davion power base and industrial empire as well. Starting with Robert and Adam, love matches took a back seat to political considerations for many Davions during the late 23rd and early 24th centuries. It was through such marriages that they defused political rivalries while strengthening their own voting bloc.

Though many important scions of First Families would crop up later in the history of the Federated Suns, including Haseks, Bulows, and even some descendants of Governor-General Varnay, none would ever have a chance of dislodging the Davions.

THE PRIME MINISTERS

Sometimes, it is hard for social historians to judge the causes and effects of change. On Arcadia, another "garden world" of the Crucis March, peace and prosperity were the watchwords even in the worst years of the Terran retreat from her former colonies. The planet's plentiful water, extensive agriculture, abundant raw materials, and booming industrial economy made Arcadia the logical heart of the new Federated Suns. But it was struggling, strife-ridden New Avalon that emerged as the lead planet of the Inner Sphere's most powerful state. Why did the poorer world succeed over the richer? Perhaps it was a matter of harsher conditions breeding tougher people, or of luxury blinding the Arcadian rulers to the need to expand aggressively. In the long run, though, New Avalon's fortunes were built solidly on the foundations laid by a single family—the Davions. If the Davions had emigrated to another world, would history have awarded New Avalon another band of visionary leaders? Or would the family have emerged on any planet they settled as the dominant influence in the politics of their region of space? That is a question historians will never be able to answer.

—From *The Dynamics of History*, by Louise Charbineau, NAIS Press, New Avalon, 3018

New Avalon's transition from single world to capital of an interstellar federation went smoothly under the leadership of the first three Davion Prime Ministers. Though differing from one another in superficial ways, each succeeding Davion contributed to the growth of New Avalon and of the Federated Suns. Between them, they also displayed the best—and worst—traits of a family whose footsteps would echo down another 700 years of peril, strife, and triumph.

LUCIEN DAVION: VISIONARY LEADER

Prime Minister of New Avalon and President of the Federated Suns, Lucien Davion stands out as the pivotal figure in his family's rise to power. Though earlier Davions had played their own important roles, Lucien was the quintessential Davion hero, whose appearance at a crucial hour of history saved New Avalon and the whole of the Crucis Reach.

The first President of the Federated Suns was quite different from the rest of his family. Indeed, there were stories—later rigidly suppressed—that Lucien was not really the son of Jerome Davion at all, but of Jason Hasek II. Though he shared the famed Hasek qualities of idealism and incorruptible honesty, these tales do not seem to contain much truth. Neither Robert nor Adam Davion was quite sterling as a character, but Lucien's father, Jerome, taught his son high standards of personal and professional conduct. Jerome Davion, the soft-spoken Ambassador, had never been well known in New Avalon society. Though he taught Lucien well, two of his other sons, Louis and Charles, were a pair of schemers who would have made their grandfather Adam proud.

For Lucien, power and prestige were a means to an end, not ends in themselves. He used his power to forge an interstellar union that he honestly believed was vital to the survival of New Avalon and the other worlds of the Crucis Reach. He seems never to have grasped that others, even members of his own family, were not equally visionary. They saw Lucien as a pawn to build a Davion dynasty rather than a folk hero come to save his world.

—From *History of the Davions*, by Enrico Rosario, Nova Press Interstellar, 3015

A DREAM FULFILLED

I see a day when the Mother of Worlds will acknowledge that her children have surpassed her, when the Children of the Stars shall claim a rightful place as equals—rather than as colonial subjects—alongside the Children of Earth. I see a time of peace and prosperity embracing not one world, but a thousand worlds. Most of all, I see a day when unity in the common cause of freedom will create an age when no world need suffer from hardships or famines or internal strife, because a thousand brother-worlds will be standing by to provide the support needed to weather the bad times or to enjoy the good.

— Lucien Davion, from a speech to the New Avalon Chamber of Deputies, 2317

Lucien was the dreamer, the statesman who forged a federation of worlds and made it work during the first critical years of growth. His diplomatic success was due as much to an admirable personality and reputation for integrity as to his natural talents as a negotiator. In his day, "the Davion wisdom" became a byword among populations who had never heard of New Avalon, but had come to love the first President of the Federated Suns for his fair and honest judgement of complicated interworld disputes.

Lucien's charm could not smooth over the difficulties of one situation, however. In 2318, the Commander of Muskegon, sovereign ruler of that planet and its colonial offshoots in the border area of the Crucis Reach, joined the Federated Suns as a full member of the Council. The colonies of Emerson and Beten Kaitos were growing restive under the impersonal government of Muskegon, however, and were refusing to obey the edicts that were part of the laws of the Federated Suns. Many of the rebellious colonial leaders believed that the two planets should fall more naturally under the protection of the Chesterton Trade Worlds, another new interstellar community forming in the shadow of Terran resurgence. The colonies declared their independence of Muskegon and signed agreements with Chesterton soon after the Commander announced his decision. The decision to turn away from the Suns in favor of another interstellar power was to cost these two worlds dearly, with far-reaching implications down to the present-day reigns of Hanse Davion and Maximilian Liao some seven centuries later.

DELEGATE FROM NEW AVALON

While we were negotiating the Crucis Pact, Lucien Davion surmised that several of the other delegates (myself included) might feel it was more democratic to hold the final meetings away from Davion's own homeworld of New Avalon. I was flattered when he suggested my homeworld, Delavan.

I had met Davion only once before and very briefly after I replaced the former delegate, who had taken ill. He had seemed pleasant enough, but I was completely unprepared for the impression he made upon arriving in the capital of Delavan. Previously, surrounded by other officials, Lucien Davion had little time to exchange even distracted courtesies. Now, he was ready to give me undivided attention, and from the moment he appeared, there was never another man, other than my husband, whose attention I wanted more.

It wasn't just that he was handsome. He had been in uniform before, looking like just another soldier, but with a little more braid on his shoulder than most. Now he was dressed like a noble, and his manner showed that this was a real individual, not just a cog in some military machine. In a room filled with dignitaries, he caught my eye just as he walked through the door. He crossed the large room, nodding to some, shaking hands with others, as he passed, but never stopping until he came up to me.

As we shook hands, he leaned over to kiss my cheek and whispered, "You're just as I remembered." What he meant by that, I had no idea, as we were both married and were there on highly official business.

As he straightened up, I noticed two things about his face. His smile was crooked—the left side of his mouth turned up farther than the right—which gave his expression a boyish quality. The other, truly charming thing about his face was that one of his eyes was green and one blue.

—From the unpublished memoirs of Duchess Ariana Fulgess, on the events of 7 November 2317

MUSKEGON

In 2163, a group of Terran colonists had set out for McHenry, a recently opened colony world. A malfunction of the colony ship's Kearny-Fuchida drive forced the colonists to desert the JumpShip in the Muskegon star system. The single marginally habitable planet of that system become the new home for nearly 100,000 people.

Conditions on Muskegon were harsh, and so only harsh measures could overcome them. Because they controlled systems vital to survival, the Captain and crew of the colony JumpShip set themselves up as leaders with absolute power over the other colonists. Within a few generations, Muskegon society had polarized into a Crew Class of aristocrats and a Colonist Class of workers. The office of Expedition Commander, now known as Commander of Muskegon, had become hereditary, and was vested with near-dictatorial powers. The people endured, however, and the colony prospered despite the severe conditions of their homeworld. Thanks to stern measures taken in the earliest days of the colony, knowledge was retained even when it was not immediately useful. The rebirth of technology waited only for industrialization.

In 2177, technicians succeeded in repairing the original colony JumpShip. A volunteer crew departed in this vessel with orders to seek out a better world on which to transplant the colony. Instead, the expedition found Emerson and Beten Kaitos, planets rich in natural resources but whose environmental conditions were not much better than Muskegon's.

A shipload of deportees and political prisoners colonized Emerson, while volunteer colonists settled Beten Kaitos. In both cases, the new colonies were ruled by the government of Muskegon, which exploited the new worlds without providing much support for the needy colonists. Raw materials from the two colonies helped build additional ships for the fleet at home, and these in turn were used to found other colonies on nearby planets. By 2190, Muskegon claimed six planets. The Commander of Muskegon was absolute ruler over these worlds through a network of governors chosen from the Crew Class.

Emerson's ties to Muskegon were weak because its first settlers had been deported criminals. When a ship from the Chesterton Trade League arrived in 2195, the Emersonians (with the tacit approval of their governor) became members of that body. For his part, the Commander of Muskegon hoped to turn Emerson into a port of entry for trade between the two interstellar groups—with high government tariffs making the venture worthwhile.

Gradually, each world in Muskegon's empire grew stronger, especially after Terra withdrew from her rebellious colonies. Trade blossomed, and Muskegon entered a period that would have been a Golden Age had it not been for the stagnation of the class system and greed of the empire's rulers.

Muskegon had been one of the worlds that Lucien Davion visited in his campaign to form the Federated Suns. Because Muskegon had become one of the strongest powers in its region of space, Commander Jehan Achmeed was undecided about whether the Crucis Pact would bolster his power or tend to dilute it. He did not sign the Pact in 2317, but had made up his mind by the following year. In 2318, the Commander of Muskegon joined the Crucis Pact.

Meanwhile, discontent with the policies of the Muskegon government on its various colony worlds had reached a fever pitch. Two of those six planets—Emerson and Beten Kaitos—went so far as to refuse to acknowledge the decision to join the Federated Suns. Declaring their independence, these two worlds invoked the support of the Chesterton Commonality States. Almost immediately, the frontier erupted in conflict.

Ironically, the decision to add Muskegon to the Federated Suns led to Achmeed losing all that he had hoped to keep intact. Even after the rebellious worlds had been conquered by the Federated Suns, they were never returned to Muskegon's rule. Two hundred years later, when a descendant of Achmeed's joined the wrong side in the Davion Civil War, the victors stripped away Muskegon's remaining possessions and reorganized its governments.

Lucien Davion may have been a shrewd diplomat, but he was a poor general. Despite the Federated Suns's superior strength, the campaign to wrest control of Emerson and Beten Kaitos from the Chesterton Commonality went poorly. The fighting lasted three years, with the two planets changing hands several times. In the end, Chesterton, supported by the Tikonov Union, retained control.

That little war ended, but other wars along the Rimward March (as the area was then known) began to escalate. Several states, including the Capellan Hegemony, the Tikonov Union, the St. Ives Mercantile Association, and the Chesterton Commonality, sought to take advantage of the Federated Suns's apparent military weakness. Disputes over claims to the important resources on the planets Redfield, Alcyone, and Daniels triggered new conflicts in which federation expeditionary troops were again defeated and driven back.

The Federated Suns's enemies soon found that it was harder to maintain their troops in forward areas than to win initial battles. Enemy garrisons along the Federated Suns border suffered from poor supply lines and growing disaffection, making them prime targets for a counterstroke. At that crucial moment, however, the Federated Suns failed to mount such an effort.

Against the background of this failure in foreign policy and the threat of further border wars, Lucien Davion decided to retire in 2332. He died a mere three months later, his health and spirit broken by what must have seemed the imminent collapse of everything he had worked to build.

Charles Davion became the new Prime Minister of New Avalon and President of the Federated Suns. Lucien was dismayed by the methods that his youngest brother was using to balance the interests of the Federated Suns against those of New Avalon. From his deathbed, Lucien wrote to his son Reynard, who was serving in an expeditionary unit on the St. Ives frontier, warning that it might be dangerous for the Davions to attempt to dominate the Federated Suns the way they controlled New Avalon. The message, if not the actual warning, proved strangely prophetic.

THE TYRANT OF MUSKEGON

When Jehan Achmeed the last Commander of Muskegon, was overthrown, it marked the end of a dynasty whose cruelty was unsurpassed anywhere outside the Bandit Kingdoms. Yet, it was not malice that led the Achmeed family to rule with an iron fist. They simply believed that it was the most pragmatic, efficient way to rule.

One of the most glaring examples of the Achmeed tendency to place efficiency above humanity occurred in 2293, during the reign of Hassan Achmeed, the 14th Commander of Muskegon. A young, unmarried mother left her baby just outside the outermost guards' station on the doorstep of the Commander's House, the seat of government. She had attached a note to the baby's bassinet, begging that the child be raised as a member of the Commander's household.

Upon hearing of this, Hassan Achmeed thundered that if he took in one baby, every unwed mother on Muskegon would appoint him her baby's godfather, and he would never have time to govern the people. He ordered the baby killed and its body placed on public display as an example to those who tried to interfere with the efficiency of his administration.

When the baby's mother learned of what Hassan Achmeed had done, she came to the Commander's House, screaming in grief and anger that she would never have left her child if she'd known that he was a man without a heart.

He made an example of her, too.

—From *Tyrannis Excelsis: Early Territorial Governments in the Inner Sphere,* by Brendan Waukonisee, Rim Worlds Press, 2913

CHARLES THE REFORMER

I agree that our new Prime Minister is more than a bit of a snake. If it takes a snake to make the federation strong, then I, for one, am ready to lie on the ground and hiss right along with him.

—From a letter by Chief Magistrate Michael Fairborn of Augusta, in a letter to his wife, 2333

Charles Davion, though handsome and charismatic, shared none of his brother's ideals. He wanted power, both for himself and for the new Davion "dynasty" that everyone denied but that all feared. Under the tutelage of Louis, the most manipulative of the Davion brothers, Charles had become a talented politician even before Lucien's retirement. After Lucien stepped down, the brothers won the Presidency for Charles by dazzling the High Council of the Federated Suns with both the Davion name and extravagant promises to restore order on the frontier. Next, they secured the Prime Ministry on New Avalon to keep their power base fully intact. Only then did Charles reveal his true nature.

Arrogant, quick-tempered, and energetic, Charles Davion transformed the loose association of Crucis Pact planets into an interstellar organization that would one day tower over the other governments in the Human Sphere. Through the force of his will, he overcame numerous obstacles to carry forward his master plan of making himself and the Davion line indispensable.

One of Charles Davion's most important actions was a complete reform of the Crucis Pact's military provisions. Until now, the Federated Suns had responded to outside aggression or internal rebellions by raising a new expeditionary force to deal with each new crisis; member worlds contributed either troops, supplies, transport, or other facilities. A vote of the Council established the force leader, and the President set strategic aims, with the advice of the Council. The outcome of the Muskegon conflict had demonstrated just how poor these arrangements were.

The memory of those defeats along the border persuaded each member world to give Charles Davion a handful of elite military units, which he used to create a Federated Suns standing army. The new Federated Peacekeeping Forces (FPF) had its own system of ranks and organization, and its own support structure to manage supplies and transport independently of the member planets.

To foster morale and identification with the new armed service, Davion ordered extensive troop training and indoctrination to focus their loyalties and pride on the Federated Suns rather than on their individual homeworlds. Many Avalonians, however, ended up in the senior ranks of the officer corps, thanks to Davion family influence. Using this kind of backroom favoritism, Charles and his brothers began to guarantee the transmission of power from Davion to Davion for several decades to come. Also during this period, a formalized bureaucracy began to supplant direct rule by the President and the Council. The Council retained control over policy matters, but routine, day-to-day operations fell to the permanent staff at the capital of the Federated Suns. Here, too, New Avalon's representation was disproportionately high.

Charles Davion did not live to see these reforms fully implemented. Like Lucien, he spent much of his term settling interworld disputes and campaigning for his ideas among member worlds. His brother Louis had been his chief deputy until 2337. When Louis died, Joseph, the fourth Davion brother, was in failing health. Thus, Reynard became Deputy Prime Minister instead. Charles had not made Lucien's mistake of failing to clearly designate an heir. Because Lucien's son was the sole offspring of the four brothers, Charles the Reformer knew that continuation of the family's power rested with Reynard.

In 2340, the second President of the Federated Suns died of a heart attack while informing the Council of his plan to send the new FPF into action on the Chesterton frontier. Though many Council members recognized the value of Charles Davion's reforms, he had not been personally popular. His bias in favor of New Avalon, in everything from political appointments to judgements in the Federation High Court, had earned him considerable opposition. Though Reynard did not come to power because of his predecessor's popularity, as had Charles, nothing could stop the Davion juggernaut.

TEMPUS TARDUS

It has taken the Davion government centuries to develop the system of checks and balances that underlies its federated system. In the early days on New Avalon, for example, the Council President could tamper with decisions of the High Court because of a short-lived legal principle known as *"tempus tardus,"* or "time slowed down." The principle came into existence during a series of cases involving colonists who had been late in taking over claims staked by their agents on other planets. These colonial developers were deliberately allowing arable lands to lie fallow while they waited offworld until the property values had risen in the food-poor colony. At that point, they planned to step in to make a killing. Once the precedent was set for the President to interfere with the High Court in *tempus tardus* cases, it allowed him to interfere in *any* civil case, which made the President *de facto* chief of the Federated Suns judicial system. It was not until 2764 that the Prince of the realm had to relinquish this sweeping judicial power.

—From *Legal Decisions through the Ages*, by Jaipur Wevensar, Teller-Burrows Corporation, 3027

Two decades under such strong leaders as Lucien and Charles Davion had been enough to bind the Federated Suns into a tightly woven net of trade and economic commitments. At every level, from the military to the bureaucracy, New Avalon's sons held the posts that kept the interstellar trade lanes open. There were regular FPF garrisons on a number of key worlds that were not yet members and few local units present to contest these strategically or economically valuable points in a showdown.

Eight years before, it had been the Davions who had kept New Avalon preeminent. Now it was the people of New Avalon, eager to retain their prominence within the Federation, who wished to keep a Davion in power. Thus, in 2340, Reynard Davion was elected Prime Minister of New Avalon and President of the Federated Suns.

THE WAR LEADER

My family has served you since the very beginning, and it will continue to serve as long as this august body is willing to accept our aid. I am not my father; I do not have his record of achievements to prove my worth to you. What I can offer you is my heart, my soul, my flesh, and my untiring devotion, not only to my own world, but to any world that seeks the protection and the freedom of these Federated Suns.

—Reynard Davion, in his acceptance speech before the Federated Suns High Council, 2340

Reynard Davion was as charming as his father Lucien, and as ruthless as any of his three uncles. Like Charles, Reynard knew how to get things done in the face of opposition, and he had few scruples about how to achieve his ends. Nevertheless, his correspondence reveals that he was as sincerely devoted to the ideals of the Crucis Pact and to the two Convenants of New Avalon as his father had been.

He was also the first of the modern Davions to show evidence that he had inherited the military prowess of his ancestors. Even the soldier Robert Davion had never been in a high enough official position to employ any military abilities that he may have inherited. Reynard was a veteran of the frontier wars, having achieved the rank of Major in the Federation Expeditionary Force of 2334 on his own merit. Prior to becoming Deputy Prime Minister, he had served as a Colonel in the FPF.

By the time Reynard came to power, the conflict along the Capellan frontier had been raging for 20 years. Lucien had fought the war against his own better judgement; Charles had kept it going to create a threat that would justify his sweeping military and political reforms. Reynard Davion resolved that his Presidency would end the conflict, but he was determined to do it from a position of strength.

In 2344, confident of his position in the federation, Reynard left New Avalon to take command of the Reserve Task Force being gathered at Achernar along the Capellan frontier. While two of the Task Forces distracted the forces around Mirach with demonstrations of strength, the Reserve Task Force penetrated to the planet Almach and severed the supply line that supported Mira and Mesartim. The sudden ferocity of the attack after more than a decade of desultory fighting overwhelmed the Tikonov forces. Moreover, the strain of meeting attacks on multiple fronts prevented the enemy from mounting an effective resistance. Perhaps no one was more surprised at these successes than the FPF itself.

Having demonstrated the muscle of the Federated Suns, Reynard now looked for the means to a peaceful settlement. Though it was the Tikonov Union that he had defeated, Davion's first negotiations were with the more important—and more powerful—Capellan Hegemony. In 2345, the two powers signed the Acala Pact, a mutual non-aggression treaty intended not only to promote peace and trade, but also to create a forum for settling future border disputes. It is likely that Davion saw the possibility of eventually absorbing the Capellan state through diplomatic means. If so, he failed to understand the Capellan character. Their motivation for signing the Pact was to *prevent* losing any of their worlds as the FPF probed deeply into their sphere of influence. Both sides apparently entered into this agreement solely to win time for military preparations.

Six months after the signing and ratification of the Acala Pact, the Federated Suns Council was called upon to approve another agreement with the Capellan leaders. This was the Almach Accord in 2346, which gave the Suns formal control over the Chesterton worlds of Mira, Mesartim, and Almach. Reynard Davion did not wait for the Council's approval before implementing the terms. It would not be the last time that a Davion assumed the prerogatives of the Council, or that the Council allowed him to get away with it.

THE VANISHING TUCAS

After much deliberation, it is the finding of this commission that Ambassador Tucas's ship was sabotaged. The only question remaining is who the saboteur may have been. In reviewing the evidence carefully, we find that the Ambassador intended to report to the Prime Minister that the Davion regime perceived weakness along the so-called Capellan March that could be exploited; that Reynard Davion intends to establish a dynasty far more extensive than any his family has yet enjoyed; that several weakly defended Capellan worlds would make profitable additions to Davion's empire; and that Reynard Davion's alleged "settlement" is a smokescreen to prevent any attempts at preemptive strikes on our part. In short, we find that Ambassador Tucas's ship was sabotaged by pro-Davion agents to prevent him from alerting us to Davion plans to annex Capellan worlds along our mutual border.

—From a confidential report by the official Capellan Commission investigating the disappearance of Ambassador Fashir Tucas, 2363

The disappearance of Ambassador Tucas is a profound embarrassment to the Federated Suns government. The Capellan government will inevitably suspect us of duplicity, despite President Davion's assurances that a final agreement has been resolved. It may be that the need for the President to renegotiate the agreement—and perhaps arrive at one less advantageous than the current version—is not the worst outcome of the Ambassador's tragic disappearance. Indeed, the worst may be the impossibility of renegotiating anything at all.

This leads us to conclude that Ambassador Fashir Tucas's JumpShip was sabotaged by anti-Davion fanatics attempting to embarrass and discredit the Prince and possibly to depose his rule.

—From the report of the official Federated Suns Commission investigating the disappearance of Fashir Tucas, 2363. Declassified 2438

After securing control of the three planets, Reynard established settlements on Mira and Mesartim, despite Mirach's claims to these worlds. He also named them independent member worlds of the Federated Suns, which gave them a higher status with their conquerors than they had enjoyed as Chesterton Trade Worlds. Almach, on the other hand, was made a Federated Suns trustee. Reynard intended to build a gigantic military staging area there against the chance of renewed war on the border.

With the border wars settled, Reynard returned to New Avalon and his normal presidential duties. The next years were mostly peaceful, with the Federated Suns expanding through political rather than military means, and devoting its energies to economic and social improvement. In 2363, hostilities again erupted on the Capellan frontier.

The events leading up to the crisis began in 2357, when Reynard occupied the Sarna Supremacy planet of Bell and demanded the immediate transfer of Chesterton and Highspire from Sarna. This set off a chain reaction within the Capellan political arena. The Capellan Prime Minister Julian Dux was forced to retire, which propelled the dynamic Seluk Tucas into his place. Tucas began immediately to do whatever necessary to unify the various disputing Capellan states against a common foe. It took him five years to carry out all his plans to secure his back, but by 2363, he was ready to confront Reynard Davion.

In March of that year, Tucas renounced the Acala and Almach Accords and demanded that Davion return all the Chesterton province worlds that had been ceded to him. He also recalled the Capellan ambassador home from New Avalon. In a final defiant gesture, Tucas upped the ante by demanding that Davion relinquish control over the Lee system. He claimed that a long-standing trade agreement with Lee was sufficient reason for his demand that the planet rejoin the Capellan fold. Originally settled by Capellan colonists from New Macao, Lee was within the region granted to Davion by the Acala Pact. This same world would continue to be the focal point of many Davion/Capellan campaigns to come.

Fashir Tucas, the Capellan ambassador to the Federated Suns, never reached home after his brother called for his return. His JumpShip vanished mysteriously on the return voyage and was never seen again, despite serious search efforts by both sides. Declaring his brother's death could only be a deliberate Davion provocation, Seluk Tucas mobilized his army. The Capellans called on member states of the Suns to depose their "false and treacherous leader" and to join the Capellan States. None took him up on the offer, and when Reynard Davion proclaimed a state of war, the vote of confidence was unanimous.

The new border conflicts went on for several years, with the Federated Suns taking most of the honors. When the FPF turned back Capellan attacks on Lee, the political repercussions toppled both Prime Magistrate Tucas and his government from power. The more the Capellans suffered internal dissension at home, the easier it was for FPF forces to continue their advances. When the Tikonov Union and the St. Ives Association were drawn into the war as well, they, too, lost worlds to the Federation. Reynard Davion conducted the strategy of the fighting, this time from the capital. When the FPF succeeded in retaking Redfield and other planets lost during the first border wars, he was satisfied that the Federated Suns had finally laid to rest the military humiliations of his father's day.

In 2367, a new government, the Capellan Confederation, was formed from the numerous smaller states that had suffered in the recent wars. Intoxicated by his military victories, Reynard Davion refused to recognize the legitimacy of the new realm. Diplomatic relations were not established, and desultory fighting continued along the frontier.

Two years later, in 2371, the 64-year-old Reynard Davion died from a recurrence of Black Marsh Fever, first contracted during his soldiering days on Gallitzin. His son, Etien Davion, was elected to both of Reynard's political posts without demur. The Davion dynasty was finally and unquestionably established.

The first three Davion Presidents had made the Federated Suns one of the major powers of the Inner Sphere. The army that Charles had forged had proven itself in battle. Meanwhile, the bureaucracy of the Federated Suns had been centralized to ensure a continuity of administration even during times of trouble.

The early Davions also left a legacy of problems. In particular, their foreign policy toward the Capellan Confederation would plague the realm for centuries to come. In fact, the Davion invasion of several Capellan worlds in 2366 foreshadowed the Age of War, during which each state of the Inner Sphere was at war with all the others.

[EDITOR'S NOTE: The creation of the Federated Suns was the first such alliance among the many worlds of the Inner Sphere. During the 24th century, ten such separate states evolved, each with a strong central government. Some came into being through the need for mutual cooperation, but others were forged through military conquest.

With its relatively large population base, easily obtained raw materials, and a tradition of many centuries of scientific superiority over the other states, the Terran Hegemony maintained technological superiority over the other states. This also bolstered its political influence, particularly when dealing with worlds lacking in such fundamental necessities as water or croplands.

As the century wore on, each state became more and more adamant about defining its borders, with border disputes becoming both frequent and intense. Struggles over strategically valuable planets escalated from small skirmishes to battlefields covering an entire world. Scientific research again turned away from improving the quality of life and toward research and development of new weapons.

In the final three decades of the 24th century, many border worlds went from being equipped with navigational aids and glorified police officers to having sophisticated ship detection equipment, fully armed K-F drive JumpShips, a variety of system defense craft, and large garrisons of crack troops.

The Age of War would soon rear its ugly head.]

THE TYRANTS

Perhaps House Davion was too successful too quickly. If they had not risen so rapidly to absolute power within the Federated Suns, their line might have produced more leaders qualified to fulfill the duties of rulership. From the very start, the Davions began to function as a dynasty, but with no provision for screening out the unfit or ensuring proper training or preparation. When fate brought forward a fool, a weakling, or a tyrant, there was no choice but for the family, the planet of New Avalon, and the Federated Suns to accept him. All too often, a ruler's flaws did not show up until after *he was in power.*

—From *The Davions: A Family Portrait,* by Johannes Bergmann, Capella Publishing House, 3008

Many people said that the marriage between Reynard Davion and Jeanne Remoix was a mistake. In later times, the superstitious even claimed that it was a cursed thing to mix New Avalon First Family blood with that of a common farmer from another world. There is no denying that Etien Davion and his two children were unstable, yet Paul, Marie, and Paul's descendants showed no taint of the so-called "demon strain" attributed to Etien's branch of the family. The truth is that the Davions, like all families, produced their share of bad apples. Etien's weak character could not sustain the rigors of supreme power, while Edmund and Edward, his twin sons, went bad because of poor upbringing. Nor was this the only time power would corrupt the line. The Davions would spawn other tyrants, just as they would produce their share of noble idealists in the years to come.

Following the death of Reynard Davion in 2371, the fortunes of his family—and of the planets that looked to the Davions for leadership—took a sharp turn for the worse. For the next half-century, the Federated Suns would find itself in difficult straits, becoming in turn the plaything of a madman, the schoolyard of a scholar, and the captive audience of twin egotists with a taste for luxury and power. Only one ruler out of Reynard's five successors was fully stable, and only two tried to provide the Federated Suns and New Avalon with satisfactory leadership.

THE MAD DAVION

Can any of you honestly claim to respect anything this man has done? How long must we go on bowing down before this family of tyrants and madmen?

—Roderick Varnay, from a speech before the High Council, 2378

Etien Davion was the eldest of Reynard Davion's three children. His mother, Jeanne Remoix, was the daughter of a landholder on the planet Lee. Reynard had met her there during a military staging operation against forces out of Chesterton and Tikonov on the Capellan frontier. Much to Charles Davion's surprise and anger, Reynard arrived back in New Avalon with the wife he had married aboard the courier ship.

Charles fought a losing battle to force his nephew to divorce Jeanne and to marry into one of the First Families. As it turned out, however, the marriage actually had some political advantages. By marrying offworld, Reynard helped to soften resentment against the pro-New Avalon bias that Charles had built up. When trouble flared again on the Capellan frontier, Reynard could claim a personal as well as a political stake in prosecuting the war to a swift and victorious conclusion.

Etien was the first Davion whose right to the Presidency was unquestioned almost from birth. Charles and Reynard had built the FPF into an army that was as loyal to the Davion name as to the government. Like Lucien before him, Reynard's solid reputation outlived him and added luster to the family name. Like his father, Etien served in a line unit during the war against the Capellans. He earned a Presidential Citation for Valor during the battles in the Lee system. A serious wound sustained while earning this honor left the younger Davion hospitalized for over a year and altered his personality forever.

Within a week after Reynard's death in 2371, the 33-year-old Etien was sworn in as Prime Minister of New Avalon. A month later, he became President of the Federated Suns. His first months in office were auspicious; a truce was called on the Capellan March, three new worlds were admitted as members of the Federation, and official emissaries were exchanged with the growing Draconis Combine. At first, Etien seemed an ideal ruler, more deferential to the Council than either Reynard or Charles had been, and more interested in the judicial work of the Presidency than either of his predecessors. Songs likening him to Lucien were common on most major worlds of his domain, and the causes of peace and growth seemed to move swiftly forward.

It was not long before Etien's beneficent image began to crack. His mood could swing sharply within the space of hours, with him becoming irritable, excitable, and prone to wild rages. Etien almost beat a courtier to death after the man failed to step out of his way during a court procession. In another instance, Etien killed one of his closest friends in a bit of temper during a casual fencing match. At first, this instability did not seem to affect his work, but soon his judgements in judicial cases became more erratic. On the advice of close advisors, Etien began to send representatives to handle this area of his responsibility. In 2373, an amendment to the Crucis Pact officially sanctioned the appointment of a Board of Magistrates to settle disputes that had previously fallen under the President's sole jurisdiction.

Gradually, Etien turned his attention to the pursuit of pleasure and left the administration of the state to the Federated Suns bureaucracy. This was a blessing because it kept the unstable President from making too many important decisions, but it also robbed the High Council of its power. Under the Articles of the Crucis Pact, the Council could not conduct any new business without the official sanction of the President or his appointed representative. Etien Davion rarely bothered to appoint a representative, and the ones that he did appoint soon ran afoul of him and were executed on a variety of spurious charges.

The rest of the Federated Suns looked to the leaders of New Avalon to curb the temper and excesses of their own Prime Minister, but 60 years of subservience to the Davion family had robbed the Avalonians of their initiative. The rest of the Davions seemed equally helpless. Etien's mother had died long before Reynard, his brother Paul was a sickly recluse with no ambition, and their sister Marie, still in her twenties, had been posted to a frontier military unit because Etien was jealous of her popularity. For a time, Etien's wife, Marion Michaels-Davion, was able to control him, but in 2376, he divorced her without cause. From then on, he deteriorated steadily, ordering arbitrary executions, indulging in the most decadent extravagances, and totally neglecting the responsibilities of government. Etien simply could not handle the strain of making decisions. While rejecting his responsibility as leader, he used his absolute power in the New Avalon Chamber of Deputies to indulge his every whim and fancy.

Not surprisingly, Etien Davion was unpopular in political circles. Sycophants might be happy to cater to his every whim, but others in important positions feared where Etien's instability might lead the realm. At least three plots were hatched against him during his eight years in office, two assassination attempts, and a plan for widespread secession from the Federated Suns. Neither bore fruit, betrayed by men eager to advance their own fortunes by protecting even a mad ruler against treachery.

In truth, Etien Davion was certifiably insane. Always inclined to melancholy, he now suffered from depression more and more frequently. The revelation of plots against him only deepened his feelings of inadequacy and grief. With the second assassination attempt in 2378, the President's emotional state hit bottom. His bodyguards killed the would-be assassin, a childhood friend and former compatriot in the Lee campaign, just as the man was about to plunge a knife in the President's back. Etien stared down at the assassin for several long minutes, then picked up his blade and said, "If Jeremiah Monroe hates me that much, then no one will hate me anymore." He stabbed himself in the heart before his guards could stop him, and died within seconds.

THE SCHOLAR

Perhaps you believe that a sick man cannot rule effectively? My brother was never ill a day in his life, but did that make him an effective ruler? Do not make the mistake of assuming that physical strength is a prerequisite of mental ability, my friends. I assure you that it is not.

—Paul Davion, from a speech before the High Council, 2378

Etien's twin sons were only 16 when he died, while his sister Marie was just 30. None were judged experienced enough to take on the responsibilities of President or Prime Minister. Indeed, the recent years under the madman had left many people doubting whether it was wise to allow any Davion to continue in power. The Davion name was still influential in some circles, however, and a clique within the High Council decided that only a Davion heir could keep factionalism from destroying the Federated Suns. They needed someone as a figurehead for their behind-the-scenes manipulation, yet they also wanted a man who projected at least the illusion of wisdom and experience. One member of the Davion family suited those requirements perfectly. It was Paul Davion, Etien's 38-year-old brother.

Paul had been a sickly child who still displayed symptoms of hypochondria and intermittent paranoia. Unable to join the military like most of his line, Paul had turned to studies in sociology, history, and some technical sciences. Throughout Etien's reign, he had happily remained far from public scrutiny. After his brother's suicide, however, the mantle of power was thrust suddenly on Paul's unwilling shoulders. Despite his attempts to refuse election to the Presidency, members of the High Council eventually persuaded him that to shirk his duty would threaten the security of the Federated Suns.

A NOBLE EFFORT

Eccentric in speech, dress, and behavior, Paul Davion was not a figure to inspire confidence among his new subjects. What he lacked in manners and breeding, he made up for with energy, goodwill, and determination. The men who had sought a puppet were startled to find that they had selected a leader with ideas of his own and the commitment to seeing them realized. Paul might have been a bit eccentric, but his peculiarities were minor and did not interfere with his work. Indeed, under Paul Davion, the Federated Suns quickly recovered from the Etien years. He promoted economic reforms that bolstered the faltering economy, helped to restore House Davion prestige, and waged a short but successful campaign on the frontier to keep the Capellans from growing too cocky.

CLOTHES MAKE THE MAN

When I met Paul Davion, he was dressed like an 1820s romantic poet, in brown flannel trousers, white ruffled shirt, brown topcoat, and with a red ascot that set off his swirl of soon-to-be-receding brown hair. This was a far cry from the standard blue, gray, or white suits and brightly colored silk shirts that were in style among businessmen in 2376. As I continued to observe President Davion in the following years, I was sometimes amazed to see him address the Council wearing garments ranging from bucklers with sashes to leather jerkins and peach-colored weskits. (He once asked me in private what I thought of a red tie that he wished to wear to a diplomatic function. On my advice, he decided against—possibly my best contribution to *haute couture* in the Crucis March.)

I hate to admit it, but there were times when he showed a real fashion flair that might have been impressive if more in step with the times.

—From *Upstairs and Downstairs: My Life as Paul Davion's Valet*, by Lorne Raskiewski, Scandalmonger Press, 2408

The new Davion ruler was a master at delegating authority, which made the Federated Suns bureaucracy even stronger than in past reigns. The Council, meanwhile, lost more of its power because of the creation of several extraordinary commissions that were given far-reaching powers in a limited sphere of responsibility. This latter practice proved so useful that many future Davion leaders would continue to appoint such commissions. Though Paul often had to oversee his government from a sickbed, he did more in one year then Etien had managed in eight. In 2378, Paul Davion married Marge Svensdottar of Arcadia, uniting the ruling houses of two key worlds in the federation. Later that year, their only child, Simon Davion, was born. With Simon added to Etien's twin sons, House Davion succession was more secure than ever.

Paul's reign was peaceful, with only a few minor skirmishes against the Capellans and a brief confrontation with the Terran Hegemony over trading rights in 2383. To settle that matter, Paul appointed Marion Davion as his deputy, and dispatched her to Terra to carry out negotiations. She was so successful that her position was made permanent in 2385, much to the dismay of Etien's sons and their ambitious mother, Marion Michaels-Davion.

Nine years later, when a real illness finally carried off the man who had struggled so long against illusory ones, Marie stepped into his shoes. She was now 46 years old and a seasoned administrator, diplomat, and soldier. No one would ever accuse her of being a figurehead, and her reign would carry on all the positive accomplishments that Paul had striven to achieve.

MAIDEN AUNT

Our worlds don't need a brilliant general or a zealous reformer. What we need today is the patient watchfulness of a maiden aunt, and who better than I to fill that need?

—Marie Davion, from a speech to the High Council, 2394

Of all the Davions who held power between the time of Reynard and Simon, Marie Davion was the soundest in mind, training, and experience. Ten years younger then her brother Etien, Marie had been her father's darling. Etien had always resented her charm and popularity, and so had kept her far from New Avalon on military postings during his reign. When Paul took power, he recalled Marie. In time, she became the frail Paul's assistant and confidante. From 2390 on, as Paul's health took a real turn for the worse, Marie Davion was virtual ruler of the Federated Suns and New Avalon.

As Paul's appointed deputy, Marie stepped into total control upon his death in 2394. Marion Michaels-Davion, ex-wife of Etien and mother of his twin sons, claimed that her children were the legitimate heirs to Davion power, but the High Council was unwilling to depose the experienced Marie in favor of the inexperienced sons of an unstable tyrant. Marion did wield just enough influence to force Marie to name her sons as principal heirs until the majority of Simon, Paul Davion's teenage son. Marie agreed to have Edmund, the elder twin, placed in an intensive training program to prepare him for a government post. Her hope, though, was to remain in power long enough to ensure that Simon followed her as President. Unfortunately, that was a futile dream.

Marie's eight-year reign was uneventful except for one domestic development and one of interstellar importance. The internal matter concerned the passage of a law that made the Prime Ministry of New Avalon a hereditary position. Henceforth, the ruling Prime Minister could freely appoint a successor without even the formality of First Family approval. Though the law merely formalized a long-established practice, it legally bolstered the future of House Davion on their home planet. From this point on, the Davions came to regard New Avalon as their personal domain, and use of the title of Prime Minster would gradually fade away.

In 2398, the second major event took place, with repercussions in the Federated Suns and in the other regions of the Inner Sphere. This year marked the start of the Age of War among the various governments in Human space. What had begun as a border dispute between the Capellan Confederation and the Free Worlds League erupted into open warfare, the first such war in over a century. Soon the whole Inner Sphere was engulfed in savage conflicts, as dozens of bloody wars were fought over frontier worlds for the next 14 years.

Although Marie Davion tried to keep the Federated Suns out of the fray, an attack on Novaya Zemlya by Capellan space forces in 2399 reignited the smoldering hatred between the two governments. The Novaya Zemlya incident was also noteworthy as the first time that a fleet of heavily armed spaceships had bombarded a world from orbit. The Capellan attack wiped out roughly 75 percent of the planet's population and crippled industrial and agricultural centers for centuries to come. The Federated Suns immediately began a crash program to build ships while keeping up more conventional FPF assaults along the Capellan frontier to tie down the opposition and prevent additional deep-penetration attacks.

Soon after the Novaya Zemlya affair, Marie Davion suffered a series of strokes and died at age 54. Rumors that her death certificate had been faked to hide poisoning were never proven, but we must note that Simon Davion was but 24 when she died, with his training in governmental affairs only barely begun. Given the military and political situation of the time, the selection of Edmund Davion as the new President was almost inevitable. A few more years might have made all the difference to the history of the Federated Suns.

TWIN TYRANTS

It is intolerable that the High Council should be dissolved on this flimsy pretext. We will not accept such a measure, and would sooner put an end to this sham of federation than to our ability to represent the views of our people in the Council forum.

—Richard Emerson Caldwell, Delegate from El Dorado, from a speech to the High Council, 2417

Etien Davion's twins were 40 years old when their aunt died. Born to luxury, they had been raised mostly by their indulgent, ambitious mother, who saw them as her personal ticket to power. That they were unmarried and childless was partly due to Marion's efforts to keep them free for a political alliance, and partly due to their personalities.

Edmund was the elder of the twins, and his character was probably the more admirable. Though a weak-willed man, especially where his mother and brother were concerned, he managed to keep his vices private enough that they harmed no one else. Like his father, Edmund suffered from severe bouts of depression that interfered with his judgement and ability. Where Etien had tended to give in to angry moods, Edmund's response was to withdraw and sulk. Marion began to groom him from an early age as the Davion heir, knowing how easily she could control him. After Marie's death, Marion's plans seemed to be proceeding smoothly.

The younger twin, Edward, was a different sort. Manipulative and cunning, he fancied himself a schemer with a talent for intrigue. In this way, he resembled his mother, though he lacked the intelligence that made her a political force under Paul and Marie. Most people saw through Edward, though Edmund never did realize that his twin brother was using him. Between mother and brother, however, Edmund Davion had little chance to make his own decisions.

Under his mother's influence, Edmund's reign began well enough. Marion Michaels-Davion was ambitious, but she was also a good, if uncompromising, decision-maker. Through her son, she implemented a number of reforms that further decreased the power of the High Council while totally eliminating First Family influence on New Avalon. The Avalonians welcomed the reforms when it became clear that the the First Families, though shorn of local power, were on their way to becoming an interstellar aristocracy within the Federated Suns.

Meanwhile, Paul Davion's example of using extraordinary commissions made up of High Council delegates was expanded and regularized, though few realized that these were actually advisory and administrative bodies rather than actual decision-making groups. Marion's deft manipulation molded a monarchy that lacked only one thing—the titular royalty that would proclaim the monarchy to all.

When Marion died in 2410, her work was only partially complete. Perhaps her greatest disappointment was that Edmund, a suspected homosexual, never gave in to her repeated requests to establish a family. His show-marriage to the daughter of the Chief Councillor of Palmyra was a complete failure. It was later proven that a daughter born to the marriage had been sired by the Commander of the Presidential Honor Guard.

Marion died knowing that the family fortunes would rest with Edward, whom she understood all too well, or perhaps with Simon Davion, Paul's son. From her deathbed, she implored Edward to see to it that her own blood remained in power after Edmund was gone, and this suited the younger twin's ambitions perfectly. His recent marriage to a prominent Arcadian socialite had already been blessed by a son, Arthur, whom Edward fully intended to see seated on the Presidential dais one day.

After Marion's death, Edward became the *eminence grise* of his brother's reign. At Edward's suggestion, Edmund began to institute further reforms that were neither as subtle nor as practical as those of his mother. Most notorious were a series of acts that placed House Davion entirely outside the laws of their planet and the Federated Suns as a whole. In the meantime, both twins became increasingly attached, like their father before them, to the trappings and pleasures of power. Before Edmund's death in 2415, the twins had twice dissolved the High Council for insulting behavior to their masters, then reconvened it with new delegates to approve some laws that the twins had requested. Government within the Federated Suns was deteriorating. Even the farthest planets were beginning to feel the effects of whimsical taxes and pointless construction projects.

THE ARES SUMMIT

In 2412, Chancellor Aleisha Liao of the Capellan Confederation invited the twin rulers to an extraordinary summit meeting to discuss limitations on the excesses of warfare between the ten rival states of the Inner Sphere. Though the Davion twins did not think much of the proposal, they sent a delegate from House Davion to the Ares Convention to advance their own public image. "Send a fool on a fool's errand" was Edward's pronouncement. They chose Simon Davion as their special envoy, and gave him negotiating power. At the time of his journey to Ares, Simon was 34, and his wide eyes and shock of tousled hair belied his political practicality and courage. He was ready and willing to represent the best interests of a government whose leaders did not care.

[EDITOR'S NOTE: Aleisha Liao convened the Ares Convention after atrocities in the Tintavel system left thousands of civilians dead. The Convention was held on the Liao world of Ares; Aleisha Liao could not possibly have been ignorant of the irony of holding a peace conference on a world named after the Greek god of war.]

The Ares Conventions were designed to limit the ravages of interstellar warfare. The accords specifically outlawed the use of nuclear weapons and the strategy of orbital bombardments of planetary targets, restricted the ability of military forces to wage war against strictly civilian targets, and set up a commission representing each of the major powers to look into possible violations of the Conventions. Any such violation would be reported to each signatory state. The unwritten but well-understood corollary was that the other governments would take united action against the violator, regardless of current political differences.

In the words of the immortal Blake, "only the Ares Conventions and its strict codes of battlefield honor saved Man from himself. Yet, these same codes also fostered the use of war to resolve every disagreement. Not since the height of the almost-forgotten Terran Roman Empire had man fought so much for so little reason."

Simon Davion signed the Ares Conventions on behalf of the Federated Suns. Although Edmund Davion refused to ratify it, the High Council defiantly voted to accept the accord, which prompted Edmund to disband it. For the next five years, it was uncertain whether the Federated Suns would adhere to this document, but Simon would ultimately see to its ratification. His attendance at the Ares meetings was the first flagstone on his road to power.

Edmund Davion died in 2415, leaving his wife's ten-year-old daughter and his brother Edward as the two closest heirs to the Davion throne. Edward moved quickly to eliminate opposition, convening a trial on grounds of high treason against Edmund's wife and her lover. Both were found guilty and executed. The daughter, Judith Davion, was quietly taken out of the public eye and then strangled by one of Edward's bodyguards. Simon escaped the purge by being away from the capital in military service on the Capellan March, but Edward sent orders to Simon's commanding officer to have his cousin killed. That officer, however, was one of the hundreds who had suffered humiliation at the hands of the two tyrants, and so he quietly made certain that the order was never recorded. He did let Simon know what was in store for him if he returned to the capital.

A FAILURE TO COMMUNICATE

It was a tense situation when Major Davion reported to my office. I had been trying to figure out the best way to break it to him that his own cousin had ordered his death.

He stood before me and saluted, reciting the formula about reporting as ordered. I stood up, walked around my desk, and handed him the order. His face turned pale, then red, then Simon Davion looked at me without a word. I could not return his gaze.

Finally, he said, "Well, sir?"

I sighed heavily, searching for the right words. "Well, Major, I guess we're just going to have to do something to upgrade our communications system. This is the fourth message this month that's been garbled in transmission."

Checking back later, I found that Corporal Stevens, who had received the message, had not recorded it. He admitted that some instinct had kept him from doing so. At my earliest convenience, I recommended Corporal Stevens for the Silver Starburst for his quick thinking under pressure.

—Colonel R. Radwell Chung, unpublished diary, ca. 2443

DEATH OF A TYRANT

Edward continued down the road to absolute power. The Presidential bodyguard, already strong during his brother's reign, now swelled even larger. He had already begun to build a secret police network prior to Edmund's death, and this too increased in size and power. Where his predecessor had been arbitrary and whimsical, Edward was ruthless and greedy. His ambitions clearly included the elimination of the entire High Council in favor of an absolute monarchy. For once, the delegates saw that threat and also managed to respond to it. The result was the so-called November Conspiracy that nearly tore the Federated Suns apart.

The Conspiracy was a widespread movement among the High Council representatives to pull out of the Federation and to demand independence from New Avalon. The leaders of the planets Augusta, Arcadia, Friesland, and El Dorado were at the heart of the plot, which was to begin when the High Council disbanded for three months starting in November 2417. FPF troops loyal to the conspirators were to seize and destroy the deep-space stations guarding New Avalon's jump points, overwhelm any JumpShips in the system, and blockade the system until the conspirators could proclaim the secession. The plotters rejected the option of assassinating Edward Davion, fearing that the Presidential Guard and the secret police would became too powerful.

Three days before the Council was scheduled to end the 2417 session, Simon Davion arrived home on New Avalon. He brought with him dispatches proclaiming that the FPF had annexed the planet Jaipur on the Capellan frontier, with a recommendation for his decoration and promotion. Though Simon knew of Edward's plans to murder him, he returned home at his own request. As he had followed a most circuitous route toward New Avalon, perhaps he had been busy rounding up support before appearing at the capital. When he did arrive, the Council welcomed him with public displays that made it impractical for Edward to arrest him.

Soon after his arrival, Simon attended a full meeting of the High Council. The moment that Edward appeared, Simon snatched a handgun from one of the Presidential Guard—no doubt a confederate who had infiltrated the ranks of the corps—and took five shots at the President, killing him instantly. Simon then threw down his weapon and put himself at the mercy of the High Council, requesting a trial for his actions. Simon and the delegates orchestrated the event so smoothly that the Presidential bodyguard was forced to accept Simon's surrender rather then avenging their leader on the spot. Showing the shrewd political instincts of many of his forebears, Simon would not only avoid punishment but would emerge as the new ruler of the Federated Suns.

It had been a period of fear and turmoil. Three of the five Davion rulers between the time of Reynard and Simon had been emotionally or mentally unstable. Even under Paul and Marie, there had been a general movement toward absolutism that could not be reversed. From a cooperative association of worlds, the Federated Suns was gradually becoming an increasingly monolithic, dictatorial state. Yet, at this moment, the Davion family was very much in danger of losing its hold entirely. Only the astute maneuverings of Simon Davion would keep the Federated Suns and House Davion united and strong.

NOVEMBER CONSPIRACY

There have been many allegations that Simon Davion's assassination of his cousin Edward was part and parcel of the so-called "November Conspiracy," and a mere sham arranged and carried out strictly to protect the conspirator from the consequences of the killing. Records of that period are extremely confused. Indeed, most have been lost, thanks to Simon's deliberate purge of Secret Police files and other documents during the Amnesty of 2418. Though the truth can now be known, Davion's behavior before, during, and after the incident strongly suggests that he was in collusion with members of the High Council. Whether he was actually involved in the entire November Conspiracy or acted to offer an alternative to it remains a key question.

Simon Davion's whole reputation was founded on claims that he knew of the conspiracy and eventually took matters into his own hands to avoid the inevitable, crippling civil war. His speeches, coins, and other propaganda endlessly echoed the selflessness of his act, that he had risked his own life and liberty to end the rule of a tyrant without the necessity of rebellion and war. If, indeed, Simon's action was part of a pre-arranged plot, it would mean that he was as cunning a schemer as any Robert, Adam, or Charles Davion. Though the official biographies portray him as "the second Lucien," it may be more accurate that Simon was motivated by simple ambition, making him no better than Edward himself. Yet Edward was reviled as a tyrant, while Simon became one of House Davion's four most revered leaders.

Most historians are inclined to doubt that Simon orchestrated the entire affair. The November Conspiracy did not require so risky an act as the outright assassination of Edward Davion right under the guns of his personal bodyguard. It was too well-planned as a secession movement to require such a gamble. More probably, Simon Davion was already planning an assassination, whether from the pure motives he claimed, or because he knew that otherwise Edward would kill him first.

It must have been during his voyage home that Simon learned of the Novembrist plot. The preparations underway at key worlds on his route would have been unmistakable to a man of his intelligence. He may well have rerouted his voyage after uncovering the plot, both to confirm it and to allow messages enough time to reach delegates on the High Council with suggestions that all cooperate. It is most likely that the assassination plot remained wholly Simon's throughout.

The question of who was using whom during the crucial days before Edward died leads to another tangled web. The High Council must have known that the welcome they extended to Simon would have major repercussions later. Did the conspirators act from a well-prepared script, or was their move a spontaneous attempt to force Edward to take some final action that would justify their secession? Was Simon Davion in on their plans, directing events on his own, or filling the role of a Judas goat to draw Edward into the last fatal outburst of his reign? And who in the Council Chamber knew that Simon Davion was going to shoot the President when he appeared that day?

The guard who provided Simon's weapon was killed in a fire before he could swear out a statement, but no one knows if his death was a convenient coincidence or perhaps part of a full-scale coverup. The guard had known links with Simon, but one of his cousins was a staff member on the Judicial Commission headed up by the Voice of Parma, a member of the inner circle of conspirators. Simon's father-in-law, Richard Caldwell, was another. It was Caldwell who saved Simon by his quick response after the assassination and Davion's surrender to the Council. The circumstantial evidence favoring collusion is formidable, but by no means conclusive.

THE FORGOTTEN DAVION

Amid the twists and turns of capital politics after the death of Edward Davion, historians tend to lose sight of one member of House Davion who was the most innocent victim of the family's intrigues and deceptions. This was Arthur Davion, son and heir of Edward, who was not yet seven years old when his father was cut down in the Council Chamber at the hands of his cousin.

Edward Davion had enjoyed a considerable reputation as a rake and a rogue. It was once said that he had slept with the wife and daughter of every prominent figure in Federated Suns politics. (The same wag went on to add that Edmund had also slept with all their husbands and sons). Almost everyone in the capital was surprised when Edward married Rachel Masters, daughter of High Council delegate Virgil Masters of Numenor, in 2309. Only Marion Michaels-Davion, Edward's mother, seemed to take it in stride. Her comment was, "That boy would have married a tree slug if he thought it would secure his inheritance." Edward openly referred to Rachel as "my senior wife" and continued to flaunt his affairs without a hint of concern for public opinion. From this union came a son, Arthur, just one month before Rachel Masters' death. Unlike Edmund's supposed daughter, there was no doubt of paternity. Arthur was the only acknowledged child out of perhaps a score that Edward was said to have sired.

After Edward's death, young Arthur was closest in line for the family succession. The High Council wanted no part of Edward's bloodline, however, nor was there a precedent for a minor becoming President of the Federated Suns. Thus did the Council eventually turn to Simon to carry on the Davion succession. Simon originally vowed that Arthur would inherit the Presidency after him, though he already had one son and there would be two more before the end of the decade.

Some have been surprised that Simon did not execute Arthur, but took the risk of allowing him to reach an age and a position to assert his claims—and his vengeance—against the man who had destroyed his father. Perhaps Simon was a good enough judge of character to realize that he could eventually win Arthur's loyalty, or he simply may have been confident enough of the political system he was already reinforcing.

In any event, Arthur Davion never posed a threat to Simon. In 2428, Arthur entered military service as an officer of the newly organized Combat Marines and served with distinction on both the Terran and Capellan frontiers. In 2447, he died in battle, holding the rank of Major and commanding a battalion during the rebellion on Alsek. His last words were reputed to have been, "Perhaps now they'll forget what my father did to them."

History would prove that hope to be vain, however. Edward would go down in history as a dangerous enemy of freedom, while Major Arthur Davion, hero of Alsek, would be no more than an obscure, half-forgotten footnote in the family annals. He never married, perhaps to reassure Simon that he harbored no plans to found his own disputing line of the dynasty. Though forgotten by time, Arthur Davion seems to have been a direct contrast to his father—a man of high character who deserved to rule, but was cheated of his right by an accident of history.

—From *A Bloody Nightmare: The Reign of Edward Davion*, by C.G. Green, Remagen Press, 2502

PRINCIPALITY

The 25th Century began in tyranny and the threat of civil war between the Davion family and the rest of the Federated Suns. It ended in a different form of tyranny and the threat of a war among the Davions. Between the accession of Simon and the rebellion of Alexander, however, there came a brief Golden Age, when a new order bound together the Crucis worlds. For a few short decades, Lucien Davion's dream of a cooperative federation became a reality. It is one of the great ironies of history that Lucien's democratic ideals came closest to fulfillment only after his descendant Simon forged a hereditary aristocracy to reign over the Federated Suns.

—From *The First Princes*, by Lydia Marsovich, Teller-Burrows Corp., 3023

For three weeks in November 2417, the Federated Suns was without a President. The assassination of Edward Davion may have ended the plan for massive secession by dissident planetary governments, but nothing else had changed. Fifteen years of rule by Etien's sons had left a sour taste with the High Council representatives. Many believed that Lucien's dream had failed, and that House Davion was finally in ruins.

It was the same man whose desperate act had struck down Edward who would step forward with a solution to the crisis, renewing not only the Federated Suns but also the fortunes of his family. Simon Davion and his descendants were destined to take up Lucien's mantle and rekindle the spirit that had carried the Federated Suns to greatness. In the end, Simon would be remembered as one of House Davion's greatest leaders.

TIME OF CRISIS

No one man can govern so vast a realm, even if that man is genius, saint, and hero all at once. And when a man of inferior or even ordinary abilities tries to exercise power over so many worlds and so many billions of people, the result is the tyranny, chaos, ignorance, greed, or hatred that we have already seen too often. Some like to hark back to the creed of the 20th century, proclaiming that all democracies are inherently good and all aristocratic oligarchies inherently evil. It is my belief, however, that only a stable aristocracy exercising authority over manageable areas of our Federated Suns can hope to blend the decisiveness of central rule with flexibility and freedom. It is not the system of government that will be good or bad, but rather the people who wield power within it. Did not my late, unlamented cousins take supreme power on the basis of a system that grew out of a democratic form of government?

—Simon Davion, from a speech to the High Council, 2417

Simon Davion faced an uncertain future in the days that followed the death of his cousin Edward. Those who had supported the fallen President—sycophants, loyal soldiers, and the uninformed masses—wanted to see Simon executed and Edward's son Arthur primed for the Presidency under the guidance of some experienced regent. A second group applauded Simon's actions but looked to the High Council to provide leadership from outside the Davion family, which they believed was now thoroughly discredited. This group saw the fall of House Davion as the ideal chance to restructure power and authority in the Federated Suns, though they lacked unity on one crucial issue. Who would rule once the Davions were no longer in the picture? Very few believed that Simon Davion would dare to lay claim to leadership of the Federated Suns.

In their plans to overthrow Edward Davion, the November Conspiracy had never entirely come to grips with the question of a successor. During those uncertain three weeks of 2417, the Novembrists were only one of the many different groups prepared to advance rival claimants. The members of the High Council knew that they must find a workable compromise that would satisfy the majority of these disparate parties, or else face an internal crisis that would tear the Federated Suns apart. There was also the constant spectre of war on several fronts, be it with the Terrans, the Capellans, or the Draconians.

Simon Davion saw the solution as well as the problem, and within those few weeks' time, he had won enough support for his plan that civil war was averted. His compromise plan would change the character of the realm forever. Though a far cry from Lucien's idealism, it was an act of sheer genius that may been the only solution to the crisis his family and his state faced.

To reconcile the demands of the various rival factions within the federation, Simon proposed to decentralize government by creating a new level of authority that would exercise local control over groups of star systems. Where each world had previously had one voice in the High Council, the new system would establish administrative units within the federation that would be answerable to the High Council, while retaining significant autonomy in matters of internal trade, defense, and so on. On one hand, the High Council would represent the will of individual member planets and would control appointments to the administrative subdivisions. Balanced against this would be smaller, interstellar governorships whose local power structures would act as a check on the federation government.

While Simon was busy lobbying for his compromise, his supporters were busy building sympathy for his cause through any propaganda means available to them. Simon had, after all, murdered his cousin, and must stand trial for it. He already had the sympathy of many High Council members, however, who believed that the Federated Suns was well rid of Edward Davion. In the course of a dramatic and highly publicized trial, Simon Davion became a folk hero on New Avalon and then throughout the Federated Suns as coverage filtered outward.

The people learned the facts about Edward Davion's nature and his excesses. The defense lawyers also managed to insert most of the more heroic details of Simon's background into the record, including his brilliant handling of the Ares Convention summit, his fine military record, and testimonials by Paul and Marie predicting that Edmund and Edward were far less suitable candidates for Presidential power than young Simon. In the end, the judge found Simon Davion guilty of justifiable homicide, but suspended his sentence on grounds that he had acted for the greater good of the Federated Suns.

A NEW ORDER

Absolved of his guilt and with his record as a diplomat and a soldier fresh in everyone's mind, Simon was ready to push his program for reform. He and his supporters continued to win support in the High Council as the representatives realized how much they stood to gain under the proposed new system. In all his discussions and meetings, Simon was careful to emphasize that the plan did not create any special role for himself or his family.

He understood better then anyone that the essentially democratic basis of the Federated Suns had to be abandoned. In practice, the concept of rule by a Council and its President was already a dead letter. Council members could not possibly represent their worlds effectively across interstellar distances, and the President's powers had become virtually unlimited. Simon's program would sweep aside the useless, outworn trappings of democratic rule. This reasoning led to the creation of an aristocracy within the Federated Suns and to a decline in the importance of the Council and the Presidency.

The new aristocracy was not, initially, a hereditary one. Rather, it was established through appointment by the High Council. The system would last, in one form or another, for nearly four hundred years. Only then would a remote descendant of Simon's begin to issue patents of nobility in his own name and without need of a vote by the Council to back him up.

The Federated Suns aristocracy would be composed of a six-tiered structure that allowed for numerous, small administrative districts with considerable local authority and a well-developed chain of command. The realm would be divided into five Principalities: the Capellan March, the Terran March, the Draconis March, the Crucis March, and the Outer March. Each Principality was further subdivided into Duchies, Margraves, Counties, and Baronies, usually on the basis of population rather than territory.

The High Council remained, but it would be modified in both composition and duties. Instead of its members being elected or appointed representatives of the various member-world governments, they would now be *actual* rulers of the member planets, and automatically granted titles of Marquess or Duke. The High Council would no longer have any true governmental function beyond the assignment of titles and as a forum of general discussion. The Princes of the five Principalities would hold co-equal powers, and within their own territories at least, were not answerable to any other. The Dukes and Marquesses on the High Council could (and frequently did) appeal to the other members of the Council to remove an unfit regional Prince, and their combined powers were sufficient to ensure that it happened with a minimum of difficulty.

This was a good system of checks and balances, but it would need some degree of centralization to keep the Federated Suns from becoming five totally independent states. Council member Richard Caldwell, Simon's father-in-law, introduced an amendment that established precedence among the five Princes. The First Prince of the Federated Suns would be the Prince of the Crucis March. It would be his job to carry out the federation-wide duties previously part of the President's role.

Now there remained only one task in redesigning the power structure of the Federated Suns—that of appointing the nobles. Among these, selection of the First Prince was the most crucial. Finding a leader from the Crucis March region acceptable to all the factions vying for power promised to be a difficult, perhaps even a dangerous, process. Some proposed Richard Caldwell, but he lacked the kind of strong personal following that would guarantee unanimous acceptance. Caldwell himself acknowledged this fact, and proposed the obvious alternative. He named Simon Davion as his choice for First Prince of the Federated Suns and Prince of the Crucis March. Though his nomination sparked a debate that lasted for three days and nights and threatened to erupt into outright violence, it eventually became clear that only a Davion of Simon's obvious ability and integrity could weld together the realm's many warring factions.

DEBATE PROCEEDINGS

MR. RODGERS: Ladies and gentlemen, I would like to remind the distinguished members of this body that my distinguished colleague Mr. Roberge is a charlatan and a pissant. . . .

MR. ROBERGE: I respectfully take exception to the distinguished delegate from New Syrtis's characterization of me, not to mention the exception I take to his character.

MS. TARKOVSKY (presiding): Gentlemen, you are both out of order!

MR. GOLDEN: Better than being out of your mind, like these two twits.

MS. TARKOVSKY: Order!

MR. ROBERGE: Madame Chairman, may I have the floor?

MS. TARKOVSKY: Chair recognizes the charming and distinguished delegate from McHenry.

MR. ROBERGE: If the accomplished and elegant delegate from New Syrtis would care to take this discussion outside, I will demonstrate that a sybaritic oaf like him cannot prevent the delegate from McHenry from personally punching him halfway from here to New Year's Eve.

MS. TARKOVSKY: Order! Order!

MS. CHEN (TO MR. GIULINI): If they do it, I'll put 20 D-Bills on the little guy.

—From *High Council Parliamentary Record*, Official Transcript of 14 March 2417

THE FIRST PRINCE

It may seem incredible that a self-confessed murderer on trial only a month before could become the foremost figure in a powerful interstellar realm. It is even more amazing that he was now named Prince by the same people who had witnessed Edward's assassination and in the same chamber where it had taken place.

Although Simon did not officially take up the reins of power until March of 2418, historians always date his rule from the previous November. His reign as First Prince was largely anticlimactic after the drama of his accession. For the most part, Simon spent his 40 years in power trying to implement the reorganization of the federation as thoroughly and peacefully as possible.

The only blight on his record occurred within the first months of his reign, when tension between the Federated Suns and the Capellan Confederation escalated to new heights. The Capellans had suffered major losses along their frontiers, and Chancellor Arden Baxter thought that the political turmoil in the Federated Suns was the ideal opportunity to win territorial concessions to balance some of the losses. When Davion refused these demands, Baxter retaliated by refusing to recognize the new Federated Suns government. Fighting erupted anew along the Capellan March, only further intensifying the bitter hatred that has marked relations between the two states through the centuries.

For some time, the Federated Suns had also been anticipating a Terran attack on their borders. When it finally came, it proved to be the major military conflict of Simon's reign. In 2431, Terran Hegemony forces won a decisive victory against the Federated Suns on Kentares IV in the Terran March. For the next eight years, the Prince of the March was unable to mount an effective counterattack. By the time he called upon First Prince Davion and the other Princes for aid, it was too late.

By 2439, the Hegemony had begun to field the first BattleMechs, and those awesome machines had made the Terran forces invincible. After the Hegemony captured Kentares, several other planets along the frontier quickly saw the wisdom of seceding from the Federated Suns and placing themselves under Terran protection. Meanwhile, the other states of the Inner Sphere were having their own problems against Terra's BattleMechs. As the century wore on, Terran leaders used the power of their armed forces to provide a balance in the ongoing Age of War. The role they sought and gradually won was that of mediator, rather than conqueror.

The deployment of these giant, man-shaped machines on the battlefield reinforced the Ares Conventions. Highly skilled warriors piloting the 12-meter high 'Mechs could fight complete battles in a small area, reducing the risk to civilian populations. Yet even the use of BattleMechs did not help make any of the battles of this era truly decisive. Most conflicts established temporary control over a small territory, and dominion could be challenged at any time by another small battle force.

Simon Davion's last act as First Prince of the Federated Suns was to approve a massive payment to the Lyran Commonwealth to purchase the most important military secret of the age. Two years before, in 2455, Lyran commandos had raided a Terran 'Mech production facility and stolen complete details on the processes of 'Mech manufacture and operation. After Davion's efficient foreign intelligence network got wind of the coup, he began to negotiate with the Lyrans to part with the data. It took two years, but the Lyrans eventually agreed that the best way to contain the Hegemony was to arm all its neighbors with the new 'Mechs. Of course, the Commonwealth leaders demanded a steep price for their "cooperation." Shortly after approving the final payment schedule that would unlock the secrets of the new technology, Simon Davion suffered a stroke and fell into a coma. He died a week later without regaining consciousness.

MILITARY SECURITY

Theoretically, there were three different layers of security at our 'Mech factories: armed guards patrolling the perimeters, a militia unit inside the electrified fences, and heavily armed elite special forces stationed at strategic points in the factory itself. There were three armored gates at each entrance, with the guards supposed to be checking the workers' IDs. There were also antiaircraft guns on the roof of each factory and more of the same inside the fences on either side of the buildings. Each of the other Houses had similar high-profile security measures at their 'Mech plants. Despite the security, however, agents of each Great House were able to steal some of the 'Mech designs of the others.

Several Davion guards were disciplined after the blueprints for an experimental 'Mech design were stolen, but the fact that the guards had been playing cards on duty and allowing the factory workers to call them by their first names was not the point. In the early days of 'Mech design and production, there was little need for either commando raids, much less full-scale military action, to breach the security of a 'Mech factory. Industrial espionage was so much easier to arrange because workers in the 'Mech factories were underpaid. No matter what the threat of punishment, there was always some worker willing to take the risks for the sake of bettering his family's lot or getting revenge against a hated supervisor.

We can infer from this that military secrets are not only the most difficult to keep, but that it becomes even more so when individuals have any pressing personal motivation to break faith with their leaders.

—From Commission on State Security Report to the High Council, 2463. Government Printing House Document 153145.88, New Avalon.

THE BLACK PRINCE

You call that lecherous pig a Prince! He's nothing but a dog, and I'd trade my life to see him dead before I'd see him hurt others the way he hurt my Cynthia.

—Raymond Sinclair, from a speech before the New Avalon High Court, 2467

Simon Davion had three children by Elizabeth Caldwell. The eldest son, James, was born before his father became First Prince, and was thirteen years older than the next child. Growing up like an only child and as the son of a noble, young James was as spoiled as Edmund or Edward had been. This made him a dangerous liability to a politician still struggling to erase the stains on the Davion name and prestige. By the time Simon died, however, James was 42 years old, and seemed to have matured enough to take up his father's work.

Though James Davion won the High Council's approval to inherit the Crucis March Principality, he was not what he appeared to be. James had always been unhappy in his marriage to Janet Solway, daughter of the Marquess of Glenmora, which Simon had arranged to suit his political needs. Though the match had produced a son, the child died in infancy. After that, the last tenuous thread holding the marriage together snapped. James and Janet were never divorced, but James became a husband in name only—and then only when the two were in the public eye. He began a series of secret extramarital liaisons, which he flaunted more and more openly, after his accession believing himself above the laws of either planet or federation.

Despite his indiscreet and profligate pursuit of the opposite sex, James Davion did prove to be an able ruler, respecting the prerogatives of subordinates and effective in political relations with his peers. That was all that mattered to the nobles of the Principalities, who merely shrugged off rumors of the Prince's infidelities. To the ordinary people of the Crucis March worlds, however, the reports of the Prince's conduct were scandalous.

Under James, the federation continued its own indecisive border wars with the Capellan Confederation and the Draconis Combine while internal trade and economic growth flourished within the realm. He also installed the talented but dangerous Richard Varnay in the twin posts of Chancellor of New Avalon and Constable of the Crucis March. Known as "The Black Prince" both for his swarthy skin and for the darker side of his personality, James might well have gone down in history as a great leader of House Davion. His career was cut short, however, when he and several aides were assassinated during a ceremonial procession.

James Davion's death brought a mixed reaction throughout the Federated Suns. Because of the scandal of his private life, many people were relieved when an assassin's bullet removed such a man from the throne on New Avalon. At the same time, few could dispute his political talents. At a time when Simon's new order was still young enough to fracture at the slightest misjudgment of authority or power, James Davion had ruled well. Luckily for House Davion and the people they ruled, the next two Princes in line were equally skilled and effective in guiding the realm.

LADY IN THE WHEELCHAIR

You say I am a cripple, unable to walk. I say one does not govern a realm from one's feet. You say I am a woman. I say I am of the same stock as Marie Davion, whom you all claim to revere. You say I am untrained. But who among you, other than those already holding a Princely title, is trained for this position? My Lords of the Council, I am the sister of James Davion, but I have never been accused of uncontrolled lust. I am the daughter of Simon Davion, but I have never been accused of murder. I will be the aunt of a new Davion heir, one with the training and the legitimacy you claim that I lack. Let me show you that I can fulfill the needs of this high office.

—Ellen Davion, from a speech to the High Council, 2467

AGAINST ALL ODDS

The assassin seeking vengeance against James had nearly extinguished the Davion family. The only survivors were Ellen Davion and Lisa Swerdlow, the pregnant wife of Ellen's younger brother Roger. For a time, however, it seemed that Ellen would not be chosen heir to the Crucis March or the Federated Suns. Even after a century and a half of Davions in power, there was no formal law guaranteeing the Davion succession. Indeed, under the new order, the High Council could choose some other qualified leader to head the Principality. Fortunately for House Davion, Ellen fought bravely for the title and power of the First Prince.

Her greatest opposition came from the Varnay family, whose importance had grown ever since Prince James installed Richard Varnay in a dual position of power in the Federated Suns government. Ellen and the Varnays waged a bitter war of words over her claim to rulership of the Federated Suns.

The odds were stacked against Ellen Davion, however. She had stayed out of the public eye while her brothers were alive, preferring service as a staff officer along the Draconis March to the political intrigues and dissolute lifestyle of Prince James's court. Because she lacked either political or governmental training or experience, many believed that she was a poor choice as the realm's new leader. Moreover, she was confined to a wheelchair because of a crippling injury sustained during the attack that killed James Davion. With the conservative shift of public mores regarding women, some even argued that her sex was yet another weakness that should prevent her ascent to power.

The Varnay faction might have kept Ellen Davion from the succession if Richard Varnay had restrained his own ambitions for even a short time. During the Council debate on the succession question, the most reasonable compromise was the creation of a Regency in favor of Roger Davion's unborn child, which medical tests had shown to be a son. The three Regents would include Varnay, Ellen Davion, and another influential Councilor. Varnay was stubborn, however, and wished to be named immediately as First Prince, with the power of the office passing into his own family line.

Though the office was not considered hereditary, the force of tradition worked against this distant descendant of New Avalon's Terran Governor. Many of the more conservative Council members could not bring themselves to break with the past by passing over a Davion heir. Some stories began to circulate suggesting that Richard Varnay had encouraged many of the excesses that marred the reign of Prince James. Labeled a panderer who wished to promote himself by dragging the Davion name through the mud, Richard Varnay saw much of his support slip away.

The Regency concept lost popularity because a Regency that ignored the Chancellor and Constable would be too weak politically to govern. It had been only 50 years since the last threat of civil war, and many feared that Varnay would take advantage of a weak government to create just such a conflict. Only a Prince could hold together the realm. Though Ellen Davion lacked experience, her promise to lean heavily on Council advice made her candidacy even stronger as Varnay's cause crumbled.

Ellen Davion was proclaimed Prince two months after the deaths of her brothers, but only with the proviso that the succession would pass to Roger Davion's child even if Ellen later had a family of her own. In view of her age and her handicap, this seemed rather unlikely.

Ellen Davion proved to be one of the most successful rulers of the Federated Suns. Common sense, stubborn determination, and a knack for drawing out the best in others helped her to overcome inexperience. Ellen knew how to delegate power without surrendering it, all the while thoroughly educating herself on any subject relevant to shaping policy. All these traits gave the new First Prince a firm grip on the reins of state.

Meanwhile, Ellen Davion's popularity was growing by leaps and bounds. "The Lady in the Wheelchair" became synonymous with courage, willpower, and self-sacrifice throughout the Federated Suns. It was Ellen who renewed the people's respect for House Davion after the excesses of the Tyrants and the mixed reputations of their successors. Unskilled politically but with an exemplary personal life, she was the antithesis of her father and brother. At that moment in history, it was integrity more than political cunning that was needed to win the hearts of her subjects.

Within five years of her accession, Ellen's power was strong enough for her to take two measures that, at least for a time, secured the Davion fortunes. First was the removal of Richard Varnay as Chancellor and Constable. In a major gamble, Ellen actually promoted her rival to fill a vacancy in the Capellan March Principality. She also installed many of her own supporters in the area to keep the new Prince under careful scrutiny. Given the importance of the Capellan frontier through modern times, this action would haunt Davion rulers three generations later.

ACT OF SUCCESSION

In 2473, after a lengthy period of debate, the Council passed the Act of Succession, which gave House Davion permanent tenure over the Crucis March Principality and the title of First Prince of the Federated Suns in perpetuity. Henceforth, the reigning Davion Prince could choose his own successor; otherwise, the succession would pass automatically to the nearest living blood-relative of the last reigning First Prince. Detailed laws of succession were included in the act, starting with the confirmation of Ellen's five-year-old nephew William as the next heir. Also established was a mechanism to create a regency for an underage Prince as well as similar provisions to regulate the succession in the other Marches of the Federated Suns.

Though Ellen Davion reigned for nearly 30 years after the Council passed the Act of Succession, that law was the crowning achievement of her career. Despite the fighting on three frontiers, the years of her rule would one day be regarded as a Golden Age.

By 2500, William Davion was 32 years old, a capable heir with broad military, governmental, and political experience. His son and two daughters secured his family succession, and the heir was popular with aristocracy and common folk alike. Ellen Davion, now 75 and feeling the full weight of her years and physical disabilities, was making plans to retire in William's favor. Already she had outlasted two Varnays, Richard and his son Thomas. The 20-year-old David Varnay, Richard's grandson, was unable to offer serious political opposition even from his Princely position on New Syrtis, and so the Davion throne seemed secure.

Before stepping down, Ellen sent her nephew on a grand tour of the Federated Suns, a two-year journey that her poor health had never allowed her to make. While William traveled, his aunt quietly went about preparing the transfer of power. When he returned early in 2502, all was in readiness. In a dignified ceremony before the Council, Princess Ellen handed over the post that she had fought so hard to win and to uphold for three and a half decades. This historic moment marked the first time that the transfer of Davion power occurred without the death of a ruler, as well as the first time that there were no disputes over the succession.

Ellen Davion lived until 2510, when she died in her sleep at age 85. Known to history as Ellen the Good, she did more then any other Davion between the time of Paul and Alexander to regain her realm's loyalty and support for House Davion.

WOMEN IN THE 25TH CENTURY

The rise and fall of individual philosophies, religions, and cultural trends did not leave many marks on the tapestry of Davion history. The NeoBaroque style of art, architecture, and music of the late 24th century meant little to the politics of the time; neither did the general collapse and rebirth of Catholicism in the same period. One of the few trends that was crucial, however, was the changing role of women between 2400 and 2500.

Marie Davion, President and Prime Minister of the Federated Suns from 2371 to 2417, was her people's ideal of late 24th century womanhood. She had taken over from her reclusive brother Paul as leader of an interstellar community, watching over it like a maiden aunt, until the new generation was ready to take up the reins of power. Like most women of her day, she had equal opportunities with men. Indeed, she was far better qualified than Paul to rule the Federated Suns.

Within the next 50 years, however, the status of woman changed drastically. By 2450, woman's main role had degenerated to that of caring for husbands and children. In an edict passed by James Davion, females were forbidden to serve in any of the purely combat arms of the Federated Suns military, though they might serve as staff officers and other support personnel. A widely circulated study purported to "prove" that women were unable to use neurohelmets to control 'Mechs because of a deep-seated psychological resistance to union of the feminine mind with any outside force. As in an earlier, less enlightened period of Human history, women were defined as the weaker sex, with more and more educational or career opportunities denied them.

When Ellen Davion ascended to the post of First Prince in 2467, the trend began to reverse itself. Though Ellen came to the job ill-prepared, she rose to meet the challenge in exemplary fashion. Under her guidance, the role of women shifted back to what it had been under Marie. She reversed several laws that excluded her sex from military and other opportunities, and funded independent research that disproved the previous "studies" on the innate inferiority of women. If anything, the new evidence demonstrated that certain feminine characteristics actually *enhanced* the effective use of neurohelmets. As a result of Ellen Davion's influence and good example, two of her grand-nieces would one day seek to grasp the ring of power for themselves.

—From *The Changing Role of Women*, by Frieda du Quoy, Sargasso Press, 2602

WILLIAM THE FAIR

He is a man of sound character, honest, and trained to serve the people. You have heard of his exploits on the Draconis frontier, as you have heard of his good work in a dozen embassies and missions on my behalf. When I promised this august body that he would be my heir, no one knew what kind of man we had chosen. By the Grace of God, we could not have chosen better.

—Ellen Davion, from a speech before the High Council, 2502

The millennium's midpoint was a transitional era throughout the Human Sphere. The Rim Worlds Republic became a hereditary fiefdom of House Amaris in 2488. The extraordinary Siriwan McAllister, who would change the course of Kuritan history, was born in 2496. In 2501, the erratic Margaret Olsen succeeded the Lyran Commonwealth's competent Steven Steiner as Archon. The first decade of the new century saw the propagation of Kalvar Lorix's *Creed,* which would serve as a code of honor first for the MechWarriors of House Liao, then for 'Mech pilots everywhere. Meanwhile, the redoubtable Albert Marik was being groomed for leadership of the Free Worlds League.

In the midst of all this, Ellen's successor came to the Davion throne in 2502 fully qualified to rule, with an honest character and a depth of experience that made him the most qualified Davion since Reynard. While Ellen Davion had been well-loved, her nephew became the darling of his realm. For a decade, he had been actively involved in the government of the Crucis March, performing state functions ranging from service in foreign embassies to his two-year inspection tour of the Federation. Unfortunately for the Federated Suns, William's reign was destined to be short. For all his ability and training, the chaos that followed him would bring the interstellar community to the brink of dissolution.

Several tragic misfortunes marked the reign of William Davion. First was the eruption of a major new war on the Capellan frontier, a conflict that cost the new Prince dearly. His only son Edward, a young field officer in a 'Mech contingent on Royalston, was killed during a hit-and-run Capellan raid in 2508. A regiment fielded by David Varnay should have been able to support Edward's unit, but they failed inexplicably to reach the proper coordinates in time to rescue Edward or his comrades from death. Subsequent investigations pronounced the commander of the unit blameless, though accusations of Varnay treachery were rife at the time. Three years later, Sondra Black Davion, Edward's widow, also died in battle when her DropShip was destroyed off Avigait. This left Edward's young son Alexander as heir-presumptive, with the youth's aunts, Laura and Cassandra Davion, as alternate choices.

In 2511, Cassandra Davion married David Varnay, youthful heir to both the Principality of the Capellan March and the Varnay aspirations of Richard's day. Because the marriage was a love-match, at least on Cassandra's part, Prince William decided that to bless the union would mend the breach in relations between these two old and respected families. William Davion was only 43 years old, leaving plenty of time for young Alexander Davion to grow into his inheritance. No one dreamed that Cassandra's marriage would one day result in a threat to the Davion right of succession.

When a virulent plague broke out on New Avalon in 2512, William was stricken. The Ten-Year Prince remained in coma for nearly a week before eventually succumbing to the disease. Just before his death, he regained consciousness only long enough to name Alexander as his heir, but he handed the signet ring of the First Prince to Cassandra Varnay and his seal to her elder sister Laura. This was taken to indicate that the two aunts would serve as the boy's Regents until he came of age. It also elevated David Varnay to a position of preeminent influence, which would have a profound effect on later Davion history.

Though there were rumors that William had died at the hands of the ambitious Varnay, the official pronouncement was death by plague. In March 2512, William Davion was buried with honor on New Avalon while the five-year-old Prince of the Federated Suns looked on.

DEATH OF PRINCE WILLIAM

Several notable works have analyzed the circumstances surrounding the death of Prince William Davion in 2512. Was it the Spotted Fever Plague or a carefully contrived murder that carried off the Prince while still in the prime of life? The question arose within days of his death, when an anonymous pamphlet entitled *The Poisoned Prince* began to make the rounds of the back streets of Avalon City.

William's death occurred as part of one of the worst plagues in the planet's long history, when the mortality rate ran 30 percent and mass quarantines restricted all contact with the heart of the Federation. Just ten days before being stricken, Prince William had insisted on visiting an overcrowded hospital in a grand gesture of sympathy with his suffering people. An autopsy proved conclusively that he was suffering from the plague when he died.

It is equally true that William's death was surprisingly convenient to David Varnay. With Edward, William's heir, dead in battle and Alexander too young to act as Prince for at least a decade, it was the perfect time to make his move.

Had William lived longer, he might have chosen another heir. Laura Davion, nearer the throne than her younger sister, might have married and brought a new faction into the picture, perhaps producing a line to rival Varnay's claim to power. Perhaps Alexander's popularity might have became strong enough to prevent the Varnay faction from acting, or the accusations of treachery against Varnay after Edward's death might have been renewed.

William's death removed many obstacles to Varnay's aspirations, and it followed suspiciously close on the heels of Varnay's marriage to Cassandra Davion. Yet, this interpretation of events is perhaps *too* obvious. David Varnay barely won the Council's approval to retain his seat as Prince of the Capellan March, and was forced to rely on Cassandra to carry out the duties of the Regency. Had it been his own plot, Varnay would surely have waited until his power over his wife was more complete. When they entered into the Regency, the two were decidedly unequal partners, with Varnay always the lesser.

Neither was Varnay ever in a position to exercise real control over the succession. Had Alexander died first, there would have been no question of a Regency and thus a fair chance of winning the Principality without further ado. David Varnay may have conspired for the throne, but the evidence suggests that it was only *after* Prince William's death opened the way. To believe that William was murdered is to suggest that David Varnay was a fool willing to execute an ill-conceived, poorly prepared plot. Few people, then or now, have ever called David Varnay a fool.

—From *History's Stepchild: The Story of House Varnay*, by Ingrid Jablonski, Interstellar Press, 2743

CIVIL WAR

By the time of Prince William, the strain of managing a multi-world, multi-interest combine had finally become too much for the Federated Suns government to handle. Each new crisis had shaped the federation for good or ill, but each solution contained the seeds for new and unexpected problems. A parallel in Terra's own remote history might be the Roman Empire, where political systems had to adapt constantly to the facade of a stable, unchanging state. As late as the reign of Rome's "Thirty Tyrants," that state was known as a republic, deferring to the propaganda but not the proclivities of its powerful First Citizens. It was much the same in the Federated Suns during the reigns of Ellen and William.

—From *Davion, Kurita and Marik: The Rival Houses of the Inner Sphere*, by Thelos Auburn, Commonwealth Historical Press, 3026

In the 88 years between the death of William Davion in 2512 and the dawn of the 27th century, the Federated Suns was governed, in name if not in fact, by one Davion: Prince Alexander. William's grandson and a pivotal figure in Davion history, Alexander inherited the Throne of the Crucis March when he was but a child of five years. Within the long reign of this brilliant ruler, the Federated Suns would teeter on the brink of dissolution as well as reach the heights of power.

The early years of his reign were a time of devastating civil wars. The ultimate outcome of his reign was peace and plenty—a prosperity that gave the Federated Suns a strong voice in the new Star League. It was quite an accomplishment for a man who spent five years of his reign a fugitive with barely enough power to raise a single company of soldiers.

THE REGENCY

It is incredible that the young Prince is still alive. The Council of Regents is planning to supplant him, but I think the Regents are divided as much against themselves as they are united in their most basic aims. I'll wager that Alexander will live only so long as he is useful. Then we shall have a tragic death and a new ruler. Pray God the Federation survives it all!

—José Estevez, Duke of New Andalusia, from a letter written three weeks before his arrest for treason, 2518

Prince Alexander's aunts, Laura Davion and Cassandra Varnay, were the heads of the Council of Regents, which would rule while Alexander was still too young to handle the affairs of his realm. Under the Act of Succession passed during the Ellen Davion era, the two aunts were each permitted to choose another Regent, with the High Council appointing a fifth member. Cassandra Varnay selected her husband, David, while Laura Davion chose the well-respected General Nikolai Rostov, a powerful military leader. After a lengthy debate, the High Council appointed Carmen Estevez Davion, William's widow, as the fifth member. Though only 43, the Dowager Princess of Davion was an invalid, crippled by strokes suffered in the wake of the sudden deaths of husband, son, and daughter-in-law. Her membership on the Council was largely honorary. Some loyalists sought to use her to offset the overweening ambition of the two Davion aunts, but she was never able to muster the energy or the support needed to resist her treacherous daughters.

On the day of the funeral ceremonies for Prince William, the backroom struggle for power had already begun among the Council of Regents. David Varnay, supported by Cassandra, put forward a motion betrothing the five-year-old Prince to David's seven-year-old niece, Cynthia. He also sought to have Cynthia named as Heir-Apparent in case of Prince William's death. He claimed that this would prevent either Davion Regent from plotting her own advancement at the Prince's expense. The latter motion was voted down, but General Rostov had backed the betrothal motion against Laura's protests. As it turned out, his was the surer instinct. As long as the Varnays were hoping for a union with the Prince, their tactics would be less brutal. This bought time for Laura and the General, who lacked the power base that the Varnays enjoyed because of their holdings on New Syrtis.

During the Regency years, the Regents openly exercised power in pursuit of personal goals. The High Council's hopes for moderation were dashed when the Dowager Princess retired into seclusion on New Andalusia, granting complete control of her vote to Laura Davion. From that point on, Rostov was the real key to the Regency, as he held the deciding vote in any dispute. When he and Laura voted together to make her Prince of the Draconis March in place of the childless Prince Vladimir Kerensky, Laura got the power base she needed to offset the Varnay Princedom in the Capellan March.

Surprisingly, the Varnays themselves proposed Rostov as Prince of the Terran March in 2515, after Prince Charles Leightan and his immediate family died during a Terran raid on Robinson. Laura could not oppose the move for fear of alienating Rostov. She could never fully trust him again, however, for he was now as much in debt to her rivals as he was to her. Moreover, ongoing problems on the Terran frontier kept the new Prince busy leading military forces instead of involved in politics on New Avalon.

By 2517, the lines were well drawn. The Prince of the Outer March, junior of the five Princes, found it expedient to steer a neutral course. He could never completely ignore the wishes of the Varnays, however, for they controlled many trading partners and key merchant routes into his territory. Meanwhile, Laura Davion was growing increasingly isolated as Rostov began to play an increasingly independent game. The last bid for sanity came from the High Council, who once again sought to use the Dowager Princess as a focal point. The effort come much too late, for she died in 2518.

Almost immediately, Varnay's ubiquitous secret service agents began to report evidence of a plot to overthrow the Regents, presumably led by the most prominent of the High Council moderates. First to be accused was José Estevez, the late Dowager Princess's cousin and the most charismatic of the moderate leaders. It did not take long for the cries of treason to become widespread throughout the government.

The Treason Trials of 2518–2520 purged the High Council of most moderates and a large proportion of the Lauraists as well, though the Varnays were careful to avoid persecuting anyone who might have connections with Rostov. With the General's power growing steadily, the two Varnays were reluctant to make him an enemy without first trying to secure him as a friend.

VARNAY VS DAVION

Aside from bloodshed and political terror, the Treason Trials left another legacy. With the Dowager Princess dead, there were now only four Regents, with equal votes in the management of the Federated Suns. The High Council had the right to name a replacement for Princess Carmen, but Varnay's political influence and Rostov's military strength kept faint-hearted Councilors from exercising their prerogative. Rather than offend one side or the other, the Council refused to name a new Regent, leaving the four survivors in contention.

Winning Rostov's permanent support was the Varnays' best hope of defeating Laura's faction. Laura also courted the bluff soldier's support. Meanwhile, she stubbornly maintained that her power to act in Princess Carmen's name had been an outright transfer of Regency authority and not, as most claimed, a simple delegation of voting power that lapsed with the Princess's death. Once again, the High Council was reluctant to take a stand until a winner emerged in this struggle for control of the state. This, in turn, only served to magnify Rostov's importance.

Rostov's power rested, not on the weak reed of politics, but on the solid foundation of his popularity with the army. In a series of campaigns on the Terran frontier, Rostov had stopped Terra's military forces cold, thanks more to brilliant strategy than to high-quality troops or equipment. In the event of open civil war, Rostov's reputation as a military leader would surely attract the widespread support of soldiers throughout the realm. David Varnay's ill-managed effort to earn equal fame on the Capellan border in 2521 only underlined Rostov's skill and influence.

In that same year, Laura won Rostov back into her camp when she proposed that he be named First Marshal of the Federated Suns, an extraordinary military command that would supersede all other military posts in all five Principalities. Because the vote on Rostov's command came up while David Varnay was still away on the Capellan front, it was Cassandra who decided, on her own, to vote against Rostov. This act of bad judgement jeopardized everything for which she and her husband had worked up until now. Her vote not only failed to block Rostov's appointment, but also created ill will.

Alexander was 15 years old as these tensions began to mount. While he was growing up in his father's palace on New Avalon, the Regents had kept him largely cut off from affairs of state. It had been the Varnays, for the most part, who had controlled access to the young Prince, and they had sought to turn him into a spoiled young tyrant with no interest in government.

Alexander, however, disliked his Aunt Cassandra and actively hated David Varnay because one of his nurses had hinted at Varnay's complicity in his father's death. He sought solace in the study of warfare, strategy, and tactics, and the memoirs of Simon, Reynard, Paul, and Ellen Davion, the family's great statesmen. By the time he was in his teens, Alexander had resolved that he would avenge his father and put his power-hungry relatives in their place. With the single-minded drive characteristic of his line, Alexander kept his plan secret, played the part of the foppish Prince, and bided his time.

Foiled in their bid to win Rostov's support, the Varnays proved that they too could play the waiting game. It was not until a border clash with the Draconis March occupied Marshal Rostov's attention in 2523 that they forced a confrontation. They kidnapped young Alexander from New Avalon and carried him off to New Syrtis, ostensibly so that he might tour the realm, as his grandfather had done before Ellen's retirement.

In fact, they never permitted Alexander to leave their capital where, early in 2524, the 17-year-old Prince was married to Cynthia Varnay. This gave Cynthia a solid claim to the Princely throne, though the Varnays looked forward to the birth of a blood heir to make their stake even stronger.

Faced with this unexpected coup, Laura Davion hastily made a separate peace with the Draconis Combine so that Rostov could mobilize for a showdown. The battle that would decide the destiny of the Federated Suns seemed at hand.

In all his calculations and schemes, David Varnay had made one serious mistake. He misjudged the relative importance of family versus spouse in the mind of his niece, Cynthia Varnay, now Davion. Unlike her aunt and uncle, Cynthia was not ambitious. Though born of strong-willed, domineering stock, Cynthia nevertheless refused to play the role that fate had handed her. In their first year of wedded life, Cynthia and Alexander discovered a love for one another, a development as unexpected as it was unwelcome to the Varnay power brokers.

Cynthia did not learn how disposable her new husband was until she reported her pregnancy in August of 2524. On that day, she discovered that the male heir she carried would, if safely delivered, be Alexander's death warrant. On that day, she also rejected her family's ambitions forever. Henceforth, she would become the Prince's most vital ally, the only one on whom he might rely through the dark years to come. Instead of being known as a pawn in the Varnays' game, Cynthia Varnay-Davion's name has become synonymous with one of history's most romantic love stories.

THREE ROSES

Alexander was away in the Capellan March, and Cynthia was at home for health reasons. I was her guest in the palace on New Avalon—invited, I suspect, to keep Cynthia company during Alexander's absence. It was the first time since their marriage that they had been separated. Cynthia, always an early riser, asked me to a sunrise breakfast on the eastern balcony. She looked wan and lacked her usual vivacity, but seemed to be in a reflective and confiding mood.

"You've heard how my family plotted against Alexander when we were first married," she said, spreading passion fruit compote on her toast. "Would you like to hear how he took the news when I broke it to him?"

"Whatever you choose to tell me I will hold in strictest confidence, " I assured her.

She ignored my remark. "As you know, my Uncle David intended for Vincent to be his opportunity to seize power. Alexander was to be killed and Uncle David appointed Royal Protector as soon as Vincent was born. I was thrilled to tell people that I was pregnant—I don't think that I was ever happier—but Uncle David's reaction! He literally wrung his hands with delight! He was saying things like, 'Congratulations. How nice for you,' but the message in his eyes said something else.

"So I knew there was something going on. I also knew I couldn't ask him outright, but I said some vague things about him helping us plan the baby's future, and his plan all came out.

"I was furious—you can imagine how I felt. I don't know what was worse, being used as a pawn in my husband's assassination, or being thought of as a political baby machine with no instincts of my own.

"That afternoon, when Alexander arrived home, I told him everything. I didn't know beforehand how I would do it, but when the time came, it all came out matter-of-factly, like I was one of his agents making a report. He listened with a surprised, hurt look on his face, and when I was finished, he sighed.

"'Well, I shouldn't be surprised,' he said finally. 'Thank you for telling me.'

"I said, 'Well, aren't you excited? Angry? Anything?'

"'Somehow,' he said, 'I can't be upset. I have you.'"

Cynthia sighed, staring at the hillsides on the horizon. "Before then, I could have forgiven Uncle David." She sipped her Chakachamna tea and sat silently for a long time. Finally, she said, "Then, Alexander gave me a gift. He'd had it jumped from Terra just for me. It was three white roses, symbolizing the three of us—him, me, and Vincent within me—and I knew that forgiveness would not be part of the picture."

—Countess Jane Pesselthalmak (2509-2604), from a letter to Zane Davion, 2553 ComStar Archives

THE REGENTS OF WAR

There are worlds besides Syrtis or fair Avalon,
And subjects still loyal despite Varnay's wrong.
Brave citizen soldiers ten thousand men strong
Will follow the Banner of Young Davion.
CHORUS
So ready the vessel, make ready to lift,
Plot coordinates to jump through the Barrier Rift.
Tonight starts the journey, 'twill ever live on
Of that proud Princely Banner of Young Davion.
Then away to Nahoni, to marsh and to mud.
'Fore I yield to the Regents, I'll shed my life's blood.
Tremble, false Varnay, you'll see more anon
Of the bold Sunburst Banner of Young Davion.

—From *The Banner of Young Davion*, as recorded in *Songs of the Crucis March*, compiled by Edward Monroe, 2896

ALEXANDER ESCAPES

Soon after discovering her uncle's plans, Cynthia Davion shared the information with her all-but-captive husband. They hatched their own plot to evade Varnay, taking advantage of Varnay's increasing concern with affairs in the Crucis and Draconis Marches. Here, Marshal Rostov, ably backed by his dynamic son Dmitri, had settled the disputes with the Federated Suns's neighbors in order to concentrate completely on internal affairs.

Late in 2524, an elite strike force of DropShips and 'Mechs landed on New Avalon as escort to Laura Davion. This show of military might cowed the High Council. Still maintaining her right to Carmen's vote as well as her own, and supported by Marshal Rostov, Laura announced that the Varnays had kidnapped the First Prince. She proclaimed them enemies of the State, Regents no longer. Those Councilors who had not already fled New Avalon loyally acclaimed these decisions and declared war on the traitors who were holding the young Prince against his will.

In answer, the Varnays assembled their own rump Council on New Syrtis. They declared that the occupation of New Avalon was itself treasonous, and in the name of the Prince, urged that a militia army be mustered against Laura and the regular army. The battle lines were drawn at last.

The first months of 2525 were a calm before the storm. Alexander, who was approaching his 18th birthday, was still more of an asset than a liability to the Varnay faction. By Federated Suns custom, he would not enter into his majority until he turned 20. In the meantime, he would serve as a useful symbol of legitimacy until Cynthia bore him a son who would be an even more potent symbol.

To counter the stories of enforced captivity on New Syrtis, Alexander was "permitted" to accompany his uncle on a tour of Capellan March worlds where troops and 'Mechs were being mustered. The fact that this would keep the young Prince away from both his wife and his aunt at the time of his son's birth was a part of Varnay's planning. Though he was unaware of Cynthia's betrayal, Varnay did know that neither she nor Cassandra were ready to see Davion killed outright. Varnay would save that part of his plan until he was sure that there was a living heir to the Princely union.

Cynthia and Alexander had not been idle, however. Though Alexander's tour would separate them, it assured his escape from certain death. Cynthia had already made contact with anti-Varnay elements in the Capellan March. As soon as the Prince's itinerary was known, she notified her contacts, who set a plan in motion.

It was at Sekulmun, third stop on the tour, that Alexander made his move. The news David Varnay had been waiting for reached them in early May 2525. Cynthia had given birth to a male child, and named him Vincent, as she and her husband had previously agreed. This good news was the signal for Alexander's escape.

Varnay could not prevent popular celebrations of the event, for the legitimacy of his cause hinged upon it. Neither could he openly turn on Alexander if that cause was to have a future. Thus, the young Prince had enough freedom of movement to be smuggled onto a ship manned by supposed members of Varnay's scout fleet. From there, he escaped offworld before the Prince of the Capellan March could react. By the time Varnay discovered what had happened, it was too late to stop Alexander, who was safely outward bound. At this same moment, Laura's forces launched an assault against Kluane that threatened to tear the Capellan March in half.

BAIT AND SWITCH

Rostov's idol was the Russian General Kutuzov, who defeated Napoleon during the latter's invasion of Russia. Using a variant of the famous General George Washington's principle of strategic retreat to prevent a less mobile army from establishing effective supply lines, Kutuzov pioneered a policy known as "scorched earth," which called for retreat into the vast Asian continent while destroying everything of value. Though this was hard on the peasant class, whose farms were burned, the maneuver kept Napoleon's armies overextended. In one of the most bitter chapters in Human history, the French army ultimately starved and froze in the severest winter in Russian history.

Rostov's favorite strategy was another variant on Washington's and Kutuzov's strategies. Named after a practice popular with dishonest merchants, the bait and switch tactic also relied on strategic retreat into vast wilderness lands to throw an opponent off balance. The object was to lay a trap for an enemy bent on easy plunder.

On the planet Talcott, for instance, Rostov pretended to flee superior Varnayist forces from nearby Salem. When the pursuers jumped into the Talcot system, they discovered that the defending forces had disappeared into the verdant farmlands of the planet. They foraged the land to replenish their food stocks and notified General Varnay that Rostov's forces had disappeared. Varnay wired back, as Rostov knew he would, that Rostov's forces were too significant to be overlooked, and ordered a thorough search of the planet. Rostov's strategy became a waiting game.

Rostov was taking a huge gamble. Knowing that his forces would be planetbound for some time, he turned off his heat-generating equipment, whose traces would have led Varnay's troops straight to him.

He knew the time had come to move when he intercepted an enemy message that Varnay himself was coming to lead the search. While the enemy was preoccupied with preparations for the arrival of their commanding general, Rostov mobilized his troops for a lightning raid on the Varnayist camp. With the element of surprise on their side and with Varnay's troops exhausted from searching, Rostov's troops easily overpowered the Varnayists, crippling their forces and destroying their newly acquired stores of provisions. Then they disappeared back into the wilderness. Varnay arrived to find his troops decimated and that Rostov had simply vanished from the face of the planet.

David Varnay was no fool, however. He realized that Rostov had found a way to make any invasion force on Talcot a sitting duck. Bloodied but unbowed, he evacuated the remainder of his expeditionary force, vowing that it would be Rostov who would fall into a trap the next time they met.

—From *Honorable Enemies: Generals of the Inner Sphere,* by Morikami Renchell, Pesht National Press, 2989

ALEXANDER IN EXILE

As the Prince and his rescuers passed through the Barrier Rift that formed the frontier between the Capellan and Crucis Marches, Rostov's veterans were cutting a wide swath through Varnay's poorly trained militia troops. During the assault on Quittacas, however, Marshal Rostov himself was killed. The death of the old war-horse of the Terran March was a shocking blow to the army's morale, which Varnay turned to his own advantage with fast thinking and typically underhanded intervention. Well-placed bribes bought him a whole series of mutinies and desertions in the ranks of Laura's main fighting force, and her offensive ground to a halt within a few weeks.

Rostov's son Dmitri, now 45 and a brilliant tactician in his own right, was cut off deep behind Varnay lines when desertion wiped out his whole line of communication. Rather than surrender, he disbanded his troops and fled with a handful of 'Mechs, seeking safety from the sudden groundswell of support for David Varnay.

This was far from the end of the Civil War. Though much of her front-line army had disintegrated, Laura still commanded the main body of Federated Suns troops and had at least nominal loyalty from all but those from the Capellan March. Varnay had won an essentially bloodless victory, but it had been costly. Varnay now lacked the resources or the trained troops to carry the war into Laura's territory. For the moment, things were at stalemate, though fighting continued in desultory fashion along the worlds lining the Barrier Rift. It was here, on the planet Nahoni, that young Prince Alexander had hidden himself in self-imposed exile. He knew that it was only a matter of time before his chance would come to avenge himself against the Regents, each of whom wished to seize Alexander's power for himself.

Alexander lay low for nearly a year and a half, celebrating his coming-of-age in the swampy encampment where he and a handful of followers had taken refuge. Both sides had declared him dead, each accusing the other of his murder. Both knew, however, that he had only disappeared, and were secretly scouring the Federated Suns for his trail. Realizing that his life was forfeit no matter which faction won the war, Alexander lived like an outlaw bandit while pondering how he could recover the throne.

VICTOR OF NAHONI

Early in 2527, elements of the 1st New Avalon Dragoons, a light armor regiment, were stationed on Nahoni while Laura mounted yet another strike into Varnay space. The Dragoons were not among her favorite troops. Colonel Gordon, their commander, had openly proclaimed that his loyalty was to New Avalon and the Crucis Prince before all else, and that his service to Laura was because of her hold on New Avalon rather than to any rights or merit of hers. Laura had stationed the Dragoons on this dreary, unimportant world and was keeping the regiment on short rations and supporting them only minimally to lure Varnay to attack and thus wipe out the regiment. In one stroke, this would eliminate an irritant while placing the enemy directly in the path of a well-planned counterstroke.

Alexander, of course, knew nothing of Laura's strategy. He had heard of the regiment's loyalties, though he knew their morale was low. Determined to test the power of the Davion name, on 24 January 2527, he set out for the Dragoon headquarters with only ten companions. The risk paid off. Gordon and his men enthusiastically hailed Alexander when he unfurled the sunburst banner on a hill overlooking the camp and proclaimed himself the rightful Prince of the Crucis March.

The Prince now had an ill-supplied but eager regiment of regular troops at his command. Within two weeks, he would also have a chance to test their mettle when Varnay finally took Laura Davion's bait. He sent two divisions of Capellan March militia, strengthened by two full regiments of BattleMechs, to occupy Nahoni.

Several factors conspired in Alexander's favor during the attack. The terrain of Nahoni, mostly vast expanses of bogs, hampered the movements of both infantry and 'Mechs. Because the Dragoons were equipped with jump infantry and hovercraft for scouting and support, they enjoyed an advantage over their opponents in this terrain. Also, Gordon's men were Regulars pitted against a mostly ill-trained militia force. Finally, the mere presence of Alexander was worth a full regiment in morale value alone, regardless of his experience as a military man.

In an inspired guerrilla campaign, Alexander and his troops withdrew deep into the marshes, striking out of the mists at isolated enemy columns floundering in the bogs. After two months of gradual retreat, Gordon broke off from the main body of Alexander's force with two hovertank companies. In a surprise attack, he encircled the advancing enemy forces and destroyed their base camp and landing facilities. The Varnay commander, General Boefers, was forced to surrender for lack of supplies.

Almost a full regiment, including four MechWarrior units, switched sides when they learned the identity of the victor of Nahoni. The remainder were held ransom while Boefers went to New Syrtis to report Alexander's terms. In exchange for the men, 'Mechs, and ships taken at Nahoni, the Prince of the Crucis March demanded his wife and his son.

If David Varnay had been on New Syrtis when General Boefers arrived with Alexander's message, it is likely that he would have refused the trade. Cassandra Varnay was another story. Under pressure from members of her Council, who had relatives captive on Nahoni, she agreed to Alexander's terms instead of responding with an ultimatum threatening the Prince's family. Cassandra decided that she could not afford to lose any more resources to the enemy. She quickly dispatched Cynthia and Vincent Davion to Nahoni with Boefers, who was instructed to retreat from the planet with everything he could salvage.

Boefers arrived only two days before the start of the Second Nahoni Campaign. Laura Davion, whose intelligence service had belatedly delivered the news of the Varnay offensive, was unaware of Alexander's presence, much less of his victory. Her plan to trap Varnay's troops thus went into motion on the assumption that the enemy was present in force.

Having concluded his trade with Boefers, Alexander had faded into the marshes before Laura's force made planetfall. While Boefers struggled to improvise some defense against these new foes, the Prince attacked one of Laura's columns and won another victory. Meanwhile, more Crucis units from Laura's army went over to the young Prince's banner. Though this gave him the nucleus of a fair-sized armed force, he was still desperately short of 'Mechs. Whether Alexander's generalship was faulty or whether pressure from his supporters outweighed his better judgement, he accepted a full-scale battle against Laura's forces on ground that gave her 'Mechs a decided advantage.

Even so, the Battle of Davion Hill might have ended in a victory for Alexander if General Boefers had not brought his troops to attack the Prince's rear, violating the terms of the trade he had effected barely two weeks before. The fact that Varnay's men were as eager to attack Laura's troops as Alexander's allowed the Prince to pull out his senior officers, 'Mechs, and 250 infantry and armor troops before his line disintegrated. Crucis DropShips in Laura's flotilla were as eager to aid their Prince as the fighting soldiers had been, which permitted Alexander and his followers to flee the Nahoni system while the other two factions continued the battle behind him.

It had been an inauspicious start, but the earlier victories at Nahoni more than outweighed the final defeat. The emergence of Prince Alexander was heralded far and wide and so were the stories of Boefers's first defeat and subsequent treachery. Alexander's cause had suffered no more than a setback. He was still a force to be reckoned with as the Davion civil war moved into a new, more devastating phase.

When Dmitri Rostov learned that the Prince was alive, he and his own band of refugees promptly transferred their allegiance from Laura to Alexander. Six months after the Prince fled Nahoni, the younger Rostov had raised the sunburst banner on Farwell. Word that the son of the Federated Suns's beloved Marshal had thrown in his lot with the Davion Prince spread like wildfire across the Crucis March, winning Alexander the support of many Regular Army units.

Meanwhile, Prince Alexander was traveling with Colonel Gordon and the other survivors of Nahoni to promote his cause from world to world. At first, they often had to dodge Laura's troops, but by the middle of 2528, Laura and her forces were pulling out of the Crucis territory. The Prince's cause was too strong in the Davion's hereditary realm.

Laura Davion withdrew into her own Draconis March to regroup and organize a renewed effort. As for the Terran March, it had split. The younger Rostov carried great authority as his father's successor to the Principality, but Laura's close ties to the Marshal were not easily forgotten. The Rostov family still lacked the lineage to command the kind of loyalty that the Davions enjoyed in the Crucis March.

STRUGGLE FOR THE THRONE

Pray God this is really the end of it all. No throne is worth what we've paid today.

—Prince Alexander Davion, after the Battle of Robinson, 2540

ALEXANDER'S CAMPAIGN

There was no longer any hope for a peaceful solution. Laura and Cassandra had each proclaimed herself rightful First Prince of the Federated Suns, ignoring Alexander's hereditary claim. From now on, there would be no more pretense of legitimacy in their campaigns. These two women realized that only naked force would bring them power, now that intrigue and politics had failed. From 2528 to 2530, there was a lull in the civil war while each faction gathered strength and cleared away internal dissension. By 2530, the war was ready to heat up again.

Fortunately for Alexander Davion, fate took a turn of its own. David Varnay, 50 years old and the most dangerous of the young Prince's opponents, was killed during the riots against the Varnays on Ashley. Though strong-willed and determined, Cassandra was no strategist, and she was overburdened with the task of holding together the remnants of the Varnay cause. With the Varnays temporarily at bay, the Prince concentrated his attention against Laura Davion, now based around Tancredi, her Draconis March capital and chief stronghold.

Alexander promoted Colonel John Gordon to the rank of General and charged him with keeping the Varnay frontier secure while the main army moved the other way. Gordon accomplished this in a series of brilliant guerrilla campaigns that were to become legend among MechWarriors. Though Gordon rarely commanded more than a handful of 'Mechs, he consistently defeated much larger forces through superior use of tactics and terrain. He raided, attacked, and harassed, but always resisted the urge to settle into a prolonged campaign on any one planet. Through these actions, Gordon kept Cassandra Varnay's forces hamstrung while his master settled accounts in the Draconis March.

LAURA'S FAREWELL

There are those who will ask why. Some will say that I despair because my forces cannot win the war, but that is not the reason. Some will also believe I cannot bear to live with the frustration of never being able to rule the Federated Suns. Others may say that I regret the loss of so many of my faithful troops in pursuit of a now-doomed cause. I assure you that although I appreciate the loyalty and courage of my troops, I hold that they entered into a bargain with me to fight in return for good government, and possibly for the rewards I would show them if victorious. They made a good bargain but a bad bet.

There may even be some who will suggest that I fear pain, imprisonment, or execution if I am captured. Those who say so simply do not know me. I despair over the direction that the Federated Suns will take without my guidance and wonder what will happen to it now.

—Laura Davion, from her 26-page suicide note, November 27, 2533

Here, much of the glory went to Dmitri Rostov. Though not the commander his father was, the Rostov name alone was enough to win support from Regular Army forces in the Draconis March. Nevertheless, the campaign for the March went slowly, due mostly to the significantly different goals held by the Prince and General Rostov. For example, Rostov's first moves were aimed at consolidating his hold on the Terran March rather than on combating Laura's main forces. Alexander needed Rostov too much to do more than plead with him to finish off Laura's force and to strike at New Syrtis while the Capellan March was in disarray.

What General Gordon might have accomplished in months took Alexander and Rostov two years to finish. In the end, however, a massive Davion strike force landed on Tancredi and decisively defeated the cream of Laura's remaining troops in 2533. Alexander's elder aunt took her own life when she saw the dust cloud raised by Alexander's 'Mechs and armored troops approaching her command post. Her senior surviving military officer, General K'trinka, surrendered on behalf of the Draconis March the following day.

ROSTOV'S PLAN

Two years of delays had been costly on the Capellan front, however. Gordon's successes had been impressive, but Cassandra Varnay was finally rallying support. More important, she had enlisted the services of a large number of mercenaries from the Capellan Confederation. Rumor had it that Cassandra offered to cede large chunks of the Capellan March to House Liao in exchange for this support, though both parties hotly denied any such bargain.

The year 2533 saw a resurgence of Varnay power. Eventually, Cassandra's forces isolated General Gordon and his small force by keeping them under siege on the planet Jaipur. Alexander had to act quickly, launching a hasty attack to relieve the pressure on his most loyal commander. Dmitri Rostov pleaded to remain in his own March to continue putting down pockets of Lauraist resistance and to keep watch on the Draconis and Terran borders. The Prince was left to rely on his own abilities and distinctly second-rate troops in the campaign to rescue Gordon.

Not surprisingly, the rescue attempt failed, and Gordon was eventually forced to surrender his troops. When brought to trial for "war crimes" by Cassandra Varnay, the young General was condemned to death and executed. Most of his troops were similarly slaughtered. The long years of civil war were breeding savagery on both sides, however, for one of Alexander's battalions retaliated by unleashing mass destruction on the Varnay planet of Smolensk. It seemed that the Ares Conventions held only for disputes between the major states, but not for wars within them.

WARTS AND ALL

The family scandal in the years before the Civil War was that Cassandra had a son out of wedlock before she married David Varnay. While Varnay refused to recognize the child (although there was no doubt as to his paternity), Terril Davion was actually the first of the Varnay pretenders.

He was not, however, ambitious, and never made any claim of his own to the seat of power. The only people who ever recognized him as being the legitimate ruler of the Federated Suns were the power brokers of House Liao, who saw him as a pliant, politically naive young man who would make a good match for their dynamic young Chancellor, Salicia Liao.

The benefits of such a state marriage for both realms were obvious, but it was not to be. First, Salicia was a connoisseur of delicate and beautiful things in all aspects of her life, and Terril had warts on his face, something Salicia could not abide. More important, the man was never really accepted as a potential leader, even within his own family. Even if he had been, a state marriage between Terril and Salicia would have embroiled the Capellan Confederation in the Davion Civil War. With everything to lose and little to gain, Terril would have had to oppose both the Davion and the Varnay factions in the fight.

Although he was given every advantage accorded to the ruling family's children and he maintained cordial relations with both sides of the family throughout his short life (no mean trick), Terril Davion has, for five hundred years, never been granted a place in the Davion family tree.

—From *Pretenders Past and Present*, by Jeana Fortunato, ComStar Publications, 3021

Cassandra Varnay's success encouraged her to muster her forces for an all-out thrust against the Crucis March. From 2534 through 2536, her troops made a slow but systematic advance, taking Crucis worlds one by one. El Dorado, one of the oldest and most important Federated Suns planets, fell late in 2536, putting Varnay in a strong position for a direct assault on New Avalon itself.

Through it all, Rostov remained concerned mainly for his own Principality, and seemed to become even more intractable as the years passed. He was a more skillful politician than his father, however, and kept just enough reinforcements flowing into Alexander's camp to keep the Prince from giving up on Rostov's support. Early in 2537, Rostov and a small honor guard finally put in a dramatic personal appearance at the Prince's field headquarters on Belladonna.

During their meeting, Alexander and Rostov could barely conceal their mutual contempt. Upon learning of the size of Rostov's force, the Prince replied irritably, "I could have done with a few more men and a lot less Rostov." For his part, the General condemned the Prince's conduct of the campaign. "Why aren't you attacking?" he demanded. "I'd be attacking all along the front. I'd be fighting anywhere one of them stood. Battles are not won by watching the enemy take your worlds." Though this bickering was an inauspicious send-off for a crucial campaign, Rostov had a plan that soon turned into a classic of modern military strategy.

The General informed Alexander that he had amassed an even more substantial army at Smolensk on the edge of the Capellan March. While Cassandra Varnay advanced slowly through the Crucis March, Rostov proposed that Alexander's army move to one flank of the Varnay line of advance while his army threatened the other. They would leave New Avalon wide open. The object was for the two Davion armies to smash through Cassandra Varnay's line of supply and to unite behind her main strike force. This would force her to turn and fight at a disadvantage, or else watch her force slowly fall apart from lack of supplies, ammunition, and spare parts. Like most modern armies, Varnay's troops could not operate without a source for equipment.

Rostov's plan was executed brilliantly. In a series of strikes, he and Alexander reclaimed all the jump routes outward from El Dorado before the Varnay forces even knew what was happening. Cassandra was forced to turn and attempt a breakout back toward her own March, but Alexander and Rostov forced her fleet to battle in space over the planet Meglan. The superior Davion forces decimated the Varnay fleet. Though some elements landed for a last-ditch fight on the planetary surface, most scattered. Cassandra Varnay escaped, but Alexander's subsequent victory on Meglan ended the Varnay threat once and for all.

In his first proclamation as undisputed master of the Federated Suns, Prince Alexander Davion gave the planet Meglan the new name of Victoria. Varnay did not even attempt to rally her forces at New Syrtis, but fled into the Capellan Confederation with the remnants of the Liao mercenaries. Though the Civil War was over for now, Cassandra Varnay would never give up hope, for she had given birth to a son only a few months before David Varnay's death in 2530. The child was now seven years old.

The task facing Alexander in 2537 was immense. The Civil War had ruined whole planets, brought the Federated Suns economy to a near standstill, and resulted in millions of deaths on over a hundred worlds. Some rebellious pockets still remained, and the High Council was only a shadow of its former self. Worst of all, though, was that the surviving Federated Suns leaders were unwilling to accept the reforms called for by this massive destruction. Many resented the fact that the young Prince—only 30 years old—was determined to take a firm hand in the reconstruction of all five Marches, not just in his own Crucis March.

ROSTOV DEFIANT

Foremost among those disputing Alexander's right to act outside the Crucis March was General Rostov, whose obsession with the power of the Terran March led him to mobilize his troops in defiance of the Prince early in 2540. The coup began tragically for the First Prince, for Rostov's first act was to order that Cynthia Davion be kidnapped during her goodwill tour of the Draconis March. The attempt was bungled, and stray laser fire killed Alexander's wife. The Prince was grief-stricken. In the campaign that followed, Alexander would act quickly, decisively, and with implacable savagery against the man responsible for Cynthia's untimely death.

As the man responsible for the triumph on Victoria, Rostov still enjoyed tremendous influence. It was Alexander, though, who represented order and stability in a Federated Suns weary of fighting and factionalism. Few worlds or army units outside Rostov's immediate sphere of influence responded to his call for action against a would-be despot. The Prince, meanwhile, gathered an overwhelming army and moved quickly on Rostov's capital of Robinson. Sheer numbers left the rebellious General with little hope of success, despite Alexander's relative lack of military ability. The outcome was not settled until the Prince himself took the controls of a BattleMech and met the General in single combat. This bravura gesture nearly cost the Prince his life, but his superior reflexes helped him win the day and turn the foolish stunt into legend, increasing his popularity. Historians, however, have suggested that an agent of Alexander's sabotaged the General's 'Mech beforehand.

After the Battle of Robinson, House Davion would never again face significant opposition from any foe within the territory of the Federated Suns. For the remaining 60 years of his life, Alexander spent his time and energy turning his military triumph into a solid political victory that would secure Davion power once and for all.

YEARS OF RECOVERY

I fear that some members of this Council are laboring under a misapprehension. For example, I often hear it said that Reynard never intended this or Simon would never have done that. My Lords, my name is Alexander, not Simon, Reynard, Paul, or Lucien. I have fought for 13 years to bring an end to the reign of terror that my two aunts imposed upon these Federated Suns. If you have not yet learned who I am, perhaps you should ask my MechWarriors. They have already shared the news with the Varnays and the Rostovs.

After the Battle of Robinson, I swore that there would never again be civil war among us. If the actions I take contradict the sacred traditions of my forefathers, it is only because that oath means more to me than any famous remark from the annals of our history. I will have my way in this, my Lords, one way or another—as you may be sure that my Aunt Laura or Prince David Varnay would have had their way if either one had survived to sit in this proud Chamber.

—Prince Alexander Davion, from a speech before the High Council, 2541

TIME TO HEAL

With the Civil War behind him, Alexander concentrated on reconstruction. The Federated Suns needed time to heal, and events beyond Alexander's borders would help the Prince guarantee that recovery period. By the mid-2500s, the Age of War was winding down as the Terran Hegemony slowly began to forge the Star League. One major power after another was joining the new body, which pledged local autonomy while promoting peace between the old rivals. Alexander Davion was reluctant to join, mostly because of weakened condition of the war-ravaged Federated Suns. He confided privately to his advisors that he believed the Star League would one day encompass the entire Human Sphere. He vowed, however, that the Federated Suns would join only when it could do so with dignity and honor. Davion wanted to see his people and their worlds recover because of their determination and inner strength, not through the assistance of outsiders who might extend economic aid purely to advance their own goals of power among men.

INNER SPHERE ECONOMICS

With all due apologies to my audience's sensibilities, nothing stimulates economy like war. There are two reasons for this. First, the need for large numbers of soldiers siphons off a percentage of the work force who would otherwise be unable to find work. Second, the civilian population remaining on the homefront are fully employed by the war effort, paying taxes, and off the welfare rolls. This stimulates the economy and keeps money (and monetary incentive) in circulation.

—From *War and Peace: Two Economic Chimeras,* by Edward Fulmos, Davion Military Press, 2888.

The fact remains that even if a state of war provides full (or almost full) employment within a society, any economic "gains" are illusory. First, though soldiers are carrying out the defense of their government, they do not actually *produce* anything in the classic sense of that term. Moreover, a wartime economy is based on destruction, rather than production. Though the workers must work much harder to reach production quotas during periods of national emergency, their products do not actually contribute to the state's wealth, for most of those war products are either destroyed or used offworld (without the recompense of interstellar trade). The workers are productive, the factories and mills are geared up to the limit, but the economy is poorer then ever.

—From *Poverty and Production: The Davion Civil War,* by Richard Finetree, Remagen Press, New Avalon, 2889

While promoting economic measures to restore the power of his realm, Alexander Davion pursued other means to promote future stability in the government. These measures also increased Alexander's personal power. The ramshackle government of the Federated Suns would need sweeping reforms to prevent a repetition of the Civil War. The Prince was intent on making those reforms, regardless of any other consequences.

His first step was to discard Simon's concept of five co-equal Marches. From now on, the First Prince would reign supreme over the entire federation. No other noble would be able to use his territory as a personal recruiting ground, for loyalties throughout the Federated Suns would trace back to the First Prince. Alexander also further reduced the powers of the High Council, realizing that the assemblage could not be relied upon for stable leadership. In other reforms, he changed the system of appointments and authority of Regents. The change excluded the Regents from any hope of succession to the throne, and also made them subject to review by a board appointed by the High Council.

THE DAVIONS SUPREME

Within five years, Alexander had enacted more laws than any one of his predecessors had put forward in a lifetime. Though he always faced opposition to his proposal, the Prince inevitably managed to get his way. He owed his throne to the military, not to the politicians, and the threat of military action was usually enough to win approval for even his most unpopular new act. Though Alexander's reforms were pointing to the creation of an autocratic society, Alexander was no tyrant. He encouraged free speech, and believed that everything he did would be justified if it freed his people from the threat of another civil war.

Where the army was concerned, Alexander was forced to tread more carefully. Most of his military reforms were designed to minimize the importance of regional loyalties in favor of the army's ties to the federation. Individual worlds continued to recruit and field their own units, such as the First New Avalon Dragoons (renamed the New Avalon Dragoon Guards and given an exceptional place in the honor roll for their part in Alexander's rise). The Prince, however, saw to it that personnel from other worlds would be integrated into such regiments, and he rarely allowed a unit to be posted to the region where it had recruited most of its men. The First Prince assumed Nikolai Rostov's former title of Marshal of the Federated Suns, placing himself at the head of the army.

BACKROOM DEALINGS

While preparing this volume, ComStar researchers have discovered documents lost in our Blessed Order's Archives since at least 2590. These documents relate to the haste with which the Federated Suns joined the Star League and the Draconis Combine's delay in doing so.

Beginning in 2551, Ian Cameron, Director-General of the Terran Hegemony, began negotiations to attract each of the five other major powers of the Inner Sphere states to join in a unified interstellar government that was to be called the Star League. To induce a like-minded, peace-loving statesman like Albert Marik to support such an idea, all it took was the promise of peace, but Cameron also offered him the position of Vice-President, an offer that the noble Albert declined. Terrence Liao, on the other hand, presided over an empire in financial trouble, and Cameron gained his support by promising favored-nation status to the Capellan Confederation. With three states out of six pledged to the League, Cameron was halfway home.

It was not difficult to induce Archon Tracial Steiner to give the Lyran Commonwealth's support to the League. After a history filled with military embarrassments, the Lyran leadership welcomed the support the Star League would offer them against an attack by a neighboring realm.

Cameron had a more difficult time persuading the two most powerful states, the Draconis Combine and the Federated Suns, to join the League. Though the current Draconis Coordinator, Hehiro Kurita, was a man of peace, his society was so steeped in the traditions of militarism that he could not join the Star League without losing support from the Draconis nobility.

The problem was different in the Federated Suns. Prince Alexander Davion had come to power only after a devastating civil war. He wanted to remain independent until the Federated Suns could enter the Star League as an equal to any of the other members, if not the preeminent power. As time went on, however, Alexander seemed to take a perverse delight in resisting Cameron's blandishments. He also seemed to be playing a waiting game with Kurita, vowing that the Suns would be the last Inner Sphere power to join the Star League.

The log jam finally broke in 2567 when Ian Cameron made Alexander Davion an offer that the Prince of the Federated Suns could no longer refuse. Cameron promised that, in the event of a Davion-Kurita war, Star League forces would stage preemptive strikes against Kuritan military targets to safeguard Davion interests. (It did not hurt that entry into the Star League would also help the Davion economy recover from both the Civil War and years of tampering perpetrated by Marik agents.) On 27 October 2567, Alexander signed the New Avalon Accords, and Ian Cameron's dream of a unified Star League was only one step from being fulfilled.

Cameron made the same offer to Hehiro Kurita. Being a canny judge of character, Cameron knew that this Kurita was no warmonger, but that he needed to give his enemies some justification for his decision to join the League. The promise of preemptive strikes against Davion gave Hehiro exactly what he needed. In 2569, the Coordinator of the Draconis Combine agreed to join the Star League.

I extend my gratitude to the researchers of ComStar who have finally managed to assemble all the pieces relating to this crucial event in Human history—the formation of the Star League.

—Anastasia Marcus, Editor

Rank and authority would remain firmly in the hands of the First Prince. From this time on, the heir to the throne would be required to serve at least five years of military training and field experience before he could legitimately be recognized as First Prince. A new rank, Field Marshal, was also created to honor talented generals, but they, too, would answer to the all-encompassing authority of the First Prince. With these military reforms, Alexander hoped to block some future general from using his popularity to challenge the legitimate government of the Federated Suns.

By 2550, Alexander's reforms were nearly complete, and the Davion family stood supreme in the realm. All significant rivals had long since been swept away, and the Terran and Outer Marches had been eliminated. Though the other Marches were now larger, they had ceased to be separate political entities. Alexander still styled himself "First Prince," but there existed no other Prince under him. Rule of the Marches was tied directly to the central bureaucracy of New Avalon, and the districts were now more military than civil divisions of the Federated Suns. Alexander's second marriage to Veronique DuVall of New Avalon in 2544 added four children to the three born of his first marriage. The succession, however, was strictly regulated by primogeniture to avoid future splits within the family.

The one threat to Alexander's position was Cassandra Varnay and her son Roger, who had gone into exile beyond the borders of the Federated Suns. In 2565, Roger Varnay, aided by a small Capellan squadron, crossed the frontier and attempted to rally supporters on Carmacks. Though he raised a fairly large militia force, they were no match for Davion regulars, particularly after the Capellan warships were annihilated in the Battle of Kigamboni late in 2565. This halted Roger's advance on New Syrtis, where he had hoped to gain massive support from longtime Varnay supporters. The Davion 'Mechs of the Eighth Syrtis Fusiliers caught him in retreat toward the frontier, destroying Varnay's force almost to a man. Though Roger Varnay escaped to keep his cause alive in exile among the Capellans, his military support was gone. With that, the threat to Davion power around New Syrtis was destroyed once and for all.

Federated Suns relations with the Capellans had never been very good, and they did not improve with the Uprising of '65. Remembering what Cassandra had done to John Gordon years before, Alexander demanded that Terrence Liao return the Varnays to the Federated Suns so that they might stand trial for war crimes. When Liao refused, Alexander imposed economic sanctions, but took no military action. It was not war but peace that Prince Alexander was hoping for. Indeed, it would not be many years before he carried out his last and greatest reform—the union of his realm with the Star League.

ALEXANDER AS MILITARY COMMANDER

Since the days of Prince Alexander, Davion admirers have tended to regard him as a paragon. They call him a brilliant ruler, a superb diplomat, and above all, a splendid soldier and leader of other soldiers in battle. Though there is no arguing Alexander's accomplishments in government or diplomacy, recent scholars are split in their opinions of his military ability.

Prince Alexander spent 15 years involved in various military operations, which gave him the reputation as one of the Great Captains of modern military history. Nevertheless, a closer examination of his campaigns suggests that he may not have been the forerunner of the Davion soldier-Prince at all. Rather, he may have been no more than a pedestrian commander whose claim to greatness was based on his choice and use of subordinates.

The Prince fought only three major campaigns completely on his own, and in each case, his overall performance was demonstrably poor. The first was the short, planet-bound campaign that led to the Second Battle of Nahoni. During the battle, Alexander completely overruled his subordinate (Colonel John Gordon) and led his forces into an uneven battle on unfavorable ground against the armies of both Laura Davion and David Varnay. His second campaign was the long retreat down the Crucis March during Cassandra Varnay's final offensive. Despite the advantages of interior lines, strong defensive positions, and short lines of communications, Davion was defeated time and again until the opposition was knocking at the very gates of New Avalon. Alexander's campaign was a well-managed example of delaying tactics, but some scholars believe that a clever strategist might have turned the tide much sooner. Instead, the final victory in the campaign was due to the intervention of General Rostov, whose reinforcements were not nearly as vital as was his brilliant plan of double envelopment on a strategic scale.

In his third campaign, fighting against Rostov on Robinson, Alexander was certainly victorious over an acknowledged master of the military arts. This was due as much to his overwhelming superiority of both numbers and political support as to any element of generalship, however. If Rostov had managed to raise any significant support before Alexander's invasion forces had reached Robinson, the outcome might well have been much different. As it was, Alexander deliberately sought out the one-on-one BattleMech engagement against Rostov for the sole purpose of enhancing an uncertain military reputation. He needed such a reputation to avoid opposition to his future plans for military reform.

Other generals dominated the rest of Alexander's campaigns. The first Nahoni campaign was a classic of guerrilla warfare, but it seems likely that Colonel Gordon exclusively managed it, with the Prince as little more than a figurehead and rallying point. Echoes of the Nahoni operations show up often in Gordon's campaigns deep in Varnay territory while Alexander and General Rostov were fighting the forces of Laura Davion. Alexander's own performance in that campaign, on the other hand, was slow and painstaking. Rostov seems to have totally dominated the Prince, who was unable to enforce his will on that determined and often duplicitous general. In the long run, that war cost Alexander dearly. It gave Cassandra Varnay the opportunity to kill General John Gordon, which deprived the Prince of this consistently brilliant and unswervingly loyal officer.

If Prince Alexander Davion lacked the military skills attributed to him by popular tradition, he did possess the one trait vital to any successful military commander. If nothing else, Alexander Davion was definitely lucky.

—From *Generals of the Inner Sphere,* by Antony Korotir, Davion Military Press, 2866

UNDER THE STAR LEAGUE

By joining the Star League, Prince Alexander brought his realm finally and irrevocably into the politics of the so-called "Inner Sphere." For centuries, the Federated Suns had remained relatively isolated from the other powers of the Inner Sphere, except for encounters that were militant in nature and inconclusive in outcome. Then, suddenly, came the dawn of a new era. The First Prince of the Federated Suns was no longer just a ruler of his own state, but now had a voice in decisions that would affect the fate of nearly the whole of interstellar Man. Isolationism had been the watchword when Lucien and his successors were in power. For the successors of Prince Alexander, however, isolationism was anathema. The struggle for political power had moved into a new arena, and the Davions, now secure on the Princely throne, would not be left out of that struggle.

—From *The Rise and Fall of the Star League*, by D. H. Rand, Tharkad Press, 2989

The Star League era was a time of peace and prosperity for the Federated Suns, as it was for the other member states. The protective umbrella of the League limited (though it did not eliminate) interstellar warfare, while promoting trade and progress. Scientific advancements reached new heights among all the member states of the League, leading to a significant improvement of the Human condition everywhere. For all this, however, the Star League contained the seeds of its own destruction.

The framers of the original League agreements had not learned from the lessons of history, perhaps because they were filled with optimism and high hopes of what the League would do for their own realms. Thus, they did not take care to build in safeguards that would prevent the same kinds of crises that had disrupted each of their own states for so many decades. By the middle of the 28th century, factionalism, disputed succession, rival economic and strategic interests, and other all-too-familiar themes would arise across the Inner Sphere, bringing down the whole structure.

REUNIFICATION WAR

No one man can rule Humanity, nor would any rational citizen welcome such a dictatorship if that man existed. The Star League gives us the best of both worlds—control over our own destiny, but with active help from our own kind in other realms. The Star League is not the path to dictatorship, as some critics tell us, but rather the road to the freedom, prosperity, and peace that Man has lacked for 5,000 years.

—Prince Alexander Davion, from a holovid address to the people of New Avalon, 2576

The Federated Suns formally joined the Star League in 2567, though it was five more years before ratification of that act. Prince Alexander's scrupulous adherence to the old, almost outdated machinery of government slowed adoption of the agreement. Some evidence shows that the Prince's exaggerated insistence on listening to "the voice of the people" was actually part of his plan to give House Davion the upper hand in its dealings with the new League. Those five years were a test of how far the Star League would go to accommodate Davion interests. Had Alexander not been satisfied, he might have stopped the whole process cold through personal appeals *against* the treaty.

As one proof of their sincerity, the League threatened to impose economic sanctions on the Capellans for harboring the Varnays in exile. Under this pressure, Chancellor Ursula Liao compromised on her position regarding the exiles. Though she would not surrender them to Alexander, she would no longer grant them any special government protection or treat their whereabouts as a state secret. Without guarantees of Liao protection, the Varnays decided to escape before government troops could arrest and deport them. Ursula Liao found it convenient to lose track of them, but the Varnays had lost their court in exile on Columbine and their fat Capellan pensions. Nevertheless, they remained in the Confederation, an ongoing thorn in House Davion's side. Alexander had won his point, though. The Federated Suns had joined the League as a major power, not as a humble supplicant.

By 2575, most of the Federated Suns troops had been integrated into the cosmopolitan Star League forces, with a small "House force" loyal directly to the Prince. At that time, Ian Cameron was attempting to persuade the four significant Periphery powers (the Taurian Concordat, the Magistracy of Canopus, the Rim Worlds Republic, and the Outworlds Alliance) to join the Star League. At this crucial moment, Cameron's diplomatic brilliance finally gave out. With his goal of universal peace within view, he became frustrated when the Periphery states adamantly rejected membership in the Star League. Despite Cameron's increasingly open threats, these distant powers held their ground. Thus it was that in 2577, the Star League declared war on the Periphery states. Davion troops were dispatched to do battle with the Outworlds Alliance and with the Taurian Concordat.

The 20-year period from 2577–2597 was dominated by the Reunification War, a savage conflict that claimed more lives than did the preceding era known as the Age of War. Davion troops played a major role in the conflict, though the Federated Suns remained largely untouched. Indeed, though this war was a drain on manpower, it proved invaluable to Alexander's ongoing program of economic recovery, allowing him to place Federated Suns production on a near-wartime footing. The war also brought House Davion gains in both prestige and territory, which Alexander used in his ongoing propaganda effort to reconcile his subjects to membership in the League.

Given the relative size of the Inner Sphere and the Periphery forces, most people in the Inner Sphere assumed that the former would make short work of the latter, but two factors peculiar to interstellar warfare interfered. The first factor was the immensity of space itself. Establishing viable supply lines across parsecs of space dwarfs the imagination of even the most capable general. Even more significant was that the relatively small populations of the Periphery worlds worked to their military advantage. Because even small planets are vast fields of war, the Star League forces were faced with the frustrating truth that it is difficult to conquer a people in hiding.

These tactics only delayed the inevitable, however. In 2597, after two decades of war and millions of deaths, the Periphery powers agreed to join the Star League.

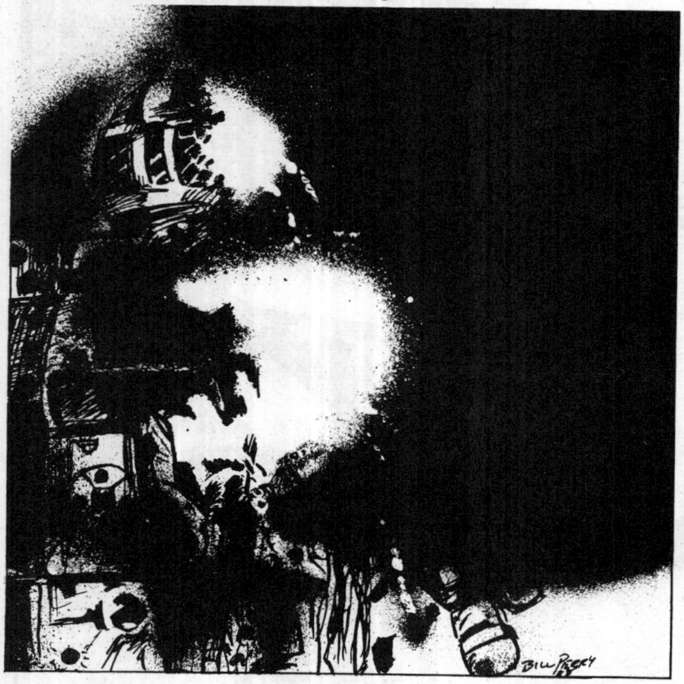

ALEXANDER'S FINAL YEARS

Prince Alexander lived another 30 years after the League Articles were signed on New Avalon. He had struggled for more than 40 years, first on the field of war and then in the political arena. Although he was 60 when the Federated Suns joined the Star League, many of his subjects still knew him as "Young Prince Davion." Though Alexander had begun to rely increasingly on his son Vincent to run the government, he was not about to abdicate the position he had fought so hard to win.

Aided by his children and then by his grandchildren, the Prince continued to hold the reigns of authority. Determined that the prerogatives he had established for the Prince of the Federated Suns would remain in Davion hands, he assigned family members to a wide range of diplomatic, political, and military functions. The wily old Prince did not wish to create any loopholes through which a new Varnay or Rostov might slip, and these practices became standard governmental policy for many centuries to come.

In 2596, Alexander dispatched an embassy of his two eldest sons, Vincent and Roger, to a meeting of the Star League on Terra. Their JumpShip was lost enroute, but no one was ever able to learn whether it was an accident or sabotage by fanatic pro-Varnay agents. The 89-year-old Prince never fully recovered from the blow. For the next four years, he experienced ever-lengthening periods of memory loss and confusion. Ian Davion, Vincent's son, governed in his grandfather's name for three of those years, but he never had the heart to declare the old man incompetent.

In 2599, Ian, too, met death unexpectedly when an assassin's bullet found his heart. Ian's heir was Zane Davion, a mere 19 years old when his father died and still serving with a frontier 'Mech unit. Ian's widow Elizabeth and various junior members of House Davion oversaw the affairs of government while Zane hastily rushed home to New Avalon.

24 January 2600 was the 63rd anniversary of the day when Alexander Davion had raised the sunburst banner before John Gordon's garrison on Nahoni. On that anniversary, the 93-year-old Alexander suddenly sat up in his bed, crying out, "Tell Gordon to bring his men up now!" Sinking back down against the pillow, he then fell into a long but fitful sleep. When he woke again, young Zane was there to see him, but the old man mistook Zane for his son Vincent. "So this is our young lad," he said, echoing his words to Cynthia when he had been reunited with his wife and newborn son so many years before. With those words, he died, making Zane Davion the First Prince of the Federated Suns. A videotaped testament, made a few days earlier during one of Alexander's last lucid periods, confirmed the succession and left Zane with some guidance and advice from the old veteran.

Of all the Davions, Alexander had experienced the longest and the most turbulent reign. Having shaped the Federated Suns into a form that would remain essentially unchanged for the next 400 years, he has become known as "the Augustus of the Federated Suns." Like the first of the ruling Caesars, his realm was in political chaos after terrible civil wars at the moment he came to power. Both Alexander and Augustus turned to one-man rule in the face of a bankrupt pseudo-democratic tradition. Finally, like Augustus, Alexander went on to become the embodiment of his people's ideals. No other Davion child would bear Alexander's name, but they would all call upon him before all others as their guiding spirit.

ALEXANDER'S HEIRS

Besides his vast political and military legacy, Alexander Davion left behind a sizeable family to carry on his name and traditions. Seven children were born to the Prince, three by his first wife and four more by his second marriage to Veronique DuVall. Prior to this, the Davion family had been small and slow to expand. After Alexander, his seven children would eventually create a sizeable aristocracy to back up the reigning Princes.

Alexander's eldest child Vincent (2525–2596) was stolid and unimaginative, but competent. He married Mary Gordon, daughter of the famous General John Gordon. From this union came Ian Davion, Prince Zane Davion, and the rest of the primary ruling line.

The second son of Alexander's first marriage was Roger (2528–2596). Roger Davion and his family settled on Victoria to govern the new colony established to commemorate the defeat of the Varnays in 2537. The Davions of Victoria would remain an important part of the family thereafter.

Melinda Davion (2530–2599), the youngest of Cynthia's children, married Edward Sanromea of El Dorado. Their line became known as the Sanromea-Davions, a family of some prominence until the eclipse of El Dorado following the economic troubles of the Second Succession War. From then on, the line continued to furnish noblemen whose distant connection to the throne was their only claim to fame.

Alexander's second wife, Veronique, gave birth to Henry Davion (2545–2614) in the same year that the Prince officially named Vincent and his progeny as his heirs to the Federated Suns throne. Henry emigrated to Royal on the Draconis frontier as part of his father's efforts to bind this Laura/Rostovist region to his regime. For similar reasons, Alexander's son Louis (2548–2609) settled on Ridgebrook in the Capellan March. While the Davions of Royal remained a prominent force for centuries, the Davions of Ridgebrook disappeared entirely during the chaos of the Succession Wars.

Veronica Davion (2550–2610) was named for her mother. Though the rumors were never proved, court gossip suggested that Veronica was not actually Alexander's daughter, but the offspring of an illicit affair between Veronique and Winston Hasek, a prominent figure in the High Council. Veronica married William Halder of Delphos in 2579. Their family, the Halder-Davions, also died out during the Succession Wars.

The youngest of Alexander's children was Lawrence (2552–2616), who settled on Lee and became a prominent local political and military figure. The Davions of Lee would be among the most famous defenders of the Federated Suns in future years, with a long and glorious history rivaling that of the main line itself.

—From *The Heirs of Alexander*, by Baxter McEnroy, University of Washington Press, 2982

CENTURY OF PEACE

When Prince Alexander died early in 2600, Zane Davion was young and poorly trained for his job as First Prince. Although secure on the throne of a peaceful and prosperous state, the new ruler would face difficulties no less troublesome than those of his predecessor.

Some commentators have noted that it was Alexander's success in putting down opposition that bred Zane's worst problems. Humanity was enjoying widespread peace for the first time since reaching deep space, and the outcome was recession and economic chaos through much of the League's farflung territories. The Federated Suns were not spared the full impact of this fiscal anarchy. Prince Zane needed a good-sized war to lift the Suns out of its depression, but wars were forbidden by the Articles of the Star League.

Ripples of discontent did lead to a few minor rebellions within Federated Suns space, the most serious of which occurred in 2617. It was set off by the arrival on New Syrtis of Josip Varnay, Roger's grandson and heir to the Varnay claim on the Capellan March. Here, Josip sent out a call that the people of the March reject membership in the Star League and embrace their former independence as the Capellan Principality. Although the magic was gone from the Varnay name, the people's frustration over their economic woes brought a large, if motley, following to Josip's standard. The Capellans did not support his cause, however, and so regular Federated Suns troops were able to make short work of the rebels. Josip escaped, but the Capellan government found it expedient to expel the family altogether.

After this, the Varnays had no hope of raising fresh efforts against House Davion. Josip died childless, as did his brother, the last of the "legitimate" Varnay heirs. In the following centuries, offshoots of the family would lay totally ridiculous claims to Federated Suns titles, but no one ever recognized any of these claims as legitimate. In an ironic twist of fate, one of these Varnay pretenders would receive a pension from the Federated Suns for leaking Kurita plans to the Prince in 2720.

To solve the economic problems of the peacetime economy, the Davion government channelled more and more capital into research and development programs, which gradually replaced warfare as the mainstay of interstellar relations. The development of hyperpulse generator technology during the first quarter of the 27th century was a significant boost to man's sprawling interstellar domain, cutting the expense and time of interstellar communications by a considerable margin.

Central to the economic recovery was the introduction of newer, cheaper methods of purifying water supplies and carrying out major terraforming efforts on worlds previously considered only marginally habitable. A massive effort to settle and explore these new worlds stimulated industry (the New Avalon shipbuilding industry tripled its profits between 2640 and 2650), promoted interstellar trade, and led to the reduction in the size of Household armies. The Century of Peace was a true renaissance of Human endeavor and achievement, with a flowering of science, technology, art, and a dozen other crucial areas of man's development. Little wonder that later generations would look back upon the Star League as the pinnacle of man's social progress, despite its chaotic beginning and disastrous end.

THE TROUBLE WITH TERRAFORMING

The marginally habitable world of Vandalia was a haven for zoological experiments in the 27th century. Many species of the planet's native wildlife seemed hearty enough to coexist with Humans, which was not always the case away from Terra. A case in point was the lizard-like creature dubbed the *Monitorens naturan* because of its resemblance to a Terran lizard. Analysis of the Vandalia Monitor's blood chemistry indicated compatibility with the Human nervous system and also proved to be similar to certain synthetic antipsychotic drugs developed by biochemists in the 2580s.

In 2616, the Davion government authorized full-scale terraforming and colonization. The desert world of Vandalia began to blossom, causing (as terraforming usually does) an ecological crisis among the wildlife adapted to the former environment. The Vandalia Monitors started to die out, with only laboratory specimens safe from the environmental changes to their former habitat.

It was in 2639 that researchers discovered the extraordinary properties of the Vandalia Monitor's blood chemistry. Preliminary findings indicated that a psychoactive chemical synthesized from the lizard's blood would either calm people prone to violence or completely prevent such impulses without impairing the individual's functioning. In other words, a person treated with an extract from the Monitor's blood remained clearheaded, able to work and play normally, free from the abnormal excitation of violent reactions as well as the fog-brained side effects of similar drugs. They were still capable of defending themselves if attacked, but felt no desire to initiate attacks. On every other level, their emotional responses were sensible and proportional. If administered to politicians, the drug would probably have prevented the warfare that would eventually ravage the Inner Sphere.

It was not to be. Although the chemical formula can be synthesized artificially, it is quite volatile, lasting only a few minutes to a few hours under laboratory conditions. Apparently, there is a still-unknown factor in the live lizard's own blood chemistry that acts as a stabilizing agent to prevent the psychoactive chemical from deteriorating when extracted from the blood.

Specimens of the Vandalia Monitor were lifted offworld when it became evident that terraforming was pushing the lizard toward extinction. They exist today on one of our Blessed Order's worlds 100 parsecs rimward of Lyran space, more than 730 light years from Terra. Attempts to adapt them to this new environment have failed because predators have prevented the Vandalia Monitors from finding an ecological niche. For nearly three generations, we have attempted to breed the lizards, but they do not take well to reproduction in captivity. If left to breed in their native environment on Vandalia, there would have been enough of the creatures to treat most of the violent criminals, politicians, and soldiers of the last four centuries, perhaps preventing hundreds of millions of deaths.

Over the last half-century, we have made periodic diplomatic initiatives to Inner Sphere governments to gain permission for breeding experiments on some desert world. All have declined. Though the House Leaders recognize the need for our Blessed Order's communications services, they seek always to minimize any other influence we may gain within their realms.

In the Periphery, the situation is different, but no more encouraging. There, the various states simply do not have the resources to spare the land areas required for this great work. Kyalla Centralla, Magistrix of Canopus, has said that if her expansion plans for the Canopian economy are successful, a world in her domain may become available for Project Monitor by 3075. By that time, however, the Canopian government will almost certainly have changed hands, and there is no guarantee that her successor will be as sympathetic to ComStar's secret experiments.

—From a report to the First Circuit of ComStar, 14 January 3024, ComStar Archives, Terra

Zane Davion and his successors steered the Federated Suns through the Century of Peace with firm, unwavering hands. The League gained 750 new colonized worlds during this period, and the Federated Suns was second only to the Capellans in colonization efforts. Some of these were new colonies on sparsely settled worlds where the Age of War and the Davion Civil War had taken their toll on the original settlements. Others, mostly in the distant Periphery, were entirely new.

Prince Alexander's line proved to be long-lived and prosperous. Zane's eldest child, Sarah Davion, succeeded him as First Prince when he died in 2659 at the age of 79. Her marriage to the common MechWarrior Mark Holt produced one son, James Holt-Davion. The succession passed to Samuel, Zane's second child, after Holt-Davion died in a fighter crash in 2663. Sarah herself confirmed Samuel as her heir after her Prince Consort died in 2678, though her devotion to Holt was so great that she continued to issue proclamations in both their names until her own death three years later at the age of 77.

Though Samuel Davion was 71 when he became first Prince, he would rule the Federated Suns for another 15 years, continuing the conservative policies that Zane and Sarah had established. His 58-year-old son Roger came to power when Samuel died in 2696, and he too maintained a policy of quiet efficiency. In the years of Zane, Sarah, Samuel, and Roger, age, wisdom, and competence ruled the Federated Suns. It is largely because of these four Princes that the era came to be known as the Century of Peace. Roger's death in 2703 signaled the succession of his 36-year-old son Joseph and an end to this great era.

BEGINNING OF THE END

No Davion ever backed down from a fight. I'm not going to be the first one to change that, not if the whole bloody Star League decides to crash the party.

—Prince Joseph Davion, on hearing of the violation of the Genoa Resolution, 2725

Joseph Davion was the youngest child of Roger Davion and Lois Chandler of Robinson. Mary Davion, their first child, had taken service in the Federated Suns Corps Diplomatique as part of her apprenticeship to the art of government. During an embassy to Luthien in 2696, Mary met and fell in love with the youngest son of Coordinator Urizen Kurita II. This was Soto Kurita, uncle of the infamous Minoru Kurita. The two were married in 2698, and Mary returned to Draconis space with her husband. She bore him three children, two of whom would distinguish themselves in the First Succession War.

Roger Davion realized that the tradition of primogeniture would one day lay Davion lands directly into the grasp of the Kuritas. Though the Davions had been willing to see the Holt-Davion line take power half a century earlier, it was only because the Davion family remained ascendant. The heirs of Mary Davion would surely be raised as Kuritas, with little regard for their mother's heritage.

In 2700, Prince Roger passed an Act of Succession that forbade Mary or her heirs from exercising any rights to the First Principality. Mary herself signed the Act in 2702, though neither Soto nor any other member of House Kurita did so.

The Kuritas made no effort to contest the succession of Mary's younger brother Joseph Davion in 2703. Though he was the youngest Prince to sit on the throne since Zane Davion more than a century before, Joseph was determined to rule with the same calm and steadfastness of his predecessors. Indeed, all was well until the death of Mary Davion-Kurita in 2715.

At that moment, Prince Joseph's son Richard was 19 years old and readying himself for the customary ceremony that would proclaim him as the Prince's heir. Just two months before the boy's 20th birthday, an embassy arrived from House Kurita with the outrageous demand that Mary's eldest son Vincent Kurita be named heir instead. In dismissing this startling claim, Joseph made the fatal error of underestimating Coordinator Takiro Kurita, Vincent's uncle. Davion believed that the Kuritas would drop their demands as soon as he showed his determination to refuse. Instead, Takiro marshalled his forces for an open confrontation.

Like most of his line, Takiro Kurita was a canny politician. Rather than attempt an act of aggression in flagrant defiance of the Star League, he made his opening salvos in the League Chambers on Terra. The Coordinator presented evidence to the other Council Lords that neither Mary nor any of the Kuritas had ever agreed to the Act of Succession of 2702. He argued that Vincent's right to inherit the Davion titles and holdings was unassailable, and he branded Joseph as a usurper. Takiro even produced a document issued in Mary's name that appointed Joseph Davion her Regent in the Federated Suns until the day her children were of age to take their rightful place. The Council had no proof until much later that this document was a forgery, and so were faced with having to decide which of the claims to the Federated Suns throne was most valid.

Until now, the League had based its power on mutual agreement and cooperation, with little dissent among the Council members on major decisions. Now the two most powerful members were at each other's throats, with neither side willing to back down or compromise. Lacking proof of either side's legitimacy, the League decided to appoint an investigatory commission, with warnings against breaking the peace. When Kurita troops attacked the Davion border world of Marduk in 2725, the League was slow to react. The clash soon escalated into a full-scale war, later known as the War of Davion Succession. This series of engagements was the first in the so-called Council Wars that would eventually bring down the Star League.

WAR OF DAVION SUCCESSION

Prince Joseph reacted with characteristic speed and recklessness to news of the attack on Marduk. Without considering all the implications, he ordered General Kessem, commander on the Draconis March frontier, to gather all available House troops for a counterstrike deep into Kurita territory. The Prince claimed that only an immediate offensive would blunt Kurita's invasion of Federated Suns territory. What he did not know was that Takiro had concentrated a large force along the Davion-Kurita border long before his actual invasion of Marduk. In retrospect, it is clear that the Combine had prepared well in advance for this action, while the Davions were hampered by a lack of supplies, transport, and other elements that a more carefully prepared campaign would have provided. An even greater obstacle was the Kuritans' numerical superiority. Though the Cameron Edict of 2650 limited the size of the Household Armies each member state could maintain, the Draconis Combine had been methodically and secretly violating that law.

General Kessem's troops reached as far as Ludwig, a fairly important Kurita administrative center, before running into significant resistance. Too late, they realized that the Kuritans were ignoring their counterattack and that the main body of the enemy force was pushing toward New Avalon. Kessem pressed on to Ludwig where he brilliantly reorganized the Davion light 'Mechs to finally break the Kurita resistance after a year-long campaign. This victory forced Kurita, whose resources were already overextended, to halt the advance on New Avalon.

Joseph's scratch Army of Crucis, which had been falling back helplessly before the Kuritans, now moved cautiously to reoccupy worlds. Meanwhile, faced with increasing pressure from the enemy, Kessem was also forced to abandon his advance. In 2729, Kessem and the Prince linked up in the Royal system, catching the main Kurita force off-guard and inflicting heavy casualties. In another rash decision, Joseph insisted on personally leading his forces in his *Marauder*. The Prince's 'Mech was disabled in the fight, and a malfunctioning escape mechanism trapped Joseph inside as the Kuritans moved in for the kill. With the death of the Prince, the Davion troops fled, and the Battle of Royal became a costly defeat for the Federated Suns.

Fortunately for House Davion, the Star League finally decided to take action. Elements of the League Regular Army—including an impressionable young MechWarrior Captain named Aleksandr Kerensky—arrived off Royal to impose a cease-fire between the two states. The League Council censured both sides for resolving the dispute by force, but House Kurita received the brunt of the criticism. All troops were ordered withdrawn to behind pre-war frontiers, and the Kurita claim to the Davion throne was refused once and for all. Because of the League's intervention, the War of Davion Succession ended in victory for House Davion, but not through any merit of the Prince or his troops. With the exception of Kessem, the Davion army leadership was outmoded in its strategy and tactics.

This war that had taken his father's life inspired a fierce determination in the heart of Richard Davion, the new First Prince, to prevent his House from ever being so vulnerable again. Battle had also claimed the life of his younger brother Leonard, who fell during the aftermath of Kessem's victory at Ludwig. The Star League's initial unwillingness to step in convinced Prince Richard that House Davion and the Federated Suns had to look out for themselves. Though he was careful not to violate Star League policies, the new First Prince set out to make his vast domain as self-sufficient as possible. Other member states obviously learned this lesson as well, for between 2730 and 2760, each of the House leaders began to increase his military preparedness and independence.

Although the Cameron Edicts limited the size of each Household military, the law specified restrictions on manpower rather than on actual materiel. This provided Richard Davion with the loophole he needed to build up his military might. In 2735, he saw to the passage of the Preparedness Act, which called for creation of a militia system similar to the Prussian *landwehr* of the 19th century. The Federated Suns House Army never exceeded the limits that Cameron's edict had imposed, but compulsory Davion military service meant that a large body of men and women would receive training. Meanwhile, arms and 'Mechs were stockpiled throughout Suns territory to permit a quick expansion of the army in time of crisis.

The break-up of the Star League was becoming inevitable as other incidents between the Council Lords put a strain on the whole clumsy fabric of the League. In 2743, Richard's son and heir Joseph was killed in a training exercise when his poorly maintained 'Mech exploded from improper heat build-up. Joseph, a well-known libertine, left behind an illegitimate son born to a young entertainer named Susan Rand. Owing to his illegitimacy, Mark Rand-Davion was excluded from the succession, but his grandfather recognized him as a member of the family and granted him large estates on Abbeville. When Richard died in 2745, the succession passed to Leonard Davion's son John, a young but promising man who would become another of House Davion's truly great leaders.

LORD JOHN

When named as the new Heir-Designate in 2743, John Davion had showed neither delight nor concern. He merely set out to prepare himself for the job—if it ever fell to him—as quickly and as efficiently as he could. Within two years, Prince Richard was dead, and John became First Prince of the Federated Suns and a Council Lord of the Star League. Though his demeanor was unassuming, underneath was that streak of Davion will. John's opponents consistently underestimated his strength of character and resolve.

When First Lord Simon Cameron died in 2751, he left his young son Richard Cameron, a child of eight, as his heir. In the face of the strained, even hostile, relations between the other members of the Star League, the League might have collapsed but for John Davion's diplomacy. Because of his eloquent arguments, the Council Lords eventually agreed to appoint General Aleksandr Kerensky, now Senior Commander of the League Regular Army, as Regent and Protector until the young First Lord reached his majority at age 18. Kerensky, who was neither ambitious nor jealous of the other Lords, should have been the ideal man for the job. As it turned out, however, military responsibilities often interfered with his political duties, and so it was that Lord John emerged as one of the leading figures of the Protectorate years.

John Davion was a complicated man. Fair-minded and tactful he might be, but above all else, he was a Davion who placed the good of his people and his House before any other considerations. As long as the Star League seemed worth preserving, John was its most stalwart supporter, but he would not be tied to that institution once its existence ceased to benefit the Federated Suns. Indeed, John Davion was as capable of intrigue as any Council Lord of the day, including the formidable Minoru Kurita. He was the man General Kerensky dubbed "the best and noblest of all the Lords" in 2764. Twenty years later, the General accused this same John of being "the traitor who destroyed all that I've worked for."

During the Regency years, Prince John was a strong supporter of Kerenky's initiatives as Protector and Regent. Yet, he did not hesitate to vote for the amendment that would reverse the Cameron Edict limiting Household troop strengths. Like Richard Davion before him, John's priority was to build Davion military strength so that his people need not fear another crisis like that of 2725–2730. When the Star League Council wished to generate revenues by raising taxes on the distant Territorial States, only John Davion and Ewan Marik opposed the measure. John understood how much resentment this tax would stir up among the Periphery worlds, but he accepted the will of the majority Council vote. Many historians believe that it was from this time on that Prince John began to believe that the Star League had outlived its usefulness.

A further problem facing the Regent was the outbreak of an inter-League territorial dispute between the Capellan Confederation and the Federated Suns. Not since the founding of the League nearly 200 years before had there been trouble along this border, but now House Liao seemed determined to harass Federation mercantile interests along the frontier.

After several incidents—some provoked by Liao, others by Davion—the Prince felt obliged to muster a sizeable draft of Household troops for a strike into Capellan space. In a bitter tripartite campaign against Tsamma, Wei, and Redfield, the Davion forces made little progress, though ultimately the Avalon Hussars broke the Liao garrison on Redfield and won that planet for the sunburst banner. This Border War of 2760–2762 sputtered to a halt more because of external political developments than because of either side's military triumphs. By 2762, the two enemy powers found they had a common interest in protecting the rights of the League member-states against the First Lord himself, who now threatened to abridge their closely guarded freedom.

EXECUTIVE ORDER 156

When Richard Cameron reached his majority in 2762, General Kerensky willingly stepped down as Regent and Protector. Having grown up in an atmosphere of political confusion and arrogance on the part of the other Council Lords, Richard took office determined to bring these unruly Lords to heel. The lessons of the War of Davion Succession and the years of secret rearming and maneuvering had not been lost on Cameron. He was still naive enough to issue Executive Order 156, however, a demand that each Council Lord completely disband his House Army. The ink was barely dry on the document before the other five Lords forced Cameron to rescind it. The young First Lord had moved too quickly and without regard for the complicated nature of Star League politics. It was through such ineptness that he lost the respect of his fellow Lords, and sowed the seeds for an even more grievous crisis to follow.

John Davion gave an eloquent speech in the debates on Executive Order 156. He admitted that an ideal League would eliminate the House Armies that could create such internal disputes as the Kurita attack on his own realm 40 years earlier. In an ideal universe, all six Houses would be united under the banner of the Star League Regular Army. He went on to say, however, that the Star League was a voluntary association of sovereign states created to promote cooperation and discourage needless conflicts. The Star League had specifically left domestic policy—including defense—in the hands of the member states, and it had no right to encroach on the sovereign rights of any member now. Davion further maintained that each state in the League needed its own armed forces for protection in case a tyrant ascended to the throne of First Lord.

Davion's was the most moderate anti-disarmament speech. The only man who spoke in favor of the measure was Stefan Amaris, ruler of the Rim Worlds Republic. History would show, of course, that his motives were purely self-seeking. When the order was rescinded, Richard Cameron turned solely to Amaris for counsel and support. No doubt it was Amaris who persuaded Richard Cameron to refuse to call a Council meeting for two years.

Meanwhile, the Periphery was growing more restive than ever. Kerensky and over 75 percent of the Regular Army were already committed to holding the Territorial States in check, and now the General needed additional troops to keep the rebelliousness from spreading like wildfire from one world to another. In 2765, when New Vandenberg and 17 other Periphery worlds revolted against Star League rule, Cameron called for the other Council Lords to contribute Household troops to support Kerensky. Again, it was only Amaris who supported the young First Lord by answering the call, but then only by pledging troops to relieve Cameron's House units garrisoning Terra and other Terran Hegemony planets. Unfortunately, the First Lord was too much under Amaris's spell to recognize the true motive behind Stefan's offer. By the end of 2767, Amaris had seized control of Terra and executed the First Lord and his entire family.

Davion tradition has it that prior to the execution of Richard Cameron, the young First Lord had named John Davion as Regent for his two-year-old daughter Amanda. Certainly, House Davion would later use the tradition as the key to its claim as rightful heir to the post of First Lord, while none of the other Houses could advance anything so significant in their own behalf. Indeed, John Davion would go down in history as "The Regent," with the story of Cameron's request receiving confirmation from several sources.

DISSOLUTION

The Amaris Coup led to full-scale civil war in the Inner Sphere. Amaris occupied his own Rim Worlds and the planets of the Terran Hegemony, while arrayed against him were Kerensky and the Star League Army. The war was slow to gather force, with five years elapsing between the coup and Kerensky's first offensive into Terran space. In that time, the Council Lords adamantly refused to enter the fray on either side. Though Kerensky fought in the name of the Star League, it was as though that body no longer existed. When Kerensky went to his former ally for assistance, John Davion replied that he must marshal his own Household forces instead, to guard against outside interference in Federated Suns affairs. When Amaris asked for support, Davion said only, "I do not recognize your right to the throne."

For seven years (2772-2779), General Kerensky advanced slowly but inexorably toward Terra, his veteran Regulars more than a match for Amaris's Household troops. Ultimately, the Usurper was taken and executed. Kerensky proclaimed himself Protector of the Terran Hegemony, then requested the Council Lords to assemble on Terra to choose a new First Lord.

Kerensky had not reckoned with the ambitions of the Council Lords, however. The only thing they could agree upon was to strip Kerensky of his duties as Protector and to appoint the Blessed Jerome Blake as Minister of Communications. It was also at this time that they began to wrangle over which one should assume the mantle of First Lord of the Star League. The clash of personalities made any further agreements impossible among them and the spectre of war inevitable.

Though no longer Protector, Kerensky was still General of the Star League Regular Army and a figure of considerable influence. While he argued for unity, the five Council Lords were already drawing up the battle lines. Each of the five Lords was determined that he would be the new leader of the Star League, and Kerensky's pleas for moderation fell on deaf ears. With Kerensky's support, Davion's claim to be Cameron's chosen Regent—and, in the absence of an heir, his preferred successor—might have carried the day. Kerensky, however, refused to consider Davion's cause unless Prince John embraced the General's plea for a return to a pre-Amaris Star League.

HEIR TO THE FIRST LORD

Jinjiro, my son,

I salute you as Coordinator-to-be!

The days are dark for our noble empire. It is a bitter irony that this Star League, which has stood in the way of our magnificent destiny for so long, is now precisely what we need to advance our interests—and yet it is now breaking down!

As mighty as our navies are, they cannot singlehandedly conquer the other five states. Thus, we have bided our time for generations, accepting that events might delay our destiny but never deny it. This breakdown of the Star League is worse than a mere delay, however. It may create a situation where our people might become the object of a combined attack by the other powers of the Inner Sphere.

There are no plans for succession should anything happen to the First Lord's family, though I doubt that any Star League member would be reckless enough to try to eliminate the Camerons. Yet I must tell you what that fool of a First Lord has had the gall to confide in me. He stated his wish that John Davion take the place of First Lord if anything were to happen to him. As if a deposed ruler's contemptible wishes could be binding on a leader secure in his people's fealty! Luckily, there was no indication that Cameron had spoken yet to Davion, nor did he mention any immediate plans to do so.

All of this indicates, my son, that war may be on the horizon. I know that your heart leaps at the prospect, yet I am concerned. Are you strong enough to take command? Reports from the palace tell me that you still spend too much time brooding upon the fate of your poor mother. Strengthen yourself! Just as the leader of the 47 ronin tended his garden, I have studied music and religion, but when the time comes to strike, I shall not fail to do so. It is your destiny as a Kurita to be strong enough to do the same.

Your father,

Kurita Minoru

[EDITOR'S NOTE: This letter passed through ComStar on 4 June 2774. The first line is a standard salutation that Minoru Kurita used in all of his letters to Jinjiro from the time the latter was thirteen years old. In the official Draconis publication of the collected letters, the references to John Davion in this letter are expurgated.]

—From the Collected Letters of Minoru Kurita, ComStar Archives, 2864.

After months of fruitless argument, the Council Lords went home and began to muster their forces for the final confrontation. Once again, Davion called on Kerensky to join him. Once again, Kerensky pleaded that Davion set aside his claim to power and help in working out a compromise. Neither man considered the other's plan acceptable. Kerensky, weary of fighting the Council Lords, decided to withdraw from the Inner Sphere once and for all.

With the bulk of the Regular Army still loyal to him, Kerensky and his men jumped out of known space in 2784. Almost oblivious to the General's last, despairing gesture, the Council Lords were busy girding up for battle. Even as Kerensky gathered his troops, transports, and warships on the Kurita world of New Samarkand, Kurita troops were pushing across the border into the Federated Suns. The First Succession War had begun.

AMARIS'S GRAVE

Stefan Amaris's death ended one problem and created another for Kerensky. The problem was that if he allowed Amaris to be buried in a marked grave, it would encourage his former followers to make pilgrimages to the gravesite to renew their devotion to his cause. If the grave were unmarked, on the other hand, it might create a myth that Amaris was still "out there somewhere," and might someday be coming back. Kerensky and his staff did not want to encourage that most perverse trait of human nature, fascination with the deeds of a great monster.

An anonymous soldier in Kerensky's command made the suggestion that the General finally adopted. Kerensky ordered a detachment of his troops to deliver Amaris's remains to the medical school of the University of New Samarkand, where they served undisclosed medical purposes.

The remains were then cryogenically preserved until 2863, when it became too costly to maintain cryogenic chambers. Amaris's remains were ultimately laid to rest in a university cemetery where Kuritan physicians bury the common criminals on whom they sometimes experiment.

—From *The End of Innocence: How the Star League Fell,* by Seth Kim and Julie Howard-Nguyen, Commonwealth Historical Press, 2958

AMARIS'S LEGACY

The death of Stefan Amaris marked the beginning of the end for both the Rim Worlds Republic and most of the hard-won influence the Periphery powers had managed to garner during the Star League era. Descendants of the Rim Worlders have compared Amaris to the Terran dictator Hitler. The comparison is faulty, however, as there were many great Germans of pre-Exodus Terra, but no one else from the Rim Worlds ever gained eminence comparable to a Beethoven, a Goethe, or a Von Braun, for example.

As a citizen of the Outworlds Alliance and a descendant of the Rim Worlders, I had hoped that in the course of our research on this volume I might find some mitigating factor to modify history's judgement of my most famous countryman. I regret to say that I have found none. As far as Amaris is concerned, the ancient slogan applies: Ig fallou blaos, dem ressensu glottuo. ("What you see is what you get," or, literally, "If you catch this, you will have to eat it.") Amaris may have been worse than even past historians have claimed.

Stefan Amaris was not a shrewd politician of the Maximilian Liao variety. In a letter to his wife written during the Civil War, he likened his quest to be dictator of the Human Sphere to a child's game called "King of the Hill". I find it significant that he chose a physical game, rather than a strategy game, for his analogy. While great gamers rarely make great leaders, any leader must understand the principles of strategy for his chosen field, be it political, military, or economic.

It is with some embarrassment that I come to the subject of my own ancestry. When Kerensky's troops slaughtered everyone in the Rim Worlds that bore the name Amaris, they overlooked maternal cousins with names like Siever, Wong, Chan, and Marcus.

With the danger of being even distantly related to Amaris, people with those names fled the Rim Worlds. My own family settled in the Outworlds Alliance. I honestly do not know if I am a descendant of Stefan the Usurper. Genealogical records that might have cleared up my ancestry were destroyed in the aftermath of the war. It is a reflection of my parents' macabre sense of humor that they named me Anastasia, after the Terran legend of Anastasia Romanov, daughter of the last Russian Czar.

Our research has resolved one matter, however. *No* conclusive link can be found between Amaris and anyone still living. With such total lack of pedigree, anyone claiming relationship to Amaris, let alone claiming leadership to any of the Rim Worlds (now the Bandit Kingdoms) on the basis of such a claim, can be assumed to be a pretender or a usurper. We can hope that the shame that my putative kinsman brought to the Periphery will never be revived, and that the carnage he brought to the Inner Sphere will never be forgotten.

—Anastasia Marcus, Historical Director, Davion Research Project, ComStar Research Archives, Terra, 3028

THE FIRST SUCCESSION WAR

If we should ever want to come back to the Inner Sphere, we will need only search for a cluster of stars burning brighter than the rest, their illumination strengthened by a thousand fires burning on a thousand planets. We'll know that is where we left the five Houses to fight themselves into extinction.

—Captain Agincort Malloy, quoted in *Reflections on the Exodus*, by Precentor Jonathan Degrassi, 2801

The departure of Kerensky and three-quarters of the Star League Regular Army in 2784 removed the final restraint binding the Council Lords to any facade of peace. Even though the Regular Army was lacking a First Lord and a central base of supply or political authority, the Star League forces had been a formidable threat to be wooed rather than opposed. With Kerensky gone, the remaining Regulars—many with valuable combat experience from the Periphery campaigns and the Amaris Coup—drifted into the various armed camps that each Lord was building to assert his claim to rulership of the League.

A DARK AGE

With the fall of the Star League, the worlds of the Inner Sphere moved into a new era, though no discontinuity was apparent to those who lived through the transition. Even today there are many who claim that the League will yet rise like a phoenix from the ashes of the Succession Wars. This is symptomatic of the Era of the Warlords.

Beginning with the Amaris Coup and stretching through the two and a half centuries since, history has become little more than a seemingly endless sequence of battles, alliances, and famous dates. In the process, however, scientific, economic, and social progress have begun to wither away. Mankind has entered an age of stagnation. Greed and ambition have so blinded the Successor Warlords that they have failed to recognize, let alone correct, humanity's slow but steady decline into darkness.

—From *Humanity at the Brink,* by Adept Armand Sieyes, ComStar General Bulletion No. 18654, Archives Publishing, Terra 3016

THE TOWNE DEBACLE

As the last of the Kerensky's transports jumped out toward the Periphery, the Federated Suns was in an extremely weak position. Because Prince John wished to behave like a proper successor to the Camerons, he had forbade his military to do anything "unseemly." For example, while the other House Lords were openly plundering the Terran state of wealth and worlds, the AFFS had to buy or secretly steal anything valuable in order not to embarrass Prince John. As a result, the Federated Suns acquired only a fraction of the supplies and military units that the other Houses gained.

This might not have been so crucial if the AFFS had been as strong as most people imagined. On paper, the Davion army was larger than any other House army, but it had several serious flaws that almost proved fatal. One of these was the resurgence of regional loyalties. During the quiet Star League era, it had become fashionable for the government to cut the AFFS budget, particularly funds for transporting units around the realm. As a result, troops became permanently assigned to a certain region for years, and began to feel more loyalty to their new homeworlds than to the Federated Suns and the Davion family. The Field Marshals of Combat Regions found their troops boastfully proud of their home region and deeply suspicious of any attempts to take them away from it. With location so predominant, cooperation between Combat Regions was poor. This, in turn, created a brittle and fragmented defense line.

Another flaw was the rampant inter-service rivalry. What had once been good-natured competition between MechWarrior, fighter pilot, and grunt had long since degenerated into bitter feuding. Active hostility between the services was common, and battlefield cooperation between them almost nonexistent.

The Towne Debacle of 2785 most clearly reveals the situation in the AFFS at that time. The Federated Suns had claimed the planet Towne, once a member of the Terran Hegemony, in 2783. While Prince John was trying to behave like a statesmanly successor to the throne of First Lord, Towne was one of the few Terran worlds that he considered valuable enough to claim openly. Not only did the planet contain many warehouses of technical machinery, but it was also one of the Federated Suns's few pathways to Terra. Because of Towne's importance, Prince John sent elements of the 56th Avalon Hussars and the 123rd Aero Interceptor Wing to garrison the planet.

Early in 2785, raiding parties from both the Draconis Combine and the Capellan Confederation made a bid to capture Towne's warehouses and industry. The planet's defenders could easily have turned back the small raiding parties, but rivalry between the Hussars and the Aerofighter Wing hampered their efforts to defend the planet. Major Wilkins of the Hussars was convinced that they must deal first with the Combine raiders, while Major Donner of the Aero Wing wanted to tackle the Capellan raiders first. Before retreating, both the Kurita and the Liao raiders were able to loot several poorly defended warehouses of all their valuables.

Seeing this weakness in the Federated Suns and anxious to exploit it, the Draconis Combine mounted a major offensive against Towne late in 2785. The Fifth Dieron Regulars, supported by seven regiments of infantry and AeroSpace Fighters, attacked the planet, forcing the Davion soldiers on Towne to beg for reinforcements. Local loyalties and a complex bureaucracy slowed the relief effort, however, and by the time AFFS JumpShips entered the Towne system, the soldiers of the 56th Hussars and the 123rd Wing were either dead or at the mercy of their conquerors. Towne was now the property of the Combine.

The slight slide in the Davion economy that had accompained the collapse of the star League now turned into an all-out crash late in 2786. Those who believed that the days of wine and roses would never end would now pay for their unwillingness to revise economic policies to meet the changing times.

KURITA OR LIAO?

Considering all that we've done in recent years, I'm not surprised that we are at war. It's a fitting punishment.
—From *The Private Journals of Prince John Davion*, NAIS Press, New Avalon, 3011

The Towne Debacle made even Prince John realize just how ineffective the AFFS had become. The idea of inheriting the title of First Lord had apparently blinded him to the needs of his realm. He was not a hard-headed Davion for nothing, however, and so the Prince gathered together his Field Marshals and girded himself for the truth. What he found was so distressing that there may be some truth to the stories that he considered sending delegates to the neighboring realms to sue for peace. What happened, however, was that Prince John announced major reforms and defensive preparations.

To stop the decay eating away at the AFFS, Prince John authorized a crash program of militarization. The government seized almost all of the Federated Suns major industries for the production of military equipment, and most other resources became the property of the AFFS. One welcome effect of this drastic policy was that it lowered the realm's high unemployment rate, easing the economic doldrums. Within six months of this industrial mobilization, the manufacture of JumpShips, 'Mechs, AeroSpace Fighters, and tanks was steadily increasing.

Prince John knew that he needed more than this to stop an opportunistic enemy, and searched about for a way to stall at least one of his "neighbors" from attacking. He ordered his Marshals to devise an offensive to spoil the plans of the realm most likely to attack the Federated Suns. In one of the great blunders of the Succession Wars, Davion's Field Marshals reported back that, between House Liao and House Kurita, the Capellan Confederation posed the most dangerous threat.

While the Davion forces were preparing their spoiling offensive against the Capellans, the Draconis Combine Mustered Soldiery was preparing its own surprise offensive to cripple the Federated Suns. Having declared himself First Lord in December 2786, Coordinator Minoru Kurita believed that the best way to assert his claim was to match his military against the only other army worthy of the effort. A decisive victory against the Federated Suns would give him virtually unlimited influence over Houses Marik, Liao, and Steiner, and thus guarantee him the First Lordship.

To design and lead the offensive, Minoru Kurita chose his unpredictable, if not insane, son, Jinjiro Kurita. Having studied the Towne Debacle closely, Jinjiro realized that the AFFS was nothing but a paper tiger. Plotting to take advantage of every flaw, every error in the Federated Suns military, he devised the most brilliant offensive that the Inner Sphere has ever seen.

When Kerensky and the Regular Army began to rendezvous around the Combine world of New Samarkand, Minoru Kurita had flooded the district with units from other regions, fearing an attack by General Kerensky's troops. Nor did the departure of the Regular Army for parts unknown allay his fears. For two years after the Exodus, the Galedon District was still on alert and busy with troop traffic moving through the region.

Federated Suns agents had grown accustomed to the activity in the district, which allowed Jinjiro to use the traffic to disguise the concentration of his troops along the Davion front. He pulled troops from all over the realm, sent them under assumed unit names through the Galedon Military District, and then secretly deployed them in positions along the border. Once in their assigned positions, these Combine units were under the strictest orders to avoid confronting AFFS units. Combine officers were ordered to lose battles rather than to tip off the Federated Suns to their strength.

A few officers within the AFFS suspected that House Kurita was up to something. Marshal E. Dryer of the 13th Avalon Hussars RCT had been warning everyone that the DCMS was planning to attack. So strident was his call to arms that he was court-martialed and demoted three ranks for physically assaulting a Field Marshal who disagreed with him. (Colonel Dryer was among the first killed in the offensive.) Though others agreed with Dryer's ill tidings, the inter-service rivalry permeating the Davion ranks kept the military strategists from getting all the facts on a possible Combine offensive.

Instead, Prince John and his Field Marshals prepared an offensive against the Capellans, massing whatever AFFS troops they could pry from their home regions. The targets of the Davion attack were the planets Mira, Mesartim, and Tikonov, which Davion agents had discovered to be major staging areas for House Liao. Successful attacks against the supplies on those planets would delay the enemy offensive for months. By the end of April, the AFFS had over 50 regiments gathered on the Federated Suns worlds of Tawas, Farwell, and Amiga.

FRONT-LINE WORRIES
5 March 2787
From: Marshal Janos Dedirth, Commander MI-3 (Combine Sector)
To: Field Marshal Timons Davion, Commander MI-2

I cannot urge strongly enough that you review the report on Draconis Combine Mustered Soldiery communiques coming in and out of the Galedon Military District. The incredible volume of communiques and the presence of numerous regiments in that region makes me quite nervous. Though the DCMS may still be in a high state of activity due to General Kerensky's rendezvous at New Samarkand in the Galedon Military District prior to the Exodus, it doesn't explain everything I've been seeing. Please give me your opinion of the report.

6 March 2787
From: Field Marshal Timons Davion, Commander MI-2
To: Marshal Janos Dedrith, Commander MI-3 (Combine Sector)

I have read your report and conclude that you are either suffering from a drug hallucination or are bucking for a promotion. If the Dragon was planning an offensive, wouldn't our operatives in the Combine have alerted us by now? If the Dragon was up to something, then why have we won the last twelve skirmishes with them? Try to calm down. It's paranoia that makes you fear House Kurita so much.

—From ComStar Research Archives, Terra, Federated Suns Folders, 26A

THE STORM BREAKS

May 1, 2787 saw the entire Kurita-Davion border erupt with fighting. Kurita units attacked every Federated Suns world on the border, intent on destroying the planets' ability to communicate with other worlds. Most of the attacks were so unexpected that many worlds went silent without sending a single distress message. Receiving only a few calls for assistance, the AFFS severely underestimated the strength of the attack. To the Field Marshals commanding the Combat Regions along the border, the sudden Kurita offensive seemed worrisome but manageable. Indeed, they made no effort to coordinate their defenses.

Three days later, Jinjiro unleashed the second phase of his offensive. Instead of launching new attacks spread out along the front, he sent a wave of Kurita regiments against the Clovis Combat Region. The regiments in this second wave pushed deep into the Region, striking at Cartago, Olancha, and Clovis. The attack on Clovis, headquarters of all AFFS forces in the region, was most crucial to Jinjiro's plans. If Clovis fell, any defense of the Region would crumble.

Field Marshal Simons, commander of the Clovis Combat Region, had guessed at Kurita's intentions and was now trying to improve the position of his headquarters when the Kurita invasion arrived. The Marshal's unit, the First Clovis Guards Regiment, along with three armored units, fought one of the most inspired defenses ever known. Their mission was to protect Gorst City, the capital of Clovis, which lay in a mountain-encircled valley. The Kurita forces attacked with five times the defender's strength, yet by the end of the first week, they could claim only a small foothold in the valley.

Throughout the fight, Marshal Simons called desperately for reinforcements from Field Marshals throughout the Draconis March. Unfortunately, he received only token support, as his neighbors were too worried about their own Combat Regions to realize how vital it was that the Clovis Combat Region remain in Davion hands.

By the end of the second week, with the Kurita troops drawing the noose tighter around Gorst City, Field Marshal Simons called for a retreat and attempted to leave the planet. As his DropShip lifted away from the planet's surface, Kurita AeroSpace Fighters attacked and destroyed the vessel. The defense of the Clovis Combat Region collapsed immediately afterward. The Arm of the Dragon now had a clear path into very the heart of the Federated Suns.

BARLOW'S FOLLY

Perched on the Davion-Kurita border, the planet Cussar had seen more than its share of raids by military forces sporting the Dragon symbol. People would jokingly put a plate at the dinner table as well as a rifle next to the front door for any Kurita soldier that might show up. Stationed on the planet was a battalion from the 38th Avalon Hussars, commanded by Colonel Michael Barlow. A native of the planet, Barlow knew its mountainous terrain well. He loved his world and believed that only House Davion would treat it with the same respect.

When the Kurita offensive began, Cussar was one of its first targets. Fight as they might, the defenders slowly lost control of the planet while the Draconis forces destroyed every building, shot every farm animal, and burned all the crops.

After receiving news that Field Marshal Simons, Commander of the Combat Region, had been killed on Clovis, Colonel Barlow's superior gave the ROW (Retreat Off-World) order. Colonel Barlow refused to abandon his homeworld. Instead, he tried to convince his comrades to remain, too. His commanding officer said this would be folly, and so all the AFFS DropShips left Cussar. Nearly every man still alive from Barlow's battalion stayed behind, however.

After the last AFFS DropShip had gone, Barlow led his battalion into the mountains circling the planet's three major cities. For the next three years, his 'Mechs attacked and harassed the Kurita forces, and Barlow's forces managed to scavenge enough parts and supplies to keep their machines up and attacking. When their 'Mechs could no longer stand, Barlow and his men took up rifles and fought on for another ten years.

After 14 years of struggle in which Barlow's men had killed countless Kurita soldiers and destroyed tons of enemy equipment and supplies, the Kurita garrison on Cussar killed the last of the guerrilla forces. To their dismay, however, the Kuritans found that the bravery of Barlow's battalion had renewed the spirit of rebellion among the citizens of Cussar. Numerous new guerrilla actions were conducted in the name of Colonel Barlow. It had been difficult enough to battle a clever enemy hidden within mountains. Now the Kuritans faced enemies hidden in the midst of every civilian crowd.

To this day, Colonel Barlow is remembered on Cussar. As the civilian population is still rebellious, three times the normal number of Draconis troops are stationed there to prevent all-out revolt. Soldiers dread being assigned to this planet because of the constant terrorism aimed at them.

Even the planet's name has changed. As the tales of Colonel Barlow's bravery spread throughout the Inner Sphere, people began to call Cussar "Barlow's Folly." The Draconis Combine itself has bowed to convention and now identifies Cussar by its new name.

—From *Heroes of the First Succession War*, by General T.J. Barker, Syrtis Military Press, New Syrtis, 3002

RETREAT

I'd sell my grandmother to be able to say that our situation wasn't critical. Trouble is, she died fighting those bastards two hours ago.

—From official AFFS transcripts, dated May 2787, AFFS Archives, Avalon City

In the third week of the offensive, Prince John, still near the Capellan border, ordered a complete and orderly withdrawal of AFFS forces from the Draconis border region. Many soldiers refused out of loyalty to their homeworlds. Others fled, leaving many tons of valuable machinery behind.

This disorganized retreat left huge gaps in the Federated Suns defensive line. Quickly discovering and exploiting these gaps, House Kurita found its invasion more than a week ahead of schedule. One of the main reasons for the swiftness of their advance was the Kuritans' willingness to bypass pockets of heavy AFFS resistance, which they left for rear-echelon units to destroy at leisure. Most pockets of resistance perished either by the sheer force of the enemy's numbers or when they ran out of supplies and food. Only a few AFFS units stationed along the border, such as McKinnon's Raiders, the Franklin Gorgons, and the Doughboys, managed to survive.

By 2792, the AFFS found itself defending Imbrial III, Kestrel, and Saunemin, each about two jumps from the capital world of New Avalon. By this time, Prince John had launched two counter-attacks. The first occurred in 2788, when John threw troops intended for the Capellan offensive against the Combine army. By striking the flank of the Combine offensive, he hoped to slow its advance. When the attempt failed at Cartago, Prince John was fortunate to escape from that world with his life.

John attempted his second counteroffensive in 2790. Gathering up most of the warships left in his navy along with virtual clouds of AeroSpace Fighters and DropShips, he assembled a taskforce to attack the Kurita invaders head-on. Their objective was not to win back worlds, but to destroy enough naval forces to force the Kuritans to slow their attacks.

This offensive had better luck than the first. Commanded by Admiral Kenneth Jones aboard the *FSS Golden Lion,* the taskforce hit over ten Kurita-occupied star systems, starting with Imbrial and ending with New Valencia, and destroyed over 20 enemy JumpShips and an equal number of DropShips. The AFFS vessels even managed to resupply the resistance on some worlds, and its AeroSpace Fighters often had time to strike at ground targets before being forced to move on.

The end for the AFFS taskforce came as they were about to head out of Kurita-held territory. Early in the war, the Federated Suns had abandoned the small one-world system of Cholame, and the Draconis Combine had reportedly not garrisoned it. Admiral Jones thought Cholame's single planet would be the perfect spot for a rest before his force made their jump to Rosamond and safety.

As the taskforce entered the system, Kurita AeroSpace Fighters immediately attacked many of the Davion JumpShips. A group of Kurita warships followed close on the heels of the fighters. The naval engagement lasted for six weeks, as Admiral Jones darted between the system jump points, hoping to buy enough time for at least a few of his ships to jump to safety. More than 100 AFFS warships, from small corvettes to the *Golden Lion* itself, had been destroyed, along with an equal number of DropShips and AeroSpace Fighters. That Admiral Jones took with him an equal number of Kurita warships was small comfort to the people of the Federated Suns.

The Capellan Confederation was not idle during battles between the Federated Suns and the Draconis Combine. Realizing that they now had little to fear from House Davion, the Capellans acquired a few Federated Suns worlds near Terra, including Addicks, Ankaa, and Small World. Even worse for the Suns was the Capellan takeovers of Salem, Victoria, Carmacks, and Sekulmun between the years of 2796–7.

Now the Federated Suns had no path to Terra. Though there was little real importance to this fact, many in the Federated Suns felt lost without a link to mankind's ancestral home.

THE PURGE

The ugliest effects of House Kurita's invasion into the Federated Suns were psychological. Among the Davion citizenry, there developed an unreasoning greed, secrecy, suspicion, and a widespread compulsion to steal supplies. A particularly wicked case in point occurred on Strawn.

In 2794, a fifth-rate petty bureaucrat, who had somehow managed to become Prime Minister of Strawn, issued an order that stripped all members of the Oriental race of their property and belongings and relocated them to camps away from sensitive government facilities.

This law struck a responsive chord among the people now so desperately afraid of the Kurita Dragon. Suddenly, anyone of Oriental lineage was a potential agent of the Draconis Combine. The Capellans were equally feared since they had cut off the Suns's bond with Terra. In the resulting hysteria, many unfortunates were accused of being an agent of both realms.

Paranoia spread like wildfire. Homes were burned, businesses bombed, and even a pasta factory was shut down because its noodles were declared "oriental". For those citizens of the Federated Suns who happened to be of Oriental ancestry, this was far from a laughing matter. Many were dragged from their homes and killed, while others watched as everything they had worked for was taken from them.

There were mixed reactions from the various planetary governments and nobility. Most were outraged by the citizens' actions and tried to stop the madness. Others took advantage of the situation, profiting from their Oriental citizens' misfortunes. A few governments and nobles actually believed in the Purge, and the history of these worlds will forever be stained by innocent blood.

After two years, the problem had so intensified that even the preoccupied Davion government could no longer ignore it. In January 2796, Prince John issued a proclamation that any form of racism was contrary to the right of freedom granted every citizen of the realm. Later, Pope Clement XX of the New Avalon Catholic Church issued a similar decree, claiming that racism could not coexist with the love of God and was therefore punishable by excommunication.

Though these actions served to calm the mania that had overtaken so many ordinary men and women, nothing could make up for the lost lives and property. To this day, there are very few willing to discuss the Purge in the Federated Suns.

—From *Atrocities Away From the Battlefield*, by Father Ryan Ramon, Unfinished Book Press, New Avalon, 3001

KENTARES IV

Remember Kentares!

—Popular rallying cry, circa 2797, quoted from *Culture Under Stress: The First Succession War*, by Duchess Greta Dewers, 3011

By 2796, the Kurita offensive had pushed its way to Saginaw, Delavan, and Odell—all worlds within one jump of New Avalon. Though heavy AFFS forces were still contesting ownership of these planets, there was little to stop the DCMS from staging an invasion of New Avalon. On the capital world, efforts to evacuate the government and civilians were already underway.

Just when victory seemed inevitable, Coordinator Minoru Kurita called a three-month halt to the offensive over strong objections from his son, Jinjiro. The Coordinator had several sound reasons for his decision, however. First, the Draconians had bypassed a large number of AFFS troops in their advance, and it was tying up too many of the Coordinator's troops to keep those areas under control. Because Kurita knew he would need many men to take New Avalon, he decided to clean up at least a few pockets of troublesome Davion resistance. Second, Minoru Kurita's supply situation was poor. Having pushed into the Federated Suns so far and so quickly meant that his supply lines were over seven jumps long. If he paused now, it would allow a buildup of supplies for the final push. Third, Coordinator Minoru needed to establish regional supply depots on conquered worlds to prevent future supply bottlenecks.

One of the worlds chosen as a regional supply station was New Rhodes III, an object of fierce fighting two months earlier. The defenders, members of the 17th Avalon Hussars, had eventually retreated offworld before the arrival of the elite second Sword of Light. With its small Type M star and quick turn-around time for DropShips, New Rhodes III could supply an entire flank of the Kurita offensive.

After the retreat of the 17th, AFFS Captain Warrent and a small band of MechWarriors had stayed behind. Hiding in the thick woods above the capital city, they watched Kurita forces land on the planet and then almost immediately ship in huge loads of supplies and equipment. Luckily for the Davions, the DCMS did not mount its usual defensive measures, such as 'Mech patrols or fighter fly-overs. Taking advantage of the Combine's laxness, Captain Warrent and his MechWarriors slipped into the supply depot and destroyed many thousand tons of supplies before being hunted down and killed.

Not only was this an important—and all too rare—military victory for the Federated Suns, it was also an important moral boost for the entire realm. So inspiring were Captain Warrent's efforts that the people of New Rhodes III became so unruly that the DCMS had to cancel its plans to build a supply depot there.

Because of the loss of supplies on New Rhodes, House Kurita's troops on nearby Kentares IV were in dire straits. The Davion presence there, led by the Seventh Crucis Lancers, had fought a long and tenacious defense, making the enemy pay for every acre of this heavily industrialized world. Now that the lack of supplies was forcing the Kuritans to slow up their efforts, the Davion soldiers redoubled their efforts and began to push back the Arm of the Dragon.

To stop this AFFS offensive, Coordinator Minoru Kurita himself arrived with his Sixth Sword of Light on Kentares IV. At first, he was almost overwhelmed by the planet's defenders, but the sheer number of Kurita 'Mechs slowly forced the Seventh Crucis Lancers back. Soon, they had been forced onto a single continent.

New Snowfield was a small town in the Carmelite Mountains of Kentares IV. It was a beautiful setting whose forests and mountains enchanted the Coordinator. Having been on the march all day, Minoru decided that this would be a good place to revive his spirit.

Above the town was a platoon of soldiers from the Seventh Crucis Lancers. Led by Sergeant Latha Pischel, the platoon had been quietly watching the coming and going of enemy forces. With the appearance of two Kurita lances in the town below, Sergeant Pischel ordered his platoon to retreat. Pischel, however, decided to stay and watch.

Down below, he saw a MechWarrior climb out from his *BattleMaster*. The MechWarrior appeared to be an old and respected one, for all the other soldiers clustered about him. After conferring with his subordinates, the man walked away with a firm gesture that none but the robed priest should follow him. As the two men strolled together and began to talk, Sergeant Pischel unslung his laser rifle and adjusted the scope. He could see that the MechWarrior was bent with age and what could only have been heavy responsibility.

For a moment, Pischel felt a flicker of pity, but he knew that war was war. Drawing a bead on the man's back, he let off a shot. The figure in his scope twitched once as though irritated by an insect sting, then collapsed. As enemy soldiers ran for their weapons and doctors rushed to the aid of the dying man, Sergeant Pischel quickly ran off to follow his men.

The man whom Sergeant Pischel had assumed to be a Colonel was, in fact, Coordinator Minoru Kurita. When the Federated Suns eventually discovered this fact, after a few days of increased enemy activity and communications, the general reaction was one of complete shock instead of joy. Though Minoru Kurita had long been despised, everyone feared the day when his son, the volatile Jinjiro, would assume control. Throughout the Davion realm, people in both high and low places shuddered to think what would happen next. Prince John ordered all AFFS forces to leave Kentares, hoping to cheat Jinjiro of any chance to vent his anger.

Jinjiro reached Kentares about a week after his father's death. Gathering his officers together, he issued a three-word command that announced months of horror. The words he spoke were: "Kill them all."

Kurita soldiers, many armed with only pistols and swords, executed over 52 million civilians on Kentares. Some of the Kurita troops, sickened at what they had been ordered to do, hid many citizens. In one instance, Talon Sergeant Tarna Oza hid ten families from execution by forcing them into a mineshaft, whose entrance she blocked with her 'Mech. She spent 20 days telling her superiors the lie that her 'Mech was damaged, until the execution parties left the area.

The Federated Suns reacted to the Kentares Massacre almost as one man. A wave of revulsion spread through the realm, fueling the cry of hate-mongers who wanted to lock up all Orientals. On the battlefield, the news made every Davion soldier fight with renewed passion. Anger and an insatiable urge for revenge now burned within the hearts of the Davion troops, sweeping away their earlier fear and depression.

"Remember Kentares!" was the common cry throughout the Federated Suns. Like the spell of a magician, the cry transformed a beaten people into grim-faced warriors. Late in 2797, Prince John hoped to take advantage of his people's renewed spirit by preparing a counterstroke, but he would not live to see the operation. In December of that year, an unknown assassin penetrated the Prince's field headquarters and struck down both the Prince and his son and heir, Joshua Davion. It was now left to Paul, Joshua's son, to stand at the head of a realm on the verge of defeat yet invigorated by a new spirit.

DOWN BUT NOT OUT

In times like these, it is the ruler's duty to look after his people. Power for power's sake is hardly worth seeking, but power to save this realm is something I must have and something I will have.

—Prince Paul Davion, in a speech before The High Council, 2800

A mere 19 years of age at the time of the assassination of John and Joshua Davion, Paul came to the Davion throne facing a critically dangerous situation and with little experience. The fury of the First Succession War was unlike anything since the terrors of the Age of War, and the Federated Suns was bearing the brunt of the fighting. The new Prince faced the worst crisis since the time of Prince Alexander, and it was fortunate for House Davion that he had inherited much of the talent, energy, and determination of his distinguished forebear. Just as Alexander's victory in the Davion Civil War had changed the face of the Federation forever, so did the policies Paul adopted to meet the common danger.

The year of the Prince's accession marked the low tide of the Federation cause. News of the Kentares Massacre was still reverberating through the realm, while the death of Prince John had thrown all plans for a Davion counteroffensive into disarray. Meanwhile, the Kuritans continued their mopping-up operations on AFFS resistance. On the other front, House Liao was still gobbling up Federated Suns worlds, slowed only by their fear of bumping heads with House Kurita.

Demonstrating a gift for strategic thought, young Prince Paul rose to the occasion. Though he might often be defeated in battle, this new Prince would one day stand among history's greatest generals for his ability to reorganize and retaliate after even the most serious reverse.

Prince Paul's first action was to give his approval to the various spontaneous offensives begun by Davion troops incensed by the Kentares Massacre. Some of the attacks failed miserably, but most succeeded, weakening the resolve of their opponents. It was in these areas that Prince Paul focused the meager reserves and supplies at his disposal. The sight of Davion soldiers attacking with an abandon supposed to be lacking in an "inferior" race shocked many Combine soldiers who already felt dishonored by the Kentares atrocities. Shaken by recent events, many DCMS soldiers laid down their arms and surrendered. These passionate victories gave the Federated Suns its first tactical, if not strategic, victories in years.

Prince Paul was quick to capitalize on this stroke of good fortune. He ensured that all prisoners were treated well and thoroughly interrogated about the Arm of the Dragon. All reports went directly to Prince Paul, bypassing Military Intelligence, which the Prince did not trust. Though the average prisoner knew only bits and pieces about Jinjiro Kurita's intentions, the pieces began to form an image of an army suddenly unsure of itself.

POWER TO THE PRINCE

Doesn't the fact that we stand on the brink of extermination make you think that perhaps we were doing things wrong?

—Prince Paul, to members of the High Council, quoted in *Political Memoirs,* by Duke Gregory Simons, NAIS Press, New Avalon, 2999

Forced to regroup and reevaluate their plan of attack, the Kuritans halted their drive. Equally content to avoid a fresh escalation, Paul took advantage of the lull to reorganize the Federated Suns defenses. During this time, he also introduced political reforms that had far-reaching consequences for House Davion and its realms. In the years from 2798 to 2801, the Federated Suns underwent a major internal overhaul, leading to yet another victory for centralization. These years and their aftermath are known today as the Reformation.

At the root of Paul's reorganization was his assumption, in 2798, of the title of Duke of New Avalon. In Star League times, the title had been granted as a mark of honor to the heir to the First Lord of the Star League, and Paul took over the title by right of his grandfather's claim to the Star League throne. During the League era, the title had accumulated a number of powers and honors that the Camerons had exercised only rarely. In Paul's hands, this courtesy title was transformed to make him the most powerful Davion ever to rule the Federated Suns.

The Duchy itself was of little importance in federation politics. During the two centuries of the Star League's existence, the Duke of New Avalon's title had slowly gathered a number of subsidiary honors that *did* make it a significant tool—in the proper hands. For example, even under Alexander and his heirs, the Federated Suns High Council had always reserved the right to create or eliminate titles and estates within the boundaries of the Federated Suns. Neither the Davions nor any other Princes could endow new members of the peerage. In 2653, the Duke of New Avalon was specifically empowered to bestow courtesy titles in the Federated Suns peerage (though not of lands or estates) to permit the First Lord and his family to honor particularly valuable followers from the Federation. In 2698, there came another measure granting the Duke of New Avalon titular ownership of 30 planets within the federation's territories. As the planets became colonized and reached a point of self-sufficiency, the Duke would grant them autonomy, and then would adopt other undeveloped planets into his keeping.

In the years following the War of Davion Succession, the practice of granting autonomy to the emergent worlds had fallen away. A number of these planets had since become totally self-sufficient. One of them, New Aberdeen, could meet all the requirements for full membership in the Federation. Having taken the title of Duke of New Avalon, Paul chose to exercise his "rights" of ownership over these worlds.

In a final power, the Duke of New Avalon was entitled to speak first at any meeting of the Federation High Council, and he had the right to summon or dismiss a High Council meeting without a vote. When the Camerons held the Duchy, these powers simply allowed the First Lord's heir to convene meetings at his convenience during his frequent visits to member states of the League. Even Alexander had not taken the power to dismiss the Council for himself.

By claiming the Duchy, Prince Paul gained all the powers needed to assume total control over the Federated Suns. He could reward followers with titles, regardless of the High Council's wishes, and many of the planets he now "owned" were significant financial assets or would make valuable endowments as estates. For the first time, the Prince of the Federated Suns became a true feudal lord. His followers owed their primary allegiance to him rather than to the federation, and it was Paul, not the government, who was in a position to reward them. To ensure perpetuation of this new order, the Prince could replenish his holdings from spoils of war or with heirless estates forfeited to the Crown to replace the lands that he handed out. In addition, the Duke's ability to summon, guide, and dismiss the High Council eventually reduced that body to little more than a rubber stamp for the Prince's will; it lacked even the few advisory powers that Alexander had left it.

Armed with these new powers, Paul Davion set out to reorganize the AFFS into a fighting force loyal to House Davion rather than to any single political or regional entity. By centralizing power and authority in his family, he made the Federated Suns better able to resist outside aggression. In 2799, he created more than 50 new noblemen, mostly distinguished officers whose services warranted exceptional notice. He also established two new military awards: the Crucis Legion (for MechWarriors) and the Order of Valor (for other, non-'Mech troops). The Prince was fond of citing a famous saying credited to the military leader Napoleon Bonaparte, "With such baubles, men are led." Paul Davion and his successors certainly proved its truth.

CREATION OF THE MIIO

One of Prince Paul's sharpest criticisms of the Davion government was that the Federated Suns civilian information gathering service was totally inadequate. Since the birth of the Star League, Military Intelligence had handled all attempts to learn the secrets of other realms. The existing civilian agency had to submit requests to MI for information, and it was up to the MI officers to decide whether to part with that intelligence. This arrangement worked well for the relatively quiet times of the Star League.

Now that whole worlds could be won or lost in a day, based on knowing or not knowing certain information, it was no longer realistic to rely totally on MI. Not only was the agency too understaffed to handle the flood of information requests, it also suffered from the factionalism that plagued the rest of the AFFS. The agency guarded its information so jealously that sometimes its people refused to part with it out for fear of weakening their own power.

Prince Paul's solution was to create the Ministry of Intelligence Investigations and Operations (MIIO) at the same time he was strengthening all his other powers during the Reformation. This Ministry's purpose was to gather information through any means necessary and to feed it back to the government as quickly as possible. To ensure the quality of the new MIIO agents, Prince Paul ordered the Ministry to encourage and reward individual initiative. Funding for the new Ministry was extraordinarily high, but the Prince's lavishness soon proved to be a good investment. Less than a year after the MIIO's creation, it was supplying the government with vital information quicker and more thoroughly than was the MI.

—From *Spies and Information for the Federated Suns: A Study*, by Ivan Navaro, Toledo Press, New Avalon 2997

ROUT OF THE DRAGON

Why should we be so amazed and honored because Dame Ilsa Liao wants to make a deal that would recognize me as First Lord? The truth is self-evident and shouldn't have to be bargained for.

—Prince Paul's remarks concerning Chancellor Liao's offer to recognize him as First Lord, in exchange for certain worlds, quoted from *Political Memoirs,* by Duke Gregory Simons, Toledo Press, New Avalon, 2999

After tending to these political matters for several years, Paul took to the battlefield once more in 2801. Leading yet another drive for Kentares IV, his strategy was not spectacular, yet the offensive gained House Davion its victory. One major reason for the success of this campaign was the pressure that the Lyran Commonwealth was placing on the Draconis Combine, drawing away key DCMS forces from the Davion front. Another important factor was the leadership of Paul's uncle, Thomas Halder-Davion, whose grasp of tactical principles was as sound as Paul's understanding of strategy. Furthermore, the efforts of McKinnon's Raiders on Kentares kept the DCMS off guard, constantly forced to spend time and resources hunting down the guerillas.

After bitter fighting, the unhappy world of Kentares, now a wasteland of corpses and ruins, was restored to Davion control. With the recovery of this planet, the Federated Suns began actively to reclaim ground lost to the Draconis Combine. Kurita generals reluctantly withdrew from several worlds deep in the Federated Suns because of the Davion forces on Kentares.

By 2808, the Federated Suns had pushed back the Draconis Combine to the border between the Crucis and the Draconis Marches. With each victory, the AFFS grew stronger and bolder. Prince Paul's military and political reforms were proving most effective, and even the most pessimistic citizens and politicians were smiling at the war news. A new confidence was in the air for House Davion, though some would later call it arrogance.

It was at this moment that Paul Davion received a totally unexpected peace proposal from the Capellan Confederation. Sensing that peace would be easier to negotiate with a realm so much on the upswing, Chancellor Ilsa Liao renounced her family's claim to the First Lordship. The proposal she sent Prince Paul promised to recognize the Davion claims to the First Lordship in exchange for control of the Chesterton region of the Federated Suns (most of which the Capellans had already wrested from the AFFS early in the war.)

Why Paul responded so coldly to this sincere peace proposal is still a matter of considerable debate. Some historians speculate that there may have been some hidden condition in the bargain to which the Prince simply could not agree. Others think that Prince Paul suspected the Capellans of sending the assassin who had killed his father. Still others believe that Paul Davion and Dame Ilsa had been secret lovers and that the Chancellor offered the peace proposal to patch up personal problems between them.

Whatever the reason, Prince Paul responded to the proposal by ordering the Third and Fifth Crucis Lancers, who especially hated the Liao forces, to renew their efforts to seize the Capellan worlds of Ulan Batar and Farwell. There were many nobles and high government officials who questioned Paul's rejection of peace with House Liao. Paul's answer was to send them or their family members into action on the frontiers.

Meanwhile, things were not going well for the DCMS. For the average Kurita soldier, indoctrinated with the idea of certain victory for the Draconis Combine, the years of slowly losing ground were disheartening. Added to this were the constant rumors about the Kentares Massacre, which troubled the sense of honor of many an average soldier.

When news reached the Kurita troops that the AFFS was planning another massive offensive in 2809, a wave of fear spread through the DCMS. The offensive was to be a simultaneous thrust from either end of the Kurita-held territory. By surrounding many Kurita-held worlds, the Davion generals hoped to trap the defending Kurita units in a pincer-like grasp.

As soon as the Davions began their offensive, panic spread among the Draconis units. Many infantrymen simply ran for the nearest DropShips, oblivious to their officers' commands. 'Mech, AeroSpace Fighter, and tank units suddenly found themselves stripped of infantry support. They, too, offered only a token defense before heading for the DropShips.

By 2818, the AFFS was fighting for Clovis, Doneval II, and Le Blanc, all planets quite near the original Davion-Kurita border. Though the initial panic had run its course, many DCMS officers were committing seppuku for having failed in their duty. This wave of suicides devastated many units, as replacement officers often lacked substantial combat experience.

The Federated Suns forces stopped for a breather near the original Kurita border, as the commanders realized that an invasion of the Draconis Combine would provide the enemy with an emotional rallying point. Moreover, they wished to avoid overextending themselves, considering the snarl caused by long supply lines and the need to continue to fight on the Capellan front.

TIME OUT

The First Succession War ended in 2821, more from sheer exhaustion than through the signing of any peace accords. None of the Great Houses was in a position to mount any new offensives, and so a semblance of peace graced the Inner Sphere for nearly a decade to follow.

Like all the Successor States, House Davion had suffered severely in this war. Its shipbuilding industries had been crippled by the funneling of key resources into the war effort and the devastation of many factory worlds. Numerous other planets had lost their production capacities because local water purification and fusion power facilities had been destroyed. Consequently, revenues for the Crown had dropped to less than a quarter of prewar levels. Casualties of the First Succession War were estimated at over a billion, most of them civilian. This number did not even begin to account for the vast numbers who succumbed to famine, plague, exposure, and other disasters related to the collapse of technology on a hundred Federated Suns worlds.

THE SECOND SUCCESSION WAR

All dressed up with nowhere to go.
—Regimental slogan for the 102nd Avalon Mixed Weapons Unit

AN UNEASY PEACE

Prince Paul used the time of relative peace in the same manner as his fellow Successor State warlords. He spurred his people to rebuild the military and to put the realm's scientific, technical, and economic resources back into a semblance of order. All the while, Paul continued to extend his personal control over many aspects of the Federated Suns government, mainly through the use of deputies appointed from the ranks of his own family. Of particular importance were his uncle, Thomas Halder-Davion, Peter Davion, the Prince's brother, and Lord William Stuart, Paul's brother-in-law.

Halder-Davion was appointed Field Marshal in 2825, and was Paul's major link to the army. Lord William Stuart, on the other hand, was a particularly canny politician. Whenever the Prince had to be absent from New Avalon, Stuart carried on Paul's work in the Court on New Avalon. As for brother Peter, appointed Duke of New Aberdeen in 2814, Paul employed him in a variety of ways, but particularly as a military leader.

With such a large and diverse family to supply officials, Paul found it easy to maintain a Davion presence in every crucial area of his far-flung realm. Although other Davion leaders had followed this same practice, Paul was the first to demonstrate just how systematic this network could become. Critics grumbled that a man could not take a step without tripping over a Davion.

Freed from many less important duties, Prince Paul turned his energies to preparations for the next war, which he and the other leading Davions were sure would come. In public, however, the Davions praised the peace and claimed that a Davion First Lord would make it last for a thousand years. This facade reassured a people who had endured too much to contemplate yet another ordeal so soon.

Most of the non-military buildups during this period of uneasy peace were directed toward worlds that had suffered most in the fighting, whether from battle damage or because their vital supplies had been threatened. The government sent major convoys of transports to these worlds. These convoys were dubbed "Red Cross Wagon Trains" for the bright red cross on the sides of the JumpShips and DropShips participating in the relief effort.

Also at this time, the Davion government participated in secret talks with the Capellans to win the release of the thousands of captured AFFS soldiers. It would have been no use attempting to negotiate with the Kuritans for their Dictum Honorium would not allow for the exchange of prisoners. Prince Davion hand-picked the negotiators to handle these delicate and often dangerous discussions. Many who served the Prince so well in this capacity later became important members of the newly created MIIO

Negotiations with the Capellan Confederation were surprisingly difficult and treacherous. The Confederation had not forgotten Prince Davion's violent reply to Chancellor Ilsa Liao's peace initiative, and they were not about to return any of Prince Paul's soldiers without making him pay dearly. A War Crimes Court put all the captured Davion officers on trial, filling every day's record with gory testimony about the crimes of the accused. The news media of both House Liao and House Davion covered this sensational trial.

All defendants were found guilty and condemned to death. Chancellor Ilsa Liao, meanwhile, was secretly informing Prince Paul that the accused were available for a price—a very high price. At first, Paul balked at the terms, but he changed his mind soon after the brutal Christmas Day execution of 20 AFFS officers, including a Marshal and two Generals. The ransom, which included tons of precious metals and minerals, paid not only for the lives of Davion officers, but for enlisted soldiers as well. So large was the ransom that it took two JumpShips of booty-laden DropShips to carry it. Paul Davion no doubt cursed the day he had so callously rejected Chancellor Ilsa's overture.

WAR HEATS UP

In 2825, the Draconis Combine launched what came to be called "Chain Gang Missions" to prevent the reconstruction efforts on border worlds of both the Federated Suns and the Lyran Commonwealth. Disgraced officers, convicts, Unproductives, and social misfits were recruited for these missions with promises of citizenship and wealth. After only the barest MechWarrior training, they were then assigned 'Mechs that were one step from the wrecking yard. The various lances and companies of these ersatz MechWarriors amounted to a little over three regiments. Once a Chain Gang Mission arrived at its target planet, the DCMS DropShips and JumpShips simply left them there. The members of the Chain Gangs did not know that the ships would never return for survivors—until it was too late.

On the battlefield, many Chain Gangs simply surrendered, hoping the enemy would treat them mercifully. Others fought bravely, but were easily cut down. Then, there were those few that fought far beyond their abilities, sowing damage and panic because planetary defenders interpreted their tenacity as an attack by frontline units. Largely through the efforts of these foolishly brave victims of the Draconis Combine, the reconstruction efforts of both the Federated Suns and the Lyran Commonwealth were thrown into disarray.

Just as there was no official end to the First Succession War, there was no official beginning to the Second. For House Davion, the skirmishing on both its borders had become so heavy that by 2828, all-out war was the only logical next step. Prince Paul and his AFFS were again eyeing the Liao frontier, especially the worlds so dear to the Capellans around Chesterton. Meanwhile, the Draconis Combine seemed to have turned its attention mainly to actions on the Lyran front, taking some of the pressure off House Davion. Most of the AFFS Marshals considered this reason enough to mount a major effort against the Capellans, advising Prince Paul to seize the opportunity before the Dragon once more turned its eyes their way.

What they did not know was that Chancellor Ilsa Liao was about to launch her own offensive against House Davion to keep the AFFS on the defensive and out of the Confederation. The main target of her thrust was the planet Orbisonia, a major staging area of the AFFS and home to the 2nd Ceti Hussars and the 14th Avalon Hussars, both 'Mech regiments.

Chancellor Ilsa Liao personally led her force of over a hundred heavy 'Mechs and lesser regiments against Orbisonia, only to discover that the outmatched Davion defenders were still too mobile for the Liao forces to be able to administer the knockout punch. On the third day of battle, the arrival of Davion's Assault Guards from the Royal Brigade sealed the fate of the Capellan effort. While fighting a rear-guard action to help her troops withdraw from the planet, Chancellor Ilsa was killed.

The AFFS was elated over this victory, which also vindicated Prince Paul's decision to concentrate his attentions against the Capellans. In early 2829, the Prince moved his field headquarters to New Syrtis to begin a full-scale campaign against the Confederation. One of his first acts was to appoint Colonel Damien Hasek, a hero of the Davion stand against Liao raiders on Demeter, as the new Duke of New Syrtis. The fiefdom of the Syrtis March was invested with all of the powers previously enjoyed by the Princes of the Capellan March, and would remain in the hands of the Colonel's descendants all the way down to the present day. The Duke was instructed to raise additional forces from his new domain to swell the ranks of the sizeable contingent that Paul had brought with him to New Syrtis.

By 2830, the Capellan campaign was in full swing while the Draconis Combine had launched a new offensive against the Lyrans. It is here, in the latter months of the year, that most historians date the start of the Second Succession War.

GENERAL MOTOCHIKA

I do my duty; no more, no less. Why? Is there anything more to a soldier's life?

—From official transcripts of the MI interrogation of General Chosokabe Motochika, 2830

In 2830, contrary to expectation, the Draconis Combine launched a series of major raids against Davion border worlds. Some attacks were aimed at border planets such as Lucerne, Franklin, and Sakhara V. More ambitious efforts were aimed at worlds deeper into the Federation interior. Though the Draconis military did not know it, the AFFS had been preparing for the invasion for some time, having broken the DCMS battle code for the operation.

General Chosokabe Motochika was the Kurita commanding officer for the raids against Lucerne and Franklin. He was a respected officer with 30 years of service and an almost religious devotion to the Kuritas and the warrior's Code of Honor. Though his battlefield performance was as dull and unimaginative as any typical Kurita officer, he had won a reputation among the AFFS as a clean and honest fighter.

THE TALON SERGEANT AND THE PROSTITUTE

Among those criminals, disgraced officers, and so-called "deviants" who were volunteered into the Chain Gang missions of 2825 were Talon Sergeant Jur Sturlaugson and Katrina Reban. Sturlaugson was a member of the "Unproductive Regiment," a Draconian euphemism for those imprisoned in a DCMS stockade. He had landed there as one of many troopers who found themselves unable to carry out the extermination order that created the infamous Kentares Massacre. Katrina Reban was an Unproductive who had answered when officials had come to her New Samarkand village calling for "true believers in the Kurita way." Whether or not she was that, believer Katrina had spent most of her 28 years as a prostitute in a DCMS brothel.

Katrina and Jur were thrown together on a Chain Gang Mission bound for the Davion world of Udibi. During the voyage, the guards allowed their charges to behave as they wished during their final hours of life. From the many physical couplings among the doomed men and women came a bond of true love between the Talon Sergeant and the Prostitute.

The raid against Udibi failed before it began. The two lances of patched and barely operating Chain Gang 'Mechs soon became scattered, and it was not long before AFFS forces began to hunt them down one by one. Talon Sergeant Sturlaugson became trapped in a desperate cat-and-mouse game with two AFFS 'Mechs while he searched desperately for Katrina Reban. This went on for two hours, until he finally came upon his love in her *Stinger*, cornered by a Davion *Marauder*. Throwing himself between Reban and the Davion 'Mech, Sturlaugson took the PPC fire meant for Reban's 'Mech.

The Davion soldiers who witnessed this act were moved by Sturlaugson's heroism and by Katrina's explanation of why he had risked his life. The soldiers brought the story to their superiors, who allowed both the media and the Prince to get wind of it. When Paul Davion learned that Sturlaugson had resisted participation in the Kentares Massacre, he immediately granted the lovers citizenship in the Federated Suns.

Jur Sturlaugson and Katrina Reban went on to serve in the AFFS and later formed their own mercenary unit, which they named the Ever-Free. The exploits of this unit, whose symbol was the broken chain, become a popular part of the legend and lore of Davion culture.

—From *True Tales and Popular Stories*, by Father Ryan Ramon, Unfinished Book Press, New Avalon, 3010

While coordinating the attacks on the two Davion worlds, the General was dismayed to learn that both planets were crawling with more AFFS troops than expected. As additional reports came in, the General understood that news of the invasion had leaked somehow and that he was now in danger of losing all his units. He gave the order to retreat.

The Coordinator himself must approve all orders to ROW (Retreat Off-World), which could create more than a bit of trouble for Kurita units in the field. On a number of occasions, the Coordinator's permission to retreat reached his men just a little too late to save anyone. To retreat without permission was a severe breach of honor, usually punishable by the deaths of the commanding officer, his staff, and a percentage of all troops that retreated. Many a Kurita commander has gone ahead anyway, pulling his troops offworld, all the while praying that the Coordinator's permission would arrive before the last soldier had retreated.

The subordinate whom General Motochika had dispatched with the request to retreat had failed to deliver it. It was not until the General's DropShips had rendezvoused with their JumpShips that he and his men discovered that they had participated in an illegal retreat that would now doom many of them to death. General Motochika had received the news in proper stoic fashion, but decided to lead his three 'Mech regiments and one tank regiment deeper into the Federated Suns instead of reporting back to Luthien.

The target of this wayward band of soldiers was Dahar IV. Dahar was one of those few fortunate worlds that had not suffered from the first war, even though it had been occupied. Because of Dahar's many deposits of metals and heavy industry, the 23rd Avalon Hussars was stationed there, along with numerous lesser regiments.

When General Motochika's force appeared above Dahar, the Davion forces immediately scattered to their heavily prepared defensive positions to wait for the Kurita DropShips and fighters to come screaming in. Instead, they watched in amazement as the Kurita DropShips slowly approached, broadcasting a desire for negotiations. Many were suspicious of General Motochika's intentions, but the MI understood from its monitoring of Kurita transmissions that something had gone wrong for the General. They assured the AFFS commanders that if Motochika allowed himself to be taken into AFFS custody, his intentions were probably sincere.

After landing on Dahar with only a regiment of 'Mechs, General Motochika and his staff did indeed allow themselves to be taken. While the MI was interrogating the General, the AFFS commander ordered his troops to stand down from their positions. At this signal, the other two Kurita 'Mech regiments left their orbiting DropShips and caught the AFFS defenders off-guard. Soon, all the major cities on Dahar were aflame, while Kurita 'Mechs roved the countryside looking for targets.

When Davion officers berated Motochika for his treachery, the General smiled and said that his troops were only following their orders. He explained that an action such as this would remove the stain of illegal retreat from their honor. His unit might now even be able to return to the Draconis Combine, assuming that they escaped the Federated Suns. With that, Motochika pulled out a small poison capsule hidden in the cuff of his uniform and swallowed it.

When Coordinator Jinjiro Kurita heard of General Motochika's actions, he gave the order that would allow General Motochika's surviving troops to return honorably to the Draconis Combine.

BATTLES FOR TIKONOV

A nice place if you're a badger or a glutton for punishment.

—Remarks by a Davion 'Mech officer during the holonews show "The Sun of Truth," 12 August 2832

With the recent flurry of Kurita border raids, Prince Paul and many of his Marshals were concerned that the Draconis Combine was turning its sights again on the Federated Suns. When several more tense months passed with only a few minor skirmishes, the AFFS decided House Kurita was not its main threat at the moment. By 2832, when all became quiet on the Draconis front, the Davions designed an offensive against the Capellan world of Tikonov, yet another prefecture capital of the Confederation.

Led by the flamboyant General H.R. "Howler" Greer, the second Ceti Hussars Regimental Combat Team spearheaded the attack. Considered a throwback to the cowboys of the American West, Howler Greer led his mobile and well-trained 'Mechs through the desert badlands of Tikonov. Urging the Hussars to attack again and again, his almost maniacal energy shattered every organized attempt to stop him.

The Howler's efforts came to an end outside the High Kremlin, the Capellan stronghold on Tikonov. A better mobile commander than siege master, the General found that he could not crack the dug-in defenders there. After six months of staring up at the walls of the Capellan fortifications, Howler Greer was forced to take his troops offworld when a Capellan relief force threatened to cut him off from his JumpShips.

The next year, however, Greer was back on Tikonov with heavier weapons and Prince Paul's threat that he would assign the General to slinging hash in an officers' mess if he did not bring Tikonov under the Federated Sunburst. In a second, more successful effort, Howler managed to seize the majority of the planet, including the High Kremlin, during the second month of the campaign.

What Howler failed to do, however, was to secure his supplies. A special Capellan strike force composed of Capellan Hussars Fighters and light 'Mechs entered the Tikonov system, striking directly at the Ceti Hussars' supply depots, with devastating results. Before the Ceti Hussars could react to this new threat, Howler learned that the rest of the Capellan Hussars, including numerous assault 'Mechs, were also on their way to Tikonov. At this blow to his cowboy pride, the General was forced to admit failure because he no longer had the supplies to fight the heavier 'Mechs. Prince Paul made good his threat. General Greer served food at an officers' mess on New Syrtis for one year.

Still determined to win Tikonov, Prince Paul decided to try again in 2834. This third attempt would use the third Deneb Light Cavalry Regimental Combat Team, commanded by General Jessica Bassner. The plan was for her swift forces to seize and hold Tikonov's major continent long enough for another RCT from the Royal Brigade of Guards to arrive and help take the rest of the planet. General Bassner and her troops held up their end of the operation quite well, maneuvering with her slower enemy so that the key cities fell into her hands. Now the regiments of the Deneb Light Cavalry awaited the relief forces.

They would never come. Just as the Second Royal Davion Guards RCT was to leave for Tikonov, Coordinator Jinjiro Kurita launched a raid against the Guards' homeworld of Sanilac. The unexpected raid caught many of the Guards units in their grounded, defenseless DropShips. Their losses were high, and by the time the 2nd RCT had fought off the Kurita raiders, there was no hope of relieving the Deneb Light Cavalry on Tikonov. General Jessica Bassner was forced to leave Tikonov to the Capellans.

PAUL'S LAST YEARS

Of course, I'm sick of all this fighting, but if I claim that I'm not going to fight anymore because I'm tired and worn out, what would stop my subjects from doing the same? It would be great if everyone on all sides could say that they were sick and tired of the war and were going to quit, but I doubt if Coordinator Kurita or Chancellor Liao will call the war on account of exhaustion.

—Prince Paul Davion, 2842

The Tikonov campaign was a puzzle for many in the AFFS. Though they understood the importance of attempting to take a capital world, the Davion armies could just as easily have bypassed Tikonov and continued the offensive elsewhere. Because of Prince Paul's seeming obsession with that planet, however, the AFFS effort against the Capellans stalled. Some of the credit for the blunted effort must go to the Capellans, of course, who had finally given up their reliance on massed artillery attacks in favor of a swifter, more flexible tactical doctrine. Strategically, the Liao military had been reorganized with new area garrisons providing a better defensive response to Davion attacks.

Another reason that the offensive against the Capellans bogged down was that Prince Paul was spending more time in the regional command headquarters on New Syrtis than on the battlefield in the late 2830s. It almost seemed that he had lost interest in the Capellan front.

What was occupying the Prince's mind was the possibility of an offensive against House Kurita. The more reports that Paul read from his Field Marshals on the Draconis front, the more he believed that the Arm of the Dragon was vulnerable. What he needed was a diversion that would allow him to move the necessary forces from the Capellan front to the Draconis front without attracting the enemy's attention. Though he wracked his brain for a solution, it was the DCMS that finally supplied Paul with his diversion.

In July 2840, the Arm of the Dragon launched a fierce raid against Robinson in an effort to freeze Davion reinforcements away from a contested area of the front. At the same time, the action on three other worlds, Tancredi, Anguilla, and Sturgis, suddenly heated up. Taking advantage of the situation, reinforcements and reserves from the Capellan front were soon streaming toward the Draconis front. With considerable glee, Prince Paul planned to trick the Draconis Combine in much the same way that it had tricked the Federated Suns at the start of the First Succession War.

In late 2841, Davion had assembled enough troops to launch a two-pronged offensive. One spearhead would strike at the worlds known as the Liao-Kurita Seam, the border between the two realms near Terra. The targets were Mallory's World, Mara, New Rhodes III, and Ozawa. The other half of the offensive aimed at the other end of the Draconis front, by passing strongly held worlds and seizing planets deeper into the Draconis Combine, such as Bryceland and Niles.

By February, Prince Paul's offensive was succeeding. Though the Kurita defenders were not the easily demoralized soldiers of the end of the first war, they could not cope with the strong and flexible attacks of the AFFS forces. Mara and Bryceland fell within three months of the opening of the offensive.

The success of this drive suddenly ground to a halt in July 2842, when unexpected news reached the front. Prince Paul was dead.

OVERTURES OF PEACE

No. You're misunderstanding what I said again. I do not think our Prince is a lesser man for actively pursuing his dream of peace. I think he is just a foolish one for not realizing that this is neither the time nor the place.

—Count Tuolov Hasek, during the FNS news program, "Federation News in Review", 8 September 2843

Prince Paul Davion died during the night of 4 July 2842. While pouring over computer reports on his latest offensive, he had been stricken with a heart attack. Early the next morning, his aides found him slumped over the computer planning board.

The realm grieved, not only because he was an admired leader and fearless defender of the realm, but because the people considered him the last of the Star League Davions. Though Paul Davion was born about the time of the League's collapse, he was the last Prince whose actions and attitude spoke of the better times of the Star League.

After Prince Davion's ornate and solemn funeral procession passed down the grand avenues of Avalon City, the reins of power passed smoothly to his son Michael. It did not take long before most people realized that Michael was the exact opposite of his father. Where Prince Paul had been obsessed with consolidating political, economic, and military power, Michael was more interested in salvaging everything that remained of prewar culture.

Following his cue, a wave of nostalgia swept the Federated Suns, generating a mania for preserving any artifact of Star League vintage. Many grumbled at the cost and expense of these salvaging efforts, but the effort would lay the foundations of information upon which Hanse Davion would one day build his institute for retrieving Mankind's lost sciences and technology.

Another significant difference between Prince Michael and his father was that of temperament. While Paul was a brilliant general who left most administrative matters to others, Michael would excel as a politician and a diplomat. Like any Davion, Michael had dutifully trained as a MechWarrior, serving as a Major in the AFFS, but he never displayed much interest or flair for war. This was a continual source of irritation, even sadness, for his father.

The biggest difference between Michael and Paul Davion, however, was that the son was far more interested in peace than war. At the root of most of the new Prince's policies was the firm belief that peace was attainable, but only if the right leader came along to see to it.

From the year of his investiture in 2843, Prince Michael spent a decade making peace overtures to the Capellan Confederation and the Draconis Combine. As his desire for truce became more passionate, Paul's peace offerings became all the more extravagant. Beginning with offers of cease-fire on certain war-ravaged worlds, the Prince had soon escalated to offering the Liaos and the Kuritas entire worlds in exchange for their recognition of his claim on the First Lordship.

The details of these peace proposals were never made public, but enough leaked out to throw the Federated Suns into turmoil throughout the first decade of Prince Michael's reign. Public opinion was dead set against any peace that would mean the loss of native soil. The nobility was also incensed at the possibility of losing their lands to a neighboring realm, and many campaigned bitterly against the idea. Not surprisingly, there were three attempts on Prince Michael's life during this time.

Despite the threats and public disapproval, Prince Michael persevered. The Capellans and the Kuritans were equally firm in rejecting all his initiatives. True to their own Code of Honor, the Combine saw offers of peace as a sign of weakness, and dismissed them as the blatherings of a defective Davion. The Capellan Confederation reacted in spite, remembering how crudely the Davions had rejected Chancellor Ilsa Liao's hopes for peace. Laurelli Liao, the current Chancellor, continued to ignore Michael's messages, though it certainly would have been to her advantage to see the Davion front cool down.

THE NOBLE ONE

Peter Davion, Paul's brother and Michael's uncle, was the successor to Field Marshal Thomas Halder-Davion, a major military figure in Paul's reign. After sustaining severe injuries in the third Tikonov campaign, Halder-Davion had been confined to a hospital for the rest of his life. Though his last seven years were spent in almost crippling pain, the loyal Field Marshal continued to serve the Federated Suns by transmitting his counsel via computer link to the New Syrtis regional headquarters. He died only a year before Paul, his friend and his Prince.

Soon after Michael Davion came to power, he elevated Peter Davion to the rank of Prime Marshal, a position created to allow a military commander to stand in for a Prince unable or unwilling (as was Michael) to participate in the war effort. As the Prime Marshal, Peter took responsibility for continuing the AFFS offensive against the Draconis Combine started in 2840. This proved more difficult than he expected because Prince Michael's policies of peace had angered so many officers in the AFFS. The 'Mech officers, in particular, were enraged at the idea of giving up worlds for which so many brave soldiers had already given their lives.

In 2846, a cabal of officers approached the 54-year-old Peter, offering to bring the bulk of the army to his banner if he would unseat the peace-lover Michael. Duke Peter Davion proved himself that rarity in politics—an honorable man. With great dignity, he refused the request of these officers and warned them that he would have to call them up on charges of treason if anyone ever breathed such a suggestion in his presence again.

Peter Davion remained unreservedly loyal to his nephew, and continued to serve the family in various capacities to the age of 91.

TISHOMINGO

In 2849, the Federated Suns was well pleased with its efforts against its enemies. On the Capellan front, the AFFS was pounding the Capellan Armed Forces, with reports of the seizure of one Liao world after another coming in. With the capture of four planets from the Kuritas, the Draconis front offensive was also starting to bear fruit. Even Prince Michael was impressed with the abilities of his AFFS.

Tishomingo, an agricultural world occupied by the Kuritans, was currently a major focus of the AFFS. Besides being an important food resource, this planet was also rich in fresh-water supplies. Because the Davion worlds in the same general vicinity were all so water-poor due to ravages of their purification facilities, Tishomingo's reliable water resources made it a tempting target.

Though Tishomingo had originally been a Davion world, the AFFS could not expect much help from the civilian population. Consistent with their policies on conquered worlds, the Kuritans obtained the loyalty of the population either through "retraining," through "restocking" it with more sympathetic citizens from other worlds, and by being sure that the pressure of their ISF (Internal Security Forces) agents kept the rest in line.

Coordinator Yoguchi Kurita had shrewdly foreseen that Tishomingo would be the Davions' next target and had ordered in his Fourth Sword of Light 'Mech regiment to reinforce the planet's defenses. Those units strengthened the well-trained members of the planet's militia units, along with three regiments of armor and three more of infantry. Leading the planet's defense would be Coordinator Yoguchi himself. He was livid at House Davion's recent military successes, and felt personally dishonored by the worlds the Combine had lost.

Despite his advance planning and strong defensive position, Yoguchi Kurita did not realize how much the AFFS coveted Tishomingo. Some 20 Davion regiments, including four 'Mech regiments, attacked Tishomingo, and the Coordinator soon found his defenses outgunned. After Davion *Marauders* overran his headquarters, he was forced to flee into the surrounding forests. The planet soon fell to the superior numbers of Davion soldiers, and the remaining Kurita regiments retreated offworld. Believing that his grasp on Tishomingo was firm, Prime Marshal Peter Davion dispatched all but a single 'Mech regiment to other hot spots on the front. He would certainly not have done so had he known his men had nearly killed Coordinator Yoguchi and that the leader was still alive somewhere on the planet.

Even the Kuritans did not know that Yoguchi had survived. Alive and well, he was secretly organizing the people of Tishomingo into a fantatic level of resistance.

Before long, the Davion regiment actually found itself under siege, with the entire garrison demoralized by the guerrilla campaign. The Davion commanders did not doubt that these were civilian efforts, but the skill behind them had the AFFS officers both puzzled and worried. They still did not know that Yoguchi was orchestrating the offensive from his forest hideout, using peasant runners to carry his commands throughout the planet.

While out on patrol one day, a Davion unit finally spotted and recognized the Coordinator. The entire garrison went on alert almost instantly, with orders to spare no effort to capture the leader of the Draconis Combine. With the people on his side, Yoguchi was spirited away into the city of Urado.

Still under the Coordinator's direction, the people of Urado rose up and killed the Davion presence in their city. Then they set about turning their city into an impromptu fortress. The Davion troops failed miserably in the first attempt to enter the barricaded city, and it would take an entire week before they were able to assemble the battalion of 'Mechs needed for the assault.

Though brave, the citizens of Urado could not hold out against Prime Marshal Peter Davion. His determination to capture Yoguchi was as fierce as the people's desire to protect him. Led by the Prime Marshal, the AFFS units did finally take the city, but it took a week of fighting against a desperate guerrilla defense. Davion had still not yet located Yoguchi when he learned that a major Kurita force had landed on the planet. The Prime Marshal was forced to retreat offworld temporarily while waiting for reinforcements, and this cheated him of the prize of Urado.

It was Roweena Kurita, the Coordinator's younger sister, who had saved Yoguchi. Like the rest of the Combine, she had believed that Yoguchi died in the initial Davion onslaught on Tishomingo. When reports of the planet's well-organized terrorist activities reached her, she grew suspicious. Playing a hunch, Roweena, as the de facto ruler of the Draconis Combine, authorized a military operation to rescue the leader of the guerrilla struggle that had the local AFFS forces so terrorized. Her instinct was a true one, for Coordinator Yoguchi was soon back in Luthien, safe and sound.

He was not safe for long, however. During Yoguchi's first night home, his concubine, the mysterious Snow Fire, quietly slit his throat with a special plasticene knife. Snow Fire then committed suicide by swallowing poison. ComStar research has proved that the beautiful Snow Fire was actually a Lyran Intelligence Corps agent, planted in the Combine several years before.

[EDITOR'S NOTE: The incredible fascination with this woman defies all explanation. Outside of the Draconis Combine, artists from every corner of the Inner Sphere have used her as the subject of countless books, plays, paintings, ballads, and legends. Indeed, the power of her image has something of a religious tinge to it.]

DID SNOW FIRE HAVE HELP?

Snow Fire is one of the most enigmatic figures in the history of the Inner Sphere. All we know of this beautiful woman is that she was born in the Lyran Commonwealth and trained by the Lyran Intelligence Corps. She was planted in the Rasalhague Military District of the Draconis Combine, where she spent several years as a low-grade Geisha secretly winning information from her clients.

She won the eye of one of the Coordinator's closest personal advisers, who bought her contract and took her to Luthien. There, she spent about a year and a half gathering information about the inner workings of the Imperial Palace, and attempting to attract the Coordinator's eye. The story of her final heroic act has been told so often that it need not be repeated here.

The question is, did the mysterious Snow Fire have help from the Federated Suns? Declassified MIIO files have yielded references to an operative known only as "The Footman," who apparently was a member of the Imperial Palace staff during the time that Snow Fire was present there.

The identity of this Footman remains unknown. He might have been Talon Sergeant Donald Baines, one of the Imperial Guards entrusted with the security of the Coordinator's bedchambers. Others speculate that the Footman was Jessica Donvel, an Imperial messenger whose keys allowed her access throughout the Imperial Palace.

It is still impossible to prove whether the Footman, as either the guard or the messenger, knew of and aided Snow Fire's mission. An agent in either role could have discovered the courtesan's true identity, either by noting the incompleteness of her security file or by breaking the code in the messages she sent to her "mother". It is also possible that the Footman could have helped Snow Fire slip her knife past the Imperial Palace's security systems.

It is interesting, even romantic, to think that Snow Fire and the Footman knew one another. Until new information comes to light, our speculations must end there.

—From "Speculations on Past Mysteries," by Precentor Ivan Fresterton, Comstar, Internal Bulletin No. 236185PF, Archives, Terra, 3001

MIYOGI'S MARATHON

One factor often overlooked in deciding whether to invade a world is the planet's emotional worth. What will happen to your enemy if you take the world? Will he shrug his shoulders and continue as before? Or, will he pull at his hair and gnash his teeth in anguish? Demoralize your enemy. Make him realize that you know what makes him happy and that you will do anything to take it away from him.

—From *A Single Slice of the Sword: Strategy in a Universe of Dwindling Resources*, Miyogi Kurita, Imperial Press, 2847

The assassination of Coordinator Yoguchi Kurita sent ripples throughout the Inner Sphere. Though none of the other states knew that Snow Fire had been a Lyran agent, every House Lord had good enough intelligence services to learn that the assassin had definitely been a foreign operative. The Kuritans, of course, knew that the assassin had laid a House Steiner regimental patch on the Coordinator's dead body, and so they turned the full fury of their hate against the Lyrans. To cover their shame that a lowly woman had felled their revered leader, the Kuritans concocted a story of Yoguchi dying heroically in a struggle against a whole band of assassins.

While the propagandists were saving the Dragon's face, the new Coordinator, Miyogi Kurita, and his military commanders were plotting their revenge. In 2853, they turned their sights on the Lyran Commonwealth's huge BattleMech production facilities on Hesperus II. Located deep in the heart of the Commonwealth, these 'Mech production facilities were the heart of the Lyrans' ability to defend themselves, as well as the largest 'Mech factories in the Inner Sphere. The huge battles that would swirl around that single rock would occupy the Draconis military for almost two years. Given the thoroughness of their strategy, it seems likely that the Kuritans had begun their planning long before Snow Fire slit their leader's throat.

During this campaign, fighting on the Davion-Kurita front tended to quiet down. With the Arm of the Dragon so focused against House Steiner, the Kuritans contented themselves, for the most part, with a defensive stance on the Davion front. In the few outright strikes against Davion planets, the Kuritans wished mainly to test the Davion defenses. In a rare act of military intervention, Prince Michael ordered his Prime Marshal to remain equally satisfied with containment, for he preferred not to heat up the situation with new offensives of his own.

One of the few decisive Davion-Kurita engagements fought during the period of the Dragon's campaign against Hesperus took place on Kentares. The Davion people and their leaders had come to regard this planet as a doomed world that was no place for ordinary civilians, and so it had never been repopulated after the Massacre of 2796–7. The AFFS had instead coverted the planet into a major resupply point.

Hoping to catch the Davions off guard during the relative lull, the DCMS launched an invasion of Kentares in 2854, planning to hold onto the world just long enough to make off with its supplies. Falling back on the tried and true strategy of harassment, Duke Peter, commanding the AFFS defenders, used raids and pinprick attacks to confuse and delay the invading Kurita forces until Davion reinforcements arrived. Kentares would remain in Davion hands, but its price was not apparent until a few months later.

THE DRAGON'S GRAND PLAN

Coordinator Miyogi, fresh from what he considered a partial victory against the Lyrans on Hesperus, decided that now was the right political moment to unleash the full power of the Dragon against both realms. The amount of manpower and machines that Miyogi would need for this Marathon Offensive must have made the Coordinator's generals grow pale. The DCMS High Command asked no questions, however, and soon the generals were setting into motion the desires of their Coordinator.

The double assault began in June 2854. The first five years of Miyogi's Marathon met with considerable success, particularly on the Lyran border, where he seized several planets. On the Davion front, the Marathon Offensive operated mainly to throw the AFFS back onto the defensive. Duke Peter had been forced to set aside any plan to win new Kurita planets in order to hold on to what he already had.

The one world the Kuritans managed to take was Robinson, the capital of the Draconis March. The planet had fallen to the Combine during the First Succession War and Davion had only managed to wrest it back within the past 20 years. Though Robinson was not a particularly valuable world, Miyogi knew that capture of a capital world represented a special psychological victory over an enemy.

The Davions had mustered a total of 13 regiments, three of them 'Mech units, to defend the planet. The Kurita forces, led by four elite Sword of Light 'Mech regiments, soon gained the upper hand, however. Field Marshal Jerome Davion, a distant cousin of the Prince and commander of the Robinson defenders, called for reinforcements.

Because so many AFFS regiments had been sent to reinforce Kentares, there were few troops near enough to aid the besieged planet. On Kentares, the DCMS forces were carrying out a series of almost suicidal invasion attempts to keep the Davion forces so tied up that they would be unable to spare any regiments to Robinson.

Coordinator Miyogi's plan worked. After a crushing pivotal defeat at Briarson's Crossroads, the Davion forces on Robinson were forced to retreat offworld in 2858. Loss of this planet hurt the morale of the AFFS, just as the Combine generals had hoped. The Davion offensive against the Combine begun so many years before, in 2840, now sputtered to a complete halt. Politically, the loss of Robinson was the final defeat of Prince Michael's policies of appeasement, and even he had to face the realization that the Inner Sphere was just not ready for peace.

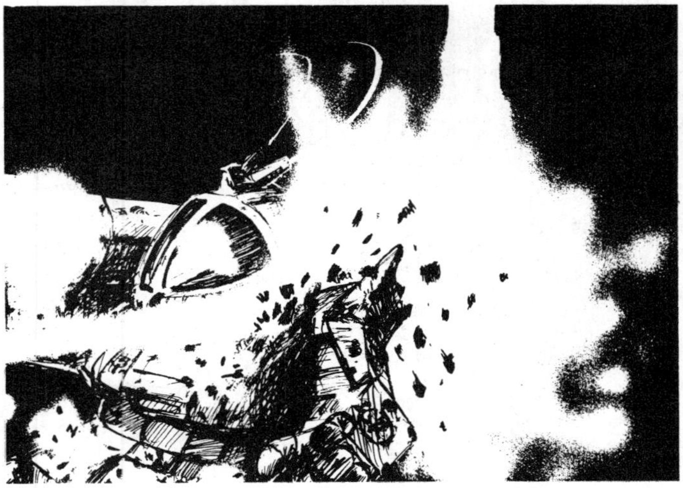

CAPELLAN JUGGERNAUT

It was inevitable that the Combine's Marathon Offensive would eventually begin to suffer from severe shortages of supplies and parts. With full-scale campaigns operating on both its borders, the supply lines became much too long and entangled. This was affecting the fighting ability and the morale of Miyogi's troops. Instead of launching immediate counterattacks, the Prime Marshal let most of his own troops rest, launching only occasional raids up and down the border to keep the Kuritans guessing.

Duke Peter Davion decided to concentrate his efforts against the Capellan front. The fall of Chesterton some 40 years earlier had begun an era of steady Davion gains against House Liao. Deprived of necessary resources, the Capellan armed forces were often forced to cut short their fighting, leaving many worlds to the Davion victors.

This steady acquisition of worlds had fostered a false impression among the Davion public that the fighting on the Capellan border was easy pickings because the Capellan soldiers were weak. Nothing could have been further from the truth. The Capellans' weak point was not cowardice, but their inadequate military industry and inefficient supply system. Indeed, the Capellans were brave and tenacious fighters, as any Davion veteran would attest.

At this moment of history, there were signs that this stubborn bravery had generated a "warrior philosophy" among the Liao infantrymen. In addition, Davion Marshals believed that the Capellan Confederation had shrunk to "its optimum defensive size." Its military was now strongly concentrated and capable of responding much more quickly than before.

Concerned with these signs of stiffening resistance, Peter Davion himself took over leadership of the invasion of the Capellan world of Cammal in 2860. According to reports, the Cammal defenders were putting up a fierce resistance to save this resource-rich planet and had managed to blunt the Davion drive.

The Prime Marshal quickly learned just how spirited was that defense when an attack by a force of Capellan *Thrush*es forced his DropShip to crashland, seriously wounding him in the process. Because of the Liao air superiority, the defenders had effectively cut off the Davion troops from outside help. With the Prime Marshal clinging to life and unable to lead, all ten AFFS regiments on Cammal, including two 'Mech units, would have to surrender.

Fate stepped in to save both the Prime Marshal and his men on Cammal. Chancellor Laurelli Liao, architect of the Capellan defensive strategy, died in October 2860, a week before Peter Davion arrived on Cammal. Her son and successor, Dainmar, was not a warrior, and he viewed the heavy action on Cammal as a losing proposition. In one of the greatest blunders in recent military history, the new Chancellor foolishly ordered his forces to retreat offworld.

Relieved that his favorite uncle had been saved, Prince Michael hoped to take advantage of the new Chancellor's weakness by authorizing a daring 'Mech raid in 2861. It was one of the few times that the Prince had taken an active role in military matters, but his timing was fortunate. The target of the Davion raid was St. Ives, the current residence of the Liao Chancellor.

The regiment assigned to carry out the raid was the elite Assault Guards of the Royal Brigade, led by a brash young officer named Colonel Rebecca Davion, Prince Michael's second child. Famous for her unpredictable wildness, Rebecca had already been reprimanded twice and demoted once for disobeying an order to retreat.

The St. Ives Raid was the perfect mission for her. Having loaded aboard the *FSS Resolve,* the Assault Guards traveled virtually unnoticed to St. Ives by using uninhabited star systems and a series of false identity transponder codes to disguise their path and destination. When they arrived at St. Ives, the Guards took the defenders so much by surprise that they made it to the planet's surface without losing a single 'Mech.

Once on the ground, Colonel Rebecca Davion split her force into two units. One engaged the Liao defenders, and the other moved deep into the heart of the planetary capital, toward the residential palace of Chancellor Dainmar. The Liao defenders were reluctant to engage in a fire fight for fear of damaging their city, but they turned back the Davion forces in a stubborn defense of the Chancellor's palace. Fortunately, the Davion troops lost only two 'Mechs in this daring raid, though both pilots were rescued. Before leaving St. Ives, Rebecca had her men put on a fearsome display of 'Mechs for Chancellor Dainmar, who was cowering in a reinforced bunker beneath the palace.

The Assault Guards had so terrified Chancellor Dainmar on St. Ives that he sued for peace with the Federated Suns in early 2862. Realizing that the Chancellor would probably sign anything at this point, Prince Michael asked that the Capellans recognize all territorial gains the Davions had made at their expense. The spineless Dainmar agreed, signing over ownership of 16 worlds to the Federated Suns.

News of the Capellan-Davion peace did not sit well with the Draconis Combine, for they would not be able to stand long against an enemy who was no longer distracted by another front. Coordinator Miyogi's grand, two-front offensive was finally showing signs of collapse because of exhaustion and because of the delays or inability to obtain supplies from the long, snarled supply lines.

Late in 2862, after observing signs of increased Davion activity in the Robinson area, the Generals of the DCMS came to Coordinator Miyogi to humbly ask that he call off the offensive before it turned into disaster. To their astonishment, the Coordinator agreed. Soon, the fighting between the AFFS and the DCMS tapered off, as both sides decided the time had come to rest their troops rather than attempt to gain extra ground. By 2864, no one would argue that the Second Succession War was over.

If the outcome of the war were measured in territorial gains, the Federated Suns was the definite winner, having taken so many planets from the Capellans. If the outcome were measured by which Successor Lord had advanced his claim to dominance over the others in the Inner Sphere, no one was victorious. If the outcome were measured by how many lives were lost and how the quality of life diminished for those who survived, then everyone lost that war.

THIRD SUCCESSION WAR

I'm glad that peace has finally returned to the Inner Sphere. Now, if logic and common sense could step in and make the peace permanent with a series of fair and equitable treaties, we could make war a thing of the past.

—From Prince Michael's Christmas Day Peace Proposal, 2863

MAN OF PEACE

Early in 2864, a group of Capellan and Free Worlds businessmen began to push for a permanent peace treaty among the five Successor States so that everyone could get on with the one truly important thing in life—business. Their efforts, dubbed "The Peace of Money Movement," received the wholehearted support of Prince Michael Davion, who had been sending out his own emissaries with feelers for a similar official peace pact. When the movement leaders arrived in New Avalon, the Prince gave them a royal welcome, and provided them with money and other means to travel through his realm with their message.

The other Successor Lords would not give the movement leaders the time of day, however, suspecting that Prince Michael's open and lavish support meant that the peace movement was just a front for the leader of the Federated Suns. That Archon Elizabeth Steiner had been making peace overtures similar to Prince Michael's was ignored, and the Peace of Money Movement disbanded early in 2866. The possibility that his own enthusiasm might have unwittingly sabotaged the Inner Sphere's chance for peace haunted Prince Michael for the rest of his life.

By 2866, it was becoming obvious that the unofficial peace would not hold. As the fighting among the other Houses resumed, Prince Michael kept the Federated Suns out of major conflicts. He reasoned that by avoiding involvement in the hostilities, the Suns could take on the role of mediator, eventually to impose a peace agreement on the other realms. In typically naive fashion, Michael further reasoned that this would lead him to ascend to the throne of First Lord.

OPENING GAMBITS

If war was as simple as Dimensional Chess, then all our Marshals would be brainy academic types. War isn't, and our Marshals definitely aren't.

—From *The Private Journals of a Loyal Marshal*, by Prime Marshal Peter Davion, Avalon Military Press, 2874

During the years of peace, the Draconis Combine became embroiled in a silent but deadly civil war. The two combatants were Coordinator Miyogi, who controlled the Combine government, and Roweena Kurita, who led the People's Reconstruction Effort, a government agency created to rebuild the war-damaged realm. By 2864, the PRE had become so powerful that it had as much influence over the general public as the regular government.

It was inevitable that the PRE and the Coordinator would end up in a struggle for control of the Draconis Combine. Instead of open warfare, which the PRE could not possibly have won, the battle would be played out with the Combine's secret police, the ISF, as its soldiers. Because the fighting between the two factions remained invisible except for the bodies of those who died, this internal struggle became known as the Shadow War.

In the effort to save face and not reveal any inner weakness, the Kuritans tried to keep news of this civil war secret from the rest of the Inner Sphere. Its efforts were in vain. The first to learn about the Shadow War was the Federated Suns, who found out in early 2865.

Marshals in the AFFS wanted to launch an immediate offensive to take advantage of the Combine's weak state. Wishing to remain a peacemaker, however, Prince Michael firmly denied any plans for invasion and forbade talk of it in his presence. This so infuriated Heir-Designate Carl Davion that he was ready to attempt to overthrow his father until Prime Marshal Peter Davion convinced him to think better of it.

The Shadow War between Miyogi and Roweena, fought through factions of the ISF, finally ended in 2865. House Kurita saw the internal war as a stain against their honorable destiny as the only proper leaders of mankind. When Coordinator Miyogi emerged victorious from the struggle with his sister, he ordered the Arm of the Dragon to purge this shame by punishing all outsiders who had learned of their troubles. The DCMS (which had been diplomatically neutral during the war) was relieved to have something to do. In 2866, in response not only to Lyran intelligence infiltration, but to Lyran economic imperialism and the vulnerability of defense plants in the Federation of Skye, the Dragon launched an offensive against the Lyran Commonwealth.

News of the Kurita offensive relieved the Marshals of the AFFS, who were afraid that Lord Kurita's revenge would be aimed at the Federated Suns. Accompanying this relief was renewed urging by the Marshals for an offensive to take advantage of the DCMS's diverted attention.

This time, Prime Marshal Peter Davion backed the Marshals. Duke Peter traveled to New Avalon from his headquarters on Kentares and personally lobbied the reluctant Prince for an offensive against the Draconis Combine. After two weeks of arguments, the Prince gave Peter Davion his blessing. Within a month after that, Golden Sunburst 'Mechs, fighters, and other vehicles were on the move.

AGENT OF THE MIIO

Ever since its creation during the First Succession War, the Ministry of Intelligence Investigations and Operations (MIIO) has been one of the most effective and creative information-gathering services in the Inner Sphere. Only the superb investigative abilities of our own Blessed Order surpass the talents of Davion agents. The reason for the MIIO's success is the strong dedication so crucial to any such agency's work. From the start, the MIIO has emphasized deep motivational and psychological training. Some of this training, such as week-long mental indoctrination to mold the recruit's spirit, is quite cruel. Nevertheless, it yields spectacular results. According to our own agents, there have been only a handful of cases where MIIO agents have broken under pressure and revealed Davion secrets.

Another important factor is the espirit de corps fostered among the recruits. All MIIO training bases are located either in the hinterland of some world or in the midst of a city whose citizens do not speak the Davion AngloFrench. This isolation forces the recruits to rely on one another for nearly everything. To further promote this spirit of cooperation, part of the training involves pairs of recruits being abandoned in the countryside or city without provisions. Their mission is to return to the base alive and unnoticed within a certain time period, but bearing some crucial bit of planted or real intelligence.

Almost as significant is the excellent education all trainees receive. Languages and the ability to lie, forge, and to disguise oneself are emphasized. Study of the long, colorful history of the MIIO is an equally vital part of the trainee's course of study. By the time he graduates, each recruit will know every important event and hero in the Ministry's history. This gives him pride in the agency's tradition and encourages a positive competitive spirit to achieve similar excellence.

One of the MIIO's most incredible feats was their discovery of the factional fighting that threatened to tear the Draconis Combine apart during the Shadow War. How the MIIO managed to learn of this conflict years before anyone in the other Successor States may be explained in several ways. With the ISF in such a state of chaos, the MIIO could have gotten wind of the trouble from some low-grade leak, such as a talkative peasant janitor, or simply by keeping their own eyes and ears open.

Another possibility is that the need for new *metsuke* to replace those being killed in the Shadow War resulted in loosened security checks for new recruits. The ISF could conceivably, but unwittingly, have recruited MIIO agents into its training program.

There is another intriguing possibility. Considerable evidence suggests that one of the ISF's top *metsuke*s, Dark Assassin Kelly Urizos, was actually a double agent for the Federated Suns. If this is true, it would mark the first and only time that one of the vaunted Dark Assassins, chief administrators and scourges of the ISF, has betrayed its allegiance to the Coordinator of the Draconis Combine.

—From "A Consideration of the MIIO," by Precentor Julius Minock, Comstar Bulletin, Archives, Terra, 3002

OPERATION PENDRAGON

ATTENTION ALL PDZs. Order code is as follows: Papa, Echo, November, Delta, Romeo, Alpha, Golf, Oscar, November 559. Personal comments to all loyal members of this AFFS mission: Like a pack of hungry hounds, we have at last been unleashed. Let us seize the opportunity to take a bite out of the Dragon.—Prime Marshal Peter Davion.

—Order for the Davion offensive against House Kurita, 2869

Because Davion Field Marshals had drawn up most of the plans for the offensive (codenamed Operation Pendragon) months earlier, the offensive was underway soon after Peter Davion got his Prince's permission. The first stage of the operation included numerous heavy raids up and down the Kurita border to discover the enemy's weaknesses and to win back the planet Robinson.

Despite the advance planning, Operation Pendragon seemed to be in trouble from the start. The Kurita defenders, few in number because so many were taking part in the offensive against the Commonwealth, were well-supplied and highly motivated. The elite Third Sword of Light regiments and its attendant lesser regiments were securely entrenched on Robinson, and the regimental commander, General Hirushi Tengwan, was an able leader.

The AFFS would soon have help for its efforts, and from an unexpected quarter. In the early days of the campaign, Coordinator Miyogi sent General Taragi Kurita to report on the progress of his forces along the Davion front. A cunning man well versed in the ways of the military bureaucracy, Taragi's ambitions would soon influence the course of the fighting.

In 2872, the Fourth Avalon Hussars were in deep trouble on Colia. Cut off from the rest of its force, the unit became isolated on a small continent where the Kuritans had concentrated many of their forces. Unable to hold a position long enough for evacuation, the 'Mech regiment had been on the run for three weeks, with two Kurita 'Mech regiments now closing in.

Trapped finally in a dead-end valley, the commander of the Hussars realized that he and his men were probably facing their last battle. They were surrounded by mountains, and they could see the dust of 'Mechs approaching through the only pass. The one odd thing was the absence of Kurita fighters making their lives hell from the air. Yet the fight never came, and the approaching columns of smoke seemed to vanish in thin air. As the Hussars cautiously moved forward, they saw half a regiment of Kurita 'Mechs standing idle while Techs worked frantically in the innards of the metal beasts.

Not ones to pass up an opportunity, the Hussars attacked immediately, leaving the Kuritans stunned and with their numbers decimated. When the Hussars interrogated some of the survivors about what had stopped their advance, they heard a true military horror story.

Apparently, a snafu in procurement had resulted in the unit receiving radiation shielding intended for tanks rather than 'Mechs. The Techs had no choice but to try to adapt the shielding for the unit's machines. The shielding proved not only inappropriate, but was of such a poor grade that it disintegrated after rough usage had compromised the 'Mechs' cooling systems, forcing them to completely shut down. A similar mixup had also kept the Kurita air support from doing its job. The Davion officers did not know what to make of this account, for the Kuritans had often turned the provision of supplies into a veritable military art.

It came out later that General Taragi Kurita had himself sabotaged the supply lines in order to make General Tengwan appear to be incompetent in his report to Coordinator Miyogi. Much to Taragi's surprise, Miyogi not only agreed with his conclusions that the commanders on the Davion front should be replaced by a single, supreme commander, he was immediately appointing his son Jon to the post, rather than Taragi himself.

PRIVATE LITTLE WAR

How in hell am I supposed to know where they went? All I know is that they've deserted, and if I ever get my hands on them, I'll slit their throats. That's assuming I live through this, of course.

—Message from Colonel Joan Didier, commander of the 45th Galax Tank Regiment, to the New Syrtis Regional Command. Published in *The Dangerous Hand: The Federated Suns and Mercenary Units*, by Duke Robert Tagart, Syrtis Military Press, 2987

In 2866, Dainmar Liao resigned as Chancellor of the Capellan Confederation, unable to face the strains of a new war. Otto Liao, his son, quietly assumed command. For the Capellan military, the change in leadership did not come a moment to soon. Under Dainmar, the Capellans had lost so many worlds to battle and to treaty that the Confederation was becoming uncomfortably thin between the Federated Suns and the Free Worlds League. The change of leadership in the Confederation did not particularly please the leaders of the Federated Suns, of course, for they had hoped to continue taking advantage of Dainmar extract a few more worlds.

The accession of Otto Liao, a veteran MechWarrior, immediately revitalized the Liao military forces. Seeing that this new Chancellor would rebuild the strength of his military, the AFFS High Command wanted to go in and seize certain Capellan worlds now, while the enemy forces were still weak.

The attack ran into immediate trouble when the Davion forces found themselves several hundred kilometers from Verlo's capital city because of flawed calculations in their drop procedure. They also quickly discovered far more and far tougher defenders on the planet than expected. Because Marshal Hasek had not wished plans for his private offensive to leak out, he had not consulted with Military Intelligence or the MIIO to determine the enemy's strength. That failure now came back to haunt him.

A year into the offensive, however, the initiative had shifted, and the Davion forces were acquitting themselves well. Though many Capellan troops were still on the planet, the Davions would certainly force their retreat offplanet before long. Though the AFFS had caught on to Field Marshal Hasek's true purposes by now, the High Command decided to wait out the offensive before deciding whether to court-martial him or award him a medal of honor.

In March 2870, a a Jump Ship appeared in the Verlo system, and sent a small courier vessel down to the planet. Believing it to be a ComStar courier, the Davion ships parked at the planet's zenith jump point let the little vessel pass unchallenged. The courier landed near the headquarters of the Davion forces on Verlo. A man and a woman dressed in unmarked overalls emerged, asking to see the commander of Clinton's Cutthroats rather than the Davion commanding officer.

For the next three days, the mysterious strangers met privately with Colonel Clinton, all the while Hasek was demanding some explanation. Clinton's only reply was a warning to the Commander to stay out of the Cutthroats' affairs.

On the fifth day, three more JumpShips arrived in the Verlo system, carrying seven DropShips, unmarked and all in pristine condition. Two weeks later, the DropShips had landed on the planet, with the apparent intention of transporting the Cutthroats offworld. The infuriated Davion commander demanded that the Cutthroats honor their contract. When Colonel Clinton refused, the Davion forces attacked the mercenaries, but it was an uneven fight because the Davion troops lacked 'Mechs. The Cutthroats left Verlo, and have never been seen or heard from since.

With the Cutthroats gone, the Davion forces on Verlo suddenly found themselves outgunned. In the battles that followed, three regiments of Davion front-line troops were slaughtered before the force could evacuate.

As for the Cutthroats, the DropShips that came to collect them were of a design not seen in the Inner Sphere since the exodus of General Kerensky. This has led to theories that the Cutthroats were "scouts" of the "Kerensky exiles" sent to monitor events in the Inner Sphere. Others have sworn that members of the Cutthroats have shown up among the mercenary Wolf's Dragoons. Regardless, the Clinton Cutthroats have joined the mysterious disappearances of the Minnesota Tribe, the Vandenberg White Wings, the Disappearing Battleship of Merope, and, of course, General Kerensky and most of the Star League Regulars.

For Field Marshal Hasek, this mystery was not as important as the fact that he must now face Prince Michael Davion and Prime Marshal Peter Davion as the man responsible for the loss of several whole regiments. Though Marshal Hasek's record was otherwise impeccable and his reconnaissance-in-force had netted the Federated Suns three important worlds, the AFFS could not simply sweep the Cutthroat Fiasco and the death of so many men under the rug. Field Marshal Hasek was court-martialed and demoted three steps in rank. As further punishment, he was transferred from active duty to an administrative post in the military bureaucracy.

Operation Pendragon, the Federated Suns offensive against the Draconis Combine, had already drawn away most of the AFFS reserves, however. There were very few front-line troops to spare for any lightning raids to seize Capellan worlds. Further, it would probably have been impossible to persuade Prince Michael to open a second risky offensive when it had been difficult enough to get his approval for Operation Pendragon.

Lack of approval for the mission did not bother Field Marshal Jerome Hasek, commander of the Sirdar PDZ. With so many of the coveted Capellan worlds within easy reach, he planned to use a tried and true method of avoiding orders to accomplish his aim. At the moment, the forces he commanded were inadequate to the job, and so Field Marshal Hasek began to hire mercenary units, who he paid out of his own pocket. When the AFFS asked why he was hiring so many mercenaries, he replied that he feared a Capellan offensive. Though highly improbable, the excuse was enough to keep the High Command off his back for the time being.

In 2869, Field Marshal Hasek hired the Clinton Cutthroats, which he considered a real stroke of luck. A 'Mech regiment with a long history of honorable service in the Free Worlds League, the Cutthroats had come to the Federated Suns because they wanted a change from the wrangling House Marik style of politics. With this veteran unit among his forces, Field Marshal Hasek was now ready to launch his private little war.

To bypass the orders not to attack the Capellans, Hasek used the infamous "Reconnaissance-in-Force" technique. He sent reconnaissance forces onto Capellan worlds, knowing that his men would be outmatched. The need to rescue these troops would trigger huge operations, full-scale offensives that would allow him to seize the Capellan worlds. The ploy was totally illegal, but as long as the Field Marshal couched his orders with words like "protective defensive actions," "short-term aggressive actions," and "minimal strategic seizure and control," the High Command could do little to stop him.

The Clinton Cutthroats was the lead unit in the Field Marshal's attempt to seize Verlo, a resource-rich world that had recently been the focus of considerable Capellan industrial growth. Hasek thought it would make an excellent addition to the Federated Suns, and he sent along seven other regiments and fighter wings to help the Cutthroats.

FALL OF DAVID

I don't give a flying fig for talk and diplomacy. Those things are for the weak-hearted and the addle-pated. Battle and bed—that's where I perform best.

—Prince Carl Davion, quoted from *My Life as a Royal Camp Follower*, by Jessica Jerloos, Risque Press of Odell, 2877

When Prince Michael Davion died in 2873, his 45-year-old son Carl succeeded him. The new Prince saw life in simple terms. To him, a Davion should be doing only one thing with his life—piloting a 'Mech. Because Carl was such a rough-mannered, sharp-tongued man, certain branches of the family actually broke off relations with the Prince, refusing to participate in the government. This would one day hurt the Federated Suns.

Meanwhile, Operation Pendragon had been under way for several years and was meeting with surprising success. The Draconis Combine's inability to properly supply its forces on the Davion front had become a chronic flaw, much to the delight of the Davion commanders.

On the Capellan front, the initiative gained by ex-Field Marshal Hasek's "reconnaissance-in-force" still existed, creating a definite bulge aimed at Sian. Gone were the days of easy victories against the Capellans, however. The Confederation had definitely shrunk to a size where it could respond easily to enemy incursions. In addition, resurgence of the Lorix Order warrior philosophy among the Liao 'Mech forces was making the Federated Suns pay dearly for its gains.

With Prince Carl at the helm, many in the Federated Suns expected the realm to enter a new phase of warfare. Indeed, Carl was soon off to the Draconis front to lead his favorite regiment, the Fourth Royal Guards. Skilled in all the military arts, he was definitely qualified to relieve the aging Prime Marshal Peter Davion of command of the AFFS.

Unfortunately, Prince Carl's military promise would never be fulfilled. In 2876, he led the first-wave assault against the planet David, but it was against the advice of his Field Marshals. Because David had been one of the first worlds to fall to the Kuritans in another, earlier war, it had important psychological as well as military value to the Davion realm.

Though the DCMS force on David was suffering from the same supply problem plaguing the whole front, they were a highly motivated force. The Kurita 'Mechs quickly contained and isolated Prince Carl's invasion points. Faced with a desperate situation, Carl decided that only a massive frontal assault against the enemy could save the day. Shouting encouragement mixed with curses, the Prince took the van of his Royal Guards, charging his *Marauder* straight into the teeth of five Kurita strongholds.

Guessing that the lunatic charging them had to be a commanding officer, the team of Kurita sharpshooters manning a twin PPC turret took aim and fired. Both salvos caught the *Marauder* full on. Screaming incoherently, Prince Carl forced his 'Mech to stumble forward. By this time, the other turrets in the stronghold had also begun to concentrate their fire on his *Marauder*. Amazingly, it withstood the punishment, and the mortally wounded Prince Carl approached the stronghold in his 'Mech through sheer force of will. Like some wounded beast, the Prince's 'Mech raised its arms and brought them crashing down upon the walls of the fortress.

Then the 'Mech collapsed, killing its maddened pilot. His attack, however, succeeded. Though the Kurita defenders would put up many months of spirited defense, the planet David would, in the end, fall to the army of the Federated Suns.

CREATION OF THE MODEL ARMY

I too enjoy the thrill of combat, but I am not foolish enough to allow myself that luxury. I know how much my people need me, and that is my greatest responsibility. My brother was a fool; I am not.

—Princess Melissa Davion, in an unpublished letter to Field Marshal Lennox, a trusted adviser and friend, 2877

Like her elder brother Carl, Melissa Davion had served long and illustriously in the AFFS, leading the Second Avalon Hussars to several key victories on both the Capellan and the Draconis fronts. Nevertheless, when she received the news of her brother's death on David in 2876, she immediately resigned her command.

As new Princess of the Federated Suns, the 47-year-old Melissa decided to make some sweeping changes in the AFFS. Though BattleMechs had dominated the battlefield for the past two centuries, the loss of production facilities and the difficulty of keeping 'Mechs in repair had given the ancient weapons of infantry, tanks hovercraft, and artillery an important role once more. Warfare was evolving and Melissa understood the need for new methods.

One of Princess Melissa's first and most important reforms was to bind together regiments from various branches of service into Regimental Combat Teams. Until now, there had rarely been any full-strength RCTs in the Davion armies, for regiments were often split up to perform various missions. An RCT was usually created solely for a single mission and then disbanded when the mission was complete. Though this system was effective enough, it did not foster much cooperation between the various service branches. Under Princess Melissa, RCTs became permanent, so that crucial cooperation between regiments could develop.

Her other reforms concerned the military bureaucracy, which had grown top-heavy with departments and agencies. By streamlining the AFFS administration, particularly the Procurement and Military Industry Liaison, she made it easier for supplies to reach the front and helped ease the tensions that sometimes strained the relations between the AFFS and the over-burdened military industry.

Well aware that some would consider her reforms an attack on tradition, Princess Melissa moved slowly but surely. For example, she often called in her Marshals and ranking members of the nobility for their counsel on some matter, whose pros and cons she presented persuasively and objectively. As often as not, their advice echoed Princess Melissa's own intentions, and everyone was glad to see the mutually agreed-upon solution become law.

By 2890, Princess Melissa had what she called her "Model Army." New Regimental Combat Teams had been deployed to face both the Kurita and Capellan enemy on the front lines. Early reports of skirmishes using the new units were encouraging. The bureaucracy was also moving supplies with greater speed and efficiency, while production was up among the military industries.

Princess Melissa felt the time had come to test her Model Army.

BATTLE FOR ROBINSON

When two armies of equal strength face off, victory will go to the side whose soldiers genuinely care for each other and will, therefore, willingly risk their lives for their comrades.

—Prince Melissa Davion in an unpublished letter to Field Marshal Gerald Lennox, 2892

That House Kurita had taken and kept the planet Robinson for so long was a constant source of anguish to the people of the Federated Suns. Over the years, the AFFS had attempted several times to win back this world that was the traditional capital of the Draconis March, but no commander had been able to penetrate the ferocious Kurita defenses.

In recent months, however, the MIIO had gathered intelligence suggesting that the Kurita forces on Robinson, led by the elite Third Sword of Light, were having supply problems. Indeed, the resistance movement on Robinson had managed to score several victories because the Kuritans seemed unwilling to waste men and materiel to combat them. Other evidence, such as a decline in Kurita raids, pointed to the severity of the supply situation. Sensing the enemy's vulnerability, Princess Melissa decided that the Second Avalon RCT should move against Robinson in an attempt to finally take back that world. The commander of the mission would be Marshal Joseph Davion II.

Joseph Davion II was the son of Prince Michael's later years. The boy was only nine when his father died, and Melissa Davion had raised him almost as her own child. Perhaps it was at her knee that he had learned to love battle as much as his father had seemed to abhor it. Many seasoned members of the AFFS were distressed at his youthful abandon when it came to military matters. Enamored of strategy and tactics as much as he loved combat, Joseph rushed into bold new adventures, yet would willingly toss out any carefully laid plan if a sudden battlefield opportunity seemed to call for it. Despite this, Joseph rose in rank like a comet. When promoted to command the Second Avalon RCT at the age of 27, he became the youngest Marshal the AFFS had ever seen.

Princess Melissa had confidence in her half-brother, however, and that was all the support he needed to plan a bold invasion of Robinson that would take advantage of his RCT's abilities. He decided that instead of dropping onto safe areas of the planet, away from the enemy and rough terrain, his units would land right in the enemy's midst. He was certain that the advantage of surprise would permit his regiments to make best use of their ability to coordinate efforts. "I'll not have a stupid dance upon some barren plain that decides nothing! Let the fight be immediate, bloody, and decisive," the Marshal reportedly told his subordinates. He got his wish.

The attack did surprise the Third Sword of Light, which had seen the AFFS DropShips coming but had expected a more conventional choice of drop zones. They fell back, but did not run. The Third Sword of Light was, after all, an elite 'Mech regiment. Mounting a counterattack, the Third had soon forced the RCT to halt its advance. Toe to toe, the two forces fought it out among the buildings of Robinson's capital city.

The vicious fighting continued for a whole month. In actions between small units, the Kurita forces were equal to the Davion units. In large-scale actions, however, where coordination between 'Mechs and other combat units was crucial, the Second Avalon RCT excelled. The Kuritans were also plagued by the difficulty with obtaining supplies, to the point that they were running out of ammunition. With no sign that he would be getting his supplies anytime soon, the Sword of Light commander decided that he had no choice but to pull his regiment out of Robinson. It would be bad enough to lose the planet, but the Kurita commander knew that neither he nor his Coordinator could bear the loss of face if he allowed one of the Combine's most prestigious units to be destroyed.

With the Third Sword of Light out of the action on Robinson, the remaining Kurita regiments were no match for the Second Avalon RCT. Soon after, the planet Robinson was liberated at last. Marshal Joseph Davion was a hero, and Princess Melissa had seen her military reforms work to perfection.

GENERAL TARAGI'S CUNNING

The final Battle of Robinson was a stunning defeat for one of the Draconis Combine's best regiments. The outcome might have been different if the elite Third Sword of Light had not been hamstrung by severe supply problems. The irony and the tragedy of the situation was that one of House Kurita's own commanders had purposely created the supply shortages to advance his ambition to become Coordinator of the Draconis Combine. The man who was willing to sacrifice his people's own military on the altar of power was Taragi Kurita.

At the time of the Hussars' invasion of Robinson, Taragi Kurita was chief aide to General Jon Kurita, commander of the Davion front and heir-apparent to the Coordinatorship. Taragi had long coveted the post of Davion Front Commander, and had been willing to disgrace his fellow generals in the battle of Colia Mountains to appear as the only one worthy of the job. When the Coordinator gave the post to his son Jon, Taragi wanted revenge on both Jon and the Coordinator himself.

Roweena Kurita, who challenged Coordinator Miyogi during the Shadow War, was Taragi's ally. She supported his efforts at sabotage as well as giving him valuable information and counsel. The wisest of this advice was that Taragi must always use finesse and not move too boldly against Jon or the Coordinator.

To strike at Jon Kurita, Taragi began again to manipulate the bureaucracy of the DCMS, throwing the supply situation up and down the Davion front into disarray. When the Federated Suns attacked Robinson, General Taragi saw his chance to disgrace Jon and the Coordinator, and he made sure that the Third Sword of Light would not receive supplies for weeks.

After the Third Sword was forced to abandon Robinson, their commanders were outraged. As the best soldiers in the DCMS, their grievances had an impact in the high military and governmental echelons of the Combine. It was not long before they were demanding that General Jon Kurita step down in favor of General Taragi.

Coordinator Miyogi was in a serious bind. If he defended his son, the military would be offended, for they despised poor commanders. If he did not defend his son, the Coordinator would be dishonored for not defending his own blood. When he turned to his generals for support, all took a neutral stand.

The final act was played out when Captain Hideyoshi Toyama of the Third Sword of Light entered the Imperial Palace with a bomb strapped to his chest. He exploded the device in the main hall of the palace, killing himself, Coordinator Miyogi, and 30 other people. How he managed to enter the palace undetected is unknown, though he must have had help from Roweena Kurita, who could still call on the secret loyalty of some ISF agents.

General Taragi arrested General Jon Kurita and made him stand trial in what was more like a kangaroo court than an act of justice. Jon Kurita, the only male offspring of Miyogi, was shot. By succession right and public support, Taragi Kurita was acclaimed as the new Coordinator of the Draconis Combine.

—From "An Analysis of the Draconis Combine Political Structure of the Early Third War," by Precentor Yvonne Rothram, Comstar Bulletin No. 65834PF, Archives, Terra 3013

PRINCE JOSEPH THE SECOND

This is how I see society. At the top are the Davions. Next come the nobility and the military. Then the government. At the bottom are the common citizens. The bottom serves the top. This is the way all societies have been since the first time a strong man declared himself king and demanded that the weaker members of the tribe serve him. Why tamper with the natural order of things?

—Prince Joseph Davion II, 2898

Not long after the victory of her forces on Robinson, Melissa Davion suddenly contracted Joshallan Fever in November of 2892. Ironically, this was a disease for which there had once been an effective treatment, but with the decline in pharmaceutical knowledge, the doctors could do little more than shake their heads sorrowfully as their ruler lay dying.

With her death, Joseph Davion II became Prince of the Federated Suns at 28 years of age. Like Carl and Paul before him, he was an active military campaigner who governed his realm from the cockpit of his silver *Battlemaster*. A charismatic man who had a way with words, Joseph made frequent trips to the front lines to boost the morale of his troops.

Unlike his immediate predecessors, Prince Joseph did not take well to advice, which meant he was surrounded either with sycophants or cronies who had no wish to cross their friend. At least once a year, Prince Joseph liked to go from New Avalon to lead an offensive on one of the fronts. During these campaigns, which might last six months or more, the government was paralyzed because the Prince refused to delegate his authority to others. It was not long before both the economic and technological decline caused by the war grew even worse because of Joseph's neglect.

Despite the new emphasis on cooperation between all branches of service with the AFFS, MechWarriors still formed an elite with special privileges. The Prince often granted landholdings and considerable local power to MechWarriors who had shown special prowess on the battlefield, thought most warriors knew nothing about administering land or governmental affairs. When these landholding MechWarriors were suddenly called to the front, their absences would seriously disrupt local government the way Prince Joseph's absences affected his non-military duties.

The one wise and constructive act of Prince Joseph's entire reign was the Armistice of Van Diemen IV. In 2903, a particularly wasteful aerospace battle for the Liao mining colony of Van Diemen IV left the Capellans anxious for a temporary cease-fire with the Federated Suns. After communicating their desire to negotiate, they waited.

High government officials tried to locate Prince Joseph, who was leading an assault on the Draconis front, far from New Avalon. When they finally were able to pass on the message from the Capellans, Joseph replied that he scorned any talk of peace that did not come from the end of his gun.

No matter what the government officials said about the advantages of a cease-fire, Joseph would not budge. When he would no longer listen to their arguments, several officials sought help from the Marshals of the High Command, who also agreed to talk to Prince Joseph. It was they who finally managed to persuade the Prince that a cease-fire on the Capellan front would allow him to conserve supplies for the wars against the Draconis Combine as well as giving him the image of a peacemaker in his people's eyes.

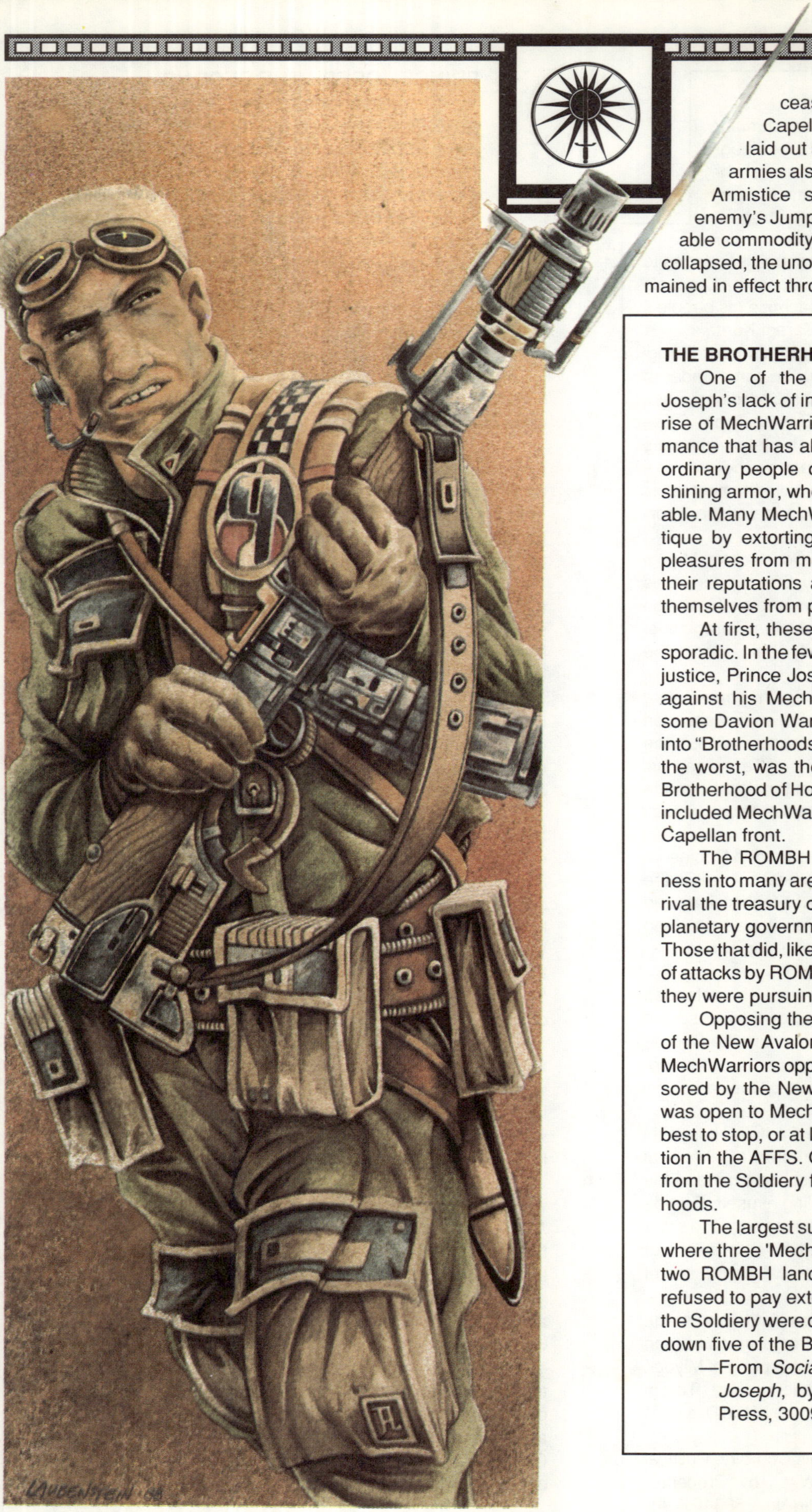

The Armistice was signed in 2905. The cease fire would remain in effect along the Capellan front for an amazing 25 years, and it laid out some rules of war that other Inner Sphere armies also unofficially adopted as time went by. The Armistice specifically forbade actions against an enemy's JumpShips, which were now a nearly irreplaceable commodity. When the Liao-Davion cease-fire finally collapsed, the unofficial ban against attacking JumpShips remained in effect throughout the Inner Sphere.

THE BROTHERHOODS

One of the cruelest consequences of Prince Joseph's lack of interest in governing his realm was the rise of MechWarrior Brotherhoods. Because of the romance that has always attached to MechWarriors, the ordinary people often looked to them as knights in shining armor, who must be both chivalrous and honorable. Many MechWarriors took advantage of this mystique by extorting money, lands, and even personal pleasures from members of the public and then using their reputations as protectors of the realm to shield themselves from prosecution.

At first, these lawless acts were only random and sporadic. In the few instances when a victim did try to get justice, Prince Joseph would not hear of such charges against his MechWarriors. Given this tacit approval, some Davion Warriors organized their illegal activities into "Brotherhoods". The first Brotherhood, and perhaps the worst, was the Royal Order of MechWarriors and Brotherhood of Honor. Formed on New Syrtis, this gang included MechWarriors from all the regiments along the Capellan front.

The ROMBH grew quickly, spreading its lawlessness into many arenas. Indeed, their income was said to rival the treasury of the entire Davion government. Few planetary governments could stand up to the ROMBH. Those that did, like Bethel and Andro, became the object of attacks by ROMBH 'Mechs, whose pilots claimed that they were pursuing Capellan infiltrators.

Opposing the ROMBH and its ilk was the Soldiery of the New Avalon Catholic Church, an association of MechWarriors opposed to the corruption. Though sponsored by the New Avalon Catholic Church, this order was open to MechWarriors of all beliefs. They did their best to stop, or at least to slow, further moral degeneration in the AFFS. On several occasions, MechWarriors from the Soldiery faced off with and fought the Brotherhoods.

The largest such battle occurred in 2921 on Hobbs, where three 'Mechs from the Soldiery attempted to stop two ROMBH lances from punishing a city that had refused to pay extortion money. The three 'Mechs from the Soldiery were destroyed, but not before they brought down five of the Brotherhood 'Mechs.

—From *Social Decay During the Reign of Prince Joseph*, by Duke Michael Getherton, Avalon Press, 3009

AID FROM AFAR

Though the armistice with the Capellans freed up troops and resources for the struggle against the Draconis Combine, Prince Joseph's offensive against House Kurita was slowing up. It had been three years since he had been able to take a world from the DCMS, but the Prince blamed everyone but himself for his army's poor performance.

There were many reasons why the Davion military was failing. For one, the AFFS was worn out after being on the offensive for more than a decade. Another reason was the wretched condition of the realm's economy. The many enemy invasions had crippled the military industries, with war supplies becoming more and more dear. In addition, insubordination among MechWarriors was becoming almost commonplace, with warriors disobeying orders when they pleased and picking fights with troopers of other services.

In 2913, the Lyran Commonwealth launched an ambitious offensive against the DCMS. Codenamed Operation Freedom, the campaign marked a surge in confidence among the Steiner fighting forces. Shinjiro Kurita, Taragi's son and the new Coordinator, believed at first that the Lyran Commonwealth's successes were a mere fluke. As the enemy offensive showed no sign of letting up, however, the Coordinator had to commit ever more resources to combat the LCCS.

Prince Joseph was elated when news came of the Commonwealth offensive. This was just the opportunity he needed to rest his forces, do a little internal head-bashing, and watch for any particular weakness that might be developing in the Draconis Combine. When both Military Intelligence and the MIIO reported no enemy activity along the Kurita front, Prince Joseph issued a partial stand-down order in 2914. Many units that had been stationed on the front lines for decades were now rotated to the rear.

Joseph felt this would also be a good time to call a halt to the growing rebelliousness among his MechWarriors. After recruiting the Soldiery of the New Avalon Catholic Church to his purposes, Prince Joseph began a grand tour of his realm. Though the official reason for his visits was to promote peace and well-being in the realm, the real reason was to break up the MechWarrior Brotherhoods. Aided by his personal guards and two companies of 'Mechs from the Soldiery, Prince Joseph managed to "persuade" most Brotherhoods of the error of their ways.

The Royal Order of MechWarriors and Brotherhood of Honor, the largest and most virulent of the warrior gangs, was not so easily cowed. In what amounted to a military campaign, Prince Joseph and his 'Mech force engaged the Brotherhood on Novaya Zemlya, Kluane, and Fortymile, where he and his men burned the headquarters of the Brotherhood to the ground. Though the action cost Joseph more than 20 warriors and twice that number of 'Mechs, the Prince did gain unexpected popularity. The public now saw him as a savior, despite his responsibility for the rise of the Brotherhoods in the first place.

For the next two years, the Kurita border was so quiet that Joseph actually began to pay attention to the less exciting affairs of his realm. He authorized some moderate policies that helped to stimulate the realm's economy, and transformed the AFFS warehouses from empty caverns into rich stores of parts and supplies.

For two years, Prince Joseph closely monitored the Lyrans' offensive against the Draconis Combine, and like many of his people, began to change his attitude toward House Steiner. Previously, the Davions and their people had regarded the Lyrans as money-hungry conmen who could not muster the courage to fight their way out of a paper bag. As Operation Freedom continued to prove the Steiner military's courage and ability, Joseph began deeply to admire the determination of the Lyrans *and* the ability of their leaders, the Steiner family.

It was for this reason that Joseph decided to renew his offensive against the Draconis Combine a year earlier than his Marshals were advising. He believed this would allow him to take advantage of the overtaxed Draconis Combine and that it would also help the Commonwealth's campaign.

In 2918, the AFFS opened its offensive with attacks on contested worlds as well as heavy raids into the Draconis Combine against the planets Lapida II, Arlington, Huan, and Capra. Davion commanders managed to catch the Combine garrisons on Lima and McComb off guard, and succeeded in pushing them off the planet within three months. The raids into Combine space were also successful as Davion 'Mechs managed to loot and destroy three major DCMS supply centers.

Though the Kuritans had their hands full on the Steiner front, the Arm of the Dragon eventually rose up to strike back at the Davion advance. In an AFFS assault on Xhosa VII, the Kuritans drove back the Davion troops, inficting heavy casualties. Though Prince Joseph was enraged, the past five years had given him the wisdom to restrain his desire to retaliate with an immediate all-out attack.

THE ARMISTICE ENDS

It's ridiculous to let the Capellans sit there and catch their breath. They're vulnerable, and we should be all over them like a bad smell.

—Field Marshal Peter Davion II, 2928

In 2928, Prince Joseph's son, Peter Davion, was 29 years old and a talented military leader in his own right. In that year, he was commander of the AFFS in the Capellan March, but that did not mean much because the Armistice of Van Diemen IV was still more or less in force. Like his father, the Prince Imperial liked nothing so much as the thrill of combat and so his current command had him feeling frustrated and restless. When Chancellor Merlin Liao abandoned his father's theory of "elastic defense" in favor of protecting certain key worlds, Field Marshal Peter Davion believed this was a golden opportunity to attack. He was anxious to seize Capellan worlds before the Chancellor returned to his senses.

For two years, he argued for a campaign. Finally, in 2930, his father gave his permission to break the armistice. The Field Marshal's operation, codenamed "Roland's Horn," would make two thrusts into the Confederation area lying near Terra. One thrust would push off from the Davion world of Addicks into Ankaa, Deneb Kaitos, Caph, Northwind, and Ruchbah. The other thrust, led by Field Marshal Davion himself, would advance from Demeter against the worlds of Halloran V and New Aragon.

In their attack on New Aragon, the First Avalon RCT achieved almost total surprise. Heavy aerospace support knocked down the Capellan air presence on the planet, and the three regiments were able to drop down virtually unopposed. Because of the Chancellor's orders to avoid losses, the Liao commanders retreated their forces offworld after only token resistance.

It was much the same with the Davion attacks on Deneb Kaitos, Ruchbah, Ankaa, and Halloran V. The only places where the Liao military put up any kind of resistance was on secondary AFFS targets such as New Hesse, Kawich, and Nopah. As Field Marshal Peter Davion assembled his forces for the second phase of Roland's Horn, news arrived from the Draconis front that his father had been seriously injured at the Battle of Xhosa VII.

XHOSA VII

'Taint nothing but a scratch.

—Prince Joseph, after his 'Mech was gunned down by a Kurita *Slayer*

The Draconis Combine had occupied the rich world of Xhosa VII ever since the First Succession War, and it was the center of supply and reinforcement for the DCMS along half the front. The planet was so heavily defended that General Sakamoto, the Kurita commander, did not expect to see any combat on this world in the near future. Like the rest of the military leaders of the Combine, he was being lulled into a false sense of security by the MIIO's vigorous disinformation campaign showing that the Federated Suns had been involved in a huge offensive against the Capellan Confederation for the past three years. While the AFFS communication bands were filled with bogus reports of Capellan actions, Prince Joseph was quietly mustering his forces around Xhosa VII.

In early 2930, he struck. Because Xhosa was a transportation hub for the Combine, Davion JumpShips and DropShips with false ID codes slipped into the system unnoticed amid the coming and going of so many other ships. The DropShips were halfway to the planet before the Kuritans identified them as Davion warships.

Prince Joseph was wounded during the initial assault. After a fierce fight for aerospace superiority, Prince Joseph believed that his fighters had won the day and that it was safe for him to land. As he left his DropShip, however, a pair of Kurita AeroSpace Fighters sneaked through the Davion defenses to bomb the Prince and his guards. One bomb exploded squarely against the Prince's *BattleMaster*, severely injuring him.

With the Davion Prince out of the fight, General Sakamoto hoped that he might be able to hold out against the invaders until the DCMS could send reinforcements. It was not to be, for the new leader, General Wallace Mickertrick, had both the skill and personal charisma to dispel the pall that had fallen over the Davion forces. It was not long before General Sakamoto found himself being pushed back.

The Kurita garrison suffered a final setback when Mickertrick killed Sakamoto in a 'Mech duel just outside of Xhosa's capital city. Now, it was the Combine's turn to suffer the loss of a valued leader. Xhosa VII fell in 2931.

PRINCE PETER

We will not win this war through attrition. Too many other warlords have tried and failed with the strategy of exhaustion. A subtler strategy will win this war, and one day the Sunburst will fly over a united League.

—Prince Peter Davion, quoted from *The New Davions: Politics Over PPCs,* by Duke Reginald Timons, Avalon Press, 2955

The explosion that had shattered Prince Joseph's *BattleMaster* had left him in a state of permanent injury. His internal organs had been badly damaged, either by the explosion or because of the radiation that leaked from his damaged 'Mech. He could spend the remainder of his life hooked up to medical life support systems, or else die.

To a man who enjoyed the thrill of battle and the air of alien worlds, the thought of a life chained to a machine was unbearable. Though he hoped to find solace in the political side of government, that had never been his pleasure or his inclination. For a warrior like him, what could compare to leading a charge of BattleMechs or to watching a vanquished enemy come forward, white flag in hand?

This trooper wears a modified uniform of the Crucis Lancers, a green and tan jerkin instead of the standard issue. She is holding an electro-plating glue gun used for minor machinery repairs.

In 2931, Prince Joseph II abdicated in favor of his son Peter. Though his body's growing deterioration prevented him from participating in the government, he lived another four years after his abdication, which was long enough to see his son's new style of leadership.

When he died, it was the end of an era. During the time of Joseph and his father, the Davions were more concerned with battle than with statesmanship. It many ways, it was a simpler, more direct age. With the accession of Peter Davion to the Princehood, things would begin to change.

Though as much a military man as his father, Prince Peter understood the desperate need for a strong government that did not try to settle every dispute through war. He saw that he must learn to take a longer view of federation affairs. Realizing that the responsibilities of government were a challenge equal to anything the battlefield had to offer, Peter began to relish the job, and rarely participated in combat during his reign. Indeed, he even made so bold as to declare in his first public address that he intended to restore the Davion government to its former heights.

Though the days of the infamous Brotherhoods were over, landholding MechWarriors were still strong enough to resist reforms from above. Paul's creation of a feudal nobility more than a century before was coming back to haunt his heirs. In Paul's day, the warrior aristocracy had owed everything to their Prince, but the hereditary descendants of those warriors owed House Davion very little. If displeased with one of these warrior aristocrats, Peter could not simply remove the noble family from their lands and titles without sparking a full-scale revolt. Resorting to legal trickery to siphon off power from the warrior nobles would have been just as risky.

Prince Peter feared that these warrior nobles would realize their political strength and try to become kingmakers within the federation. As the head of House Davion, he was not about to allow that to happen. The situation was inherently unstable, and Peter would devote most of his reign to seeking a solution. While the war continued along the frontiers, Peter reformed the government where he could, and found ways to placate the nobility when reforms frightened them. Knowing that he could not directly attack the power of the MechWarrior aristocrats, he attempted to raise another faction of his own as a bulwark against the warrior nobles' intervention in government. Peter was only partially successful, and it would ultimately cost him his life.

KNIGHTS OF THE VOID

I've always found fighting in an AeroSpace Fighter much more honest than fighting in a 'Mech. In a fighter, you maneuver through space, cutting and slicing at your opponent any way you can. And if your fighter should get hit, you tumble straight into the black void. To me, that has always been much closer to the truth than a 'Mech fight.

—General Sylvia Nathan, quoted in *The Scythe of the PPC*, by Duchess Fredrica Davion, Addicks Press, 3021

During his rule, Prince Peter Davion lavished enormous sums, special military honors, and patents of nobility on members of the Federated Suns AeroSpace Fighter arm. The Knights of the Void soon became his special favorite among the armed services, and he never ceased to promote loyalty and a high *esprit de corps* among their ranks.

Peter's special concern for his fighter pilots coincided with Chancellor Ingrid Liao's new offensive strategy for her armed forces. To bolster her ground forces, she had begun to use large groups of AeroSpace Fighters. In 2952, the new strategy paid off when her fighters intercepted a raid by House Marik. Greatly

Major Jeffrey Bannson wears a special urban camouflage uniform. The gray, brown, and black helmet, boots, and jerkin are issued to troops detailed for fighting in city streets. Because Major Bannson is attached to the Royal Brigade of Guards, his parade uniform would be trimmed with blue, white and red. Troopers in Brigade units are allowed to wear MechWarrior gloves as a sign of their elite status.

The grim minor official of the federal government pictured here is about to use a data probe. His red tie is an official symbol worn by all members of the Federated Suns government, as are the white wrist cuffs. The data probe is used for a variety of tasks, such as inventories on cargo coming into and out the Federated Suns. The probe is attached to the small computer on the man's hip.

encouraged, Chancellor Ingrid decided to launch a major offensive composed of almost three hundred AeroSpace Fighters against the Davion world of Lee. After the Davion Military Intelligence and the MIIO learned of her intentions, Prince Peter moved to reinforce the world. For the first time in years, Peter Davion entered a battle to direct the planet's defense.

In a three-day battle on and around Lee, the Davion defenders crushed the Capellan attack. In the air, the AFFS AeroSpace Fighters crippled the Capellan air strike force, and succeeded in knocking out 40 of them. This left the Capellan 'Mechs, tanks, and infantry on the ground without the full weight of the massive air support they needed. At that point, Prince Peter led his Davion 'Mechs in an attack that easily pushed the Capellans off Lee, with heavy casualties. The Capellans had lost so many fighters in "the Great Lee Turkey Shoot," as the battle came to be known, that it took them years to regain their former strength.

Though he had spent the entire battle slugging it out among the rest of the MechWarriors, Prince Peter especially praised the AeroSpace Pilots, whom he credited with the victory. Not even the dimmest MechWarrior could miss seeing which way the wind was blowing. Not long after the Great Lee Turkeyshoot, the Warrior's Cabal was formed.

HEROES OLD AND NEW

The MechWarrior, mighty in his towering suit of mechanical armor, has been but one focus of romantic idealism in the Inner Sphere. In other times, other warriors also embodied the virtues of honor, honesty, and goodness. During the Star League era, for example, it was the common trooper of the Regular Army who captured the popular imagination. Later, naval officers, tank commanders, and even VTOL pilots also had their day as heroes of the public's fantasies. When the image of the MechWarrior began to capture people's hearts and minds, it was usually because of the exploits of certain elite units such as the Sword of Light regiments in the Draconis Combine or the Royal Brigade of Guards in the Federated Suns.

Not surprisingly, in the dark days when MechWarriors were either stealing from the public or trying to manipulate the Davion government during Prince Joseph's reign, the AeroSpace Fighter Pilot was considered the upholder of honor and truth. This became even more so when Prince Peter Davion ennobled the Knights of the Void.

—From *Romanticism in Common Culture*, by Rebecca Greenburg, New Syrtis Press, 2999

THE MECHWARRIOR ELITE

The death of Prince Peter is, by contemporary standards, a recent event. Several noted sociopolitical theorists have advanced the theory that the events leading up to it could arise again in the coming years. Official Davion propaganda claims that the so-called "Cabal" was a conspiracy of a few disgruntled Warriors who resented Prince Peter's efforts to curtail their power. That is certainly part of the story, but probably only the tip of the iceberg.

The prominent sociologist R.N. Herrick drew numerous comparisons between post-Star League society and the deteriorating conditions of Imperial Rome during the so-called "Age of Thirty Tyrants." Like Rome in that era, today's Successor States are held together by the personal authority of warlords who maintain their power so long as they can maintain the support and loyalty of their armies. As the Romans learned only too well, an army with that much political power rarely hesitates to use it when and if its best interests demand it.

There are, of course, points of divergence between the Roman model and our present society. The MechWarriors involved in the Cabal are more similar to the powerful land-owning barons and knights of Medieval England and France than to the leaders of Rome's legions. It is true, however, that the breakdown of Roman government was directly linked to the increasing autonomy of ambitious field commanders. Once the Roman dynastic succession was broken, anyone of reasonably noble birth became eligible for the throne. During the Age of the Thirty Tyrants, a number of comparatively low-born soldiers took the purple simply because they had distinguished themselves as loyal soldiers. This parallels the situation that Peter Davion faced.

By virtue of their training and ownership of increasingly scarce BattleMechs, MechWarriors are vital to the survival of the Successor States. Modern warfare has demonstrated that 'Mechs are essential to victory, no matter what infantry and armor enthusiasts claim to the contrary. Warlords cannot wage war without MechWarriors, and so they grant them lands, titles, and political power to win their loyalty.

Bestowing such power on often combative and ambitious men and women makes it that much more likely that those MechWarriors will fail in their loyalty if forced to choose between their Lord and themselves. In such a situation, the Warrior will attempt either to influence a Lord's policies or to replace him with someone else.

Prince Peter's solution was to create another powerful group to offset the power of the MechWarriors. Though the faction helped Prince Andrew defeat the disorganized Cabalists, this solution triggered the very crisis it was to prevent. Moreover, the two factions may find enough common ground to cooperate against the government, or one faction may find it convenient not to intervene at a moment of crisis, leaving the Prince high and dry.

In the past decade, Prince Hanse Davion has initiated another solution. The creation of the New Avalon Institute of Science suggests the possibility of reform through the conversion of an essentially feudal army into a professional one. Warriors molded by institutions and holding allegiance to the State rather than to personal overlords should ultimately become the backbone of a new kind of fighting force. Similar academies exist in other Successor States as well, and so the trend is likely to take hold everywhere.

A similar transition from feudalism to nationalism also revolutionized Europe and led to the fall of the hereditary monarchies. If similar changes occur in our own society, what will happen to the Successor Lords? Will the restoration of the Star League become as foolish a concept as the revival of the Roman Empire in the face of clearcut national and linguistic boundaries? Would the most talented Successor Warlord be any more successful in uniting a formerly feudal Inner Sphere than Napoleon was in uniting post-feudal Europe?

—From *The Inner Sphere and the Modern Feudalistic State,* by Natalie Wessick, Remagen Press, 3022

This Corporal's white uniform shows that she is a member of an AFFS security force used to protect military bases. She holds a large stun stick, a common weapon of security personnel. The members of the efficient Davion police force also wear this type of uniform.

HAVE IT YOUR WAY

One of the strangest mutinies in history occurred in 2970, when mercenaries from Beaufort's Cossacks went on a shooting spree in a suburb of New Defiance City on the planet Moravian. A clause in the unit's contract specified that twice a week the government would pay for a free meal for all members of the unit at a local Triple-F restaurant, the most popular chain in the Federated Suns. The fact that Capellan naval forces blockaded Moravian shortly after the unit's arrival should not have made any difference, as Moravian produced its own wheat, beef, corn, sugar, milk, and potatoes. The only offworld elements of Triple-F cuisine were pickles, sesame seeds, and whalemeat. Nevertheless, when the troops could not get exactly what they ordered, Colonel Regnus accused Triple-F of defaulting on the unit's food supply contract.

The manager of the outlet, afraid of losing the most lucrative contract he had ever had, was quick to point out that the blockade had prevented supplies from arriving. He went on to say that he had in good faith fulfilled the contract as best he could, and that even if he could have anticipated the blockade, he did not have facilities to refrigerate a whale. Upon hearing this, the Cossacks, against orders from Colonel Regnus, went on a rampage, commandeering civilian motor vehicles, screaming out the windows at the natives, and discharging their weapons. They stormed the Triple-F franchise, chanting "Where's the pickle? Where's the lettuce?" Their chant made no sense, as Triple-F had never used lettuce in any of its products. (Perhaps the troops were confusing the eatery with a similar, competing chain.)

The manager and staff, frightened by the outburst, fled the restaurant. After a few minutes more of chanting, the troops began to feel foolish and left.

Colonel Regnus took no disciplinary action. He remarked to one of his aides, "We've got to get these soldiers into a real fight. When they start letting off steam like this, the cutting edge is definitely off the rose."

Despite all the commotion and the shooting, not a single civilian was hurt during the soldier's outburst.

—From *A Child's History of Mutinies,* by Henrietta Chabriole, Subversive Press, 3021

THE CABAL

Prince Peter had entered the Battle of Lee mainly to boost his military prestige so that he could continue his reforms. In the wake of that victory, he pushed for further attacks on the Capellans to maintain pressure against House Liao and to keep his warriors preoccupied.

Because of the glowing performance of the Davion aerospace pilots at Lee, Prince Peter believed that he could now accelerate his plans to be rid of certain MechWarriors who had consistently opposed his reforms. In a purge that some likened to a bureaucratic lynching, MechWarriors who had spoken against the Prince suddenly found themselves cashiered out of the AFFS as well as stripped of their lands, money, and 'Mechs.

As the purge gained speed, the Warrior's Cabal collected more and more members who feared for their existence as MechWarriors. Though Prince Peter's advisors urged him to move more cautiously, the Prince believed that the high stature now enjoyed by his AeroSpace Pilots would protect him.

In 2961, the Prince had gone to the Draconis front to tour the newly conquered world of Breed. Though he was ostensibly there to tour the planet's reconstruction efforts, the Prince had actually come to Breed to personally dismiss three dissident officers. One of these, Colonel Jeffrey Dempsey, had threatened the Prince's life if he dared to strip him of his rank and honor. As Prince Peter's entourage approached Colonel Dempsey's headquarters, Dempsey made good his threat. Five 'Mechs burst from the forest to either side of the base and opened fire on the Prince's vehicle. The Prince was killed instantly.

Though Colonel Dempsey and his conspirators were immediately arrested, the other members of the Cabal agitated for their release. The dissidents had managed to win so much sympathy for their cause on Breed that the Prince turned the capital city into an armed fortress to keep the people of Breed or the other Cabalist officers from liberating the Colonel from his keep. The one error in Colonel Dempsey's calculations was that he did not foresee what effect Prince Peter's death would have on his son Andrew. A Major in the Royal Brigade and a well-respected MechWarrior, it was Andrew who announced the news of his father's assassination to the Federated Suns, displaying such grief and yet so much dignity that the Cabal's hopes of winning public support were dashed. The fact that Prince Peter had a son to avenge his death convinced many other MechWarriors to stay away from joining the Cabal.

Instead of a civil war in the Federated Suns, a small mutiny broke out in the Draconis March, where the Cabal had been the strongest. When five 'Mech regiments refused to take the customary oath of fealty to the new First Prince, Andrew gathered up the Royal Brigade and a large aerospace force for a short and savage campaign to wipe out the rebellion. Fortunately for the Suns, the Draconis Combine economy had fallen on hard times, and so the Kuritans were in a poor position to exploit the chaos in the Draconis March.

Prince Andrew attacked so swiftly that the mutineers had neither time to flee nor to receive aid. In the course of a campaign that culminated in the destruction of the second Avalon Borderers on Deshler, all the rebellious MechWarriors had either surrendered or been killed in combat.

Countess Cordelia Spenser of Torrence, Lady-in-Waiting to Yvonne Davion and mistress of Field Marshal Ran Felsner, wears a dress in the latest Royal Court fashion.

REGROUPING

The mutiny of the Cabalists and the murder of Prince Peter were evils that did result in some good. Because the malcontents in the army were exposed, Prince Andrew was able to strip them of their powers and their holdings.

The vast majority of AFFS units remained loyal to the Davions during the mutiny and were later eager to prove that loyalty by accepting the military reforms that had cost Prince Peter his life. Andrew also helped things along by sugarcoating the major reforms in showy honors and larger retirement and benefit plans. Though this caused the Federated Suns some economic hardship, it was worth it to bolster the loyalty of the army and to encourage longer terms of service.

During Prince Peter's reign, there had been times of near quiet along both Marches. As news of the MechWarrior Cabal leaked out, however, both the Draconis and the Capellan commanders hoped to take advantage of House Davion's misfortunes by escalating the war along both fronts. Though both realms were too late to capitalize much on the unrest within the Federated Suns, they nevertheless persevered in their attacks.

Facing attacks on both fronts taxed the AFFS and prevented Prince Andrew from launching counterattacks, even though the Capellan offensive was sporadic and inept. Forced into a defensive stance, AFFS Marshals used the tactic of brief skirmishes and raids to harass enemy buildups. The new Prince found it necessary to balance resources and strategies to avoid giving either one of his foes an opening. Time and again, he demonstrated his skill in deploying his meager forces, particularly at Lee, Rio, and Kasai. Border worlds continued to change hands regularly on both fronts, but on the whole, Prince Andrew proved that he was fully capable of holding his own in the Succession Wars.

When House Liao completed the ten-year reorganization of its aerospace force in 2992, the Marshals of the Federated Suns realized that they had lost another chance to defeat House Liao. With the Liao fighters better organized and better trained than ever before, they could easily hold their own against Davion pilots, as AFFS losses on St. Andre in 2992 proved.

In 2999, Andrew Davion died of natural causes, leaving behind three children. The eldest was Marie Davion-St. Claire, the illegitimate daughter of Andrew and his mistress, Simone St. Clair. Marie would later become the focal point for a dynastic crisis, despite the fact that she had been specifically excluded from the succession. Andrew's two sons, Ian and Hanse, were the children of the Prince's official marriage to Jennifer Campbell. At the moment of Prince Andrew's death, the Federated Suns was poised on the thin edge between success and collapse. It would fall to these two sons, who would one day be known as "the Hound" and "the Fox," to steer the realm through the most critical phase of the post-League era.

THE HOUND

They call our new Prince "The Hound," and I must say that the label is not far from the mark. He has the kind of expression you might associate with a faithful old hunting dog, with those large, sad eyes and a nose just a little too big for his face. Of course, the real source of that nickname is his character. Ian Davion has the tenacity and determination of a bloodhound on the trail, and I don't think he'd give up an idea if all the 'Mechs from here to Luthien stood in his way.

—From the diary of Naomi Gavin Rollings, Countess of Trethany Valley, Galax, 4 September 2999

Nelitha Smith-Fantod, Assistant Minister of Exploration, is shown here wearing her business suit. The broad-shoulder cape and the exaggerated wrist and ankle cuffs are signs of her importance. It is said in New Avalon that if a person coming into a room has to turn his shoulders to get through the door, then he must be a Minister. Nelitha carries her Minister's Baton, another sign of her high status.

Prince Ian Davion took up the leadership of the Federated Suns in mid-2999. Though only 19 years old, Ian had already received extensive training as a warrior, but had not yet undergone training for the complexities of statecraft. Neither did he have any great interest in that. Everyone in the Court of New Avalon had long since discovered that fate had played House Davion a fickle hand by making the pugnacious Ian the elder, and the wiser, more level-headed Hanse the younger of the family.

Of course, Prince Andrew was not bound to select his eldest son as heir, but several factors conspired to make it necessary. Foremost among these was the attitude of the army, which preferred a trained MechWarrior on the throne over a youngster who had not yet seen combat. In addition, Hanse was little more than a boy, which was the last thing the realm needed at its helm.

On his deathbed, Andrew confirmed the selection of Ian as Prince Imperial. Even then, many court advisors already recognized Hanse Davion's potential and they urged Ian to groom him for a greater role than that of commanding marshal of the AFFS, as their father had wished. Luckily for the future of the Federated Suns, Ian was more than happy to take the advice.

In many ways, Ian resembled Prince Joseph II. He was happiest when leading soldiers and 'Mechs into battle, and preferred to leave the problems of politics to others. Nor did he appreciate being told what to do, and so he often disregarded the experience, abilities, and suggestions of others when it came to the issues of government. Before long, the realm began to suffer.

Luckily, there were plenty of battles to occupy Ian, and so direction of the government passed into the hands of a small group of talented men. One of these was Benjamin Sandoval, Duke of Robinson and the leader of the Draconis March. Duke Sandoval's talents were matched by long experience. As good as he was, however, he lacked one major qualification for the exercise of authority in the Prince's absence. He was not a blood relative of the Davion family, who would always hesitate considerably before giving any non-Davion too much power.

The other two notable political figures of Ian's reign were George and Michael Hasek. George Hasek was Duke of New Syrtis, and hence had authority similar to Duke Sandoval's. His son Michael married Marie Davion, the illegitimate daughter of Prince Andrew, in 3003. This gave Michael the family connection that the other two lacked, and the weak-willed Marie was soon under the spell of her clever, ambitious husband.

RISE OF HASEK-DAVION

Decline? Why must it be a decline? Couldn't it be that the Capellans have improved their fighting skills? Why blame our soldiers for a supposed decline in their skills? Why blame me?

—Duke Michael Hasek-Davion, from official court transcripts, 10 September 3005

Michael Hasek (who took the name Hasek-Davion after the marriage) was even more clever than either his father or Duke Sandoval. By the time he was 27, his influence was already as great as theirs. From 3005 through 3013, Michael Hasek-Davion held considerable authority in political and diplomatic operations, leading three separate ministries while at the zenith of his power. Hasek-Davion was both ambitious and unscrupulous, however, and his methods offended those who believed in ethics in government. One of these was Hanse Davion, who disliked Michael from the first time they met. Hanse was kept busy with training, however. When he was given a military command on the Draconis

Duke Thomas Dryden of Grosvenor wears a style of suit currently popular with Federated Suns aristocracy. The spurs on his boots show that he was once a MechWarrior, which he wants no one to forget.

March in 3007, those duties removed him as a major factor in court politics for the rest of Ian's reign.

It was during Hasek-Davion's period of political authority that House Liao managed to lure Wolf's Dragoons out of Federated Suns employ. This was probably not Hasek-Davion's intention, but it is symptomatic of how his misuse of authority was taking its toll. At that time, he had been encouraging minions at various levels of the administration to divert funds and supplies to nobles from whom he was seeking favors. The Dragoons must have felt such diversion of supplies and funds keenly, and it would have contributed to their desire to move on. Indeed, this kind of corruption within House Davion's administrative ranks led to a general decline in military prowess, despite the efforts of Prince Ian, who did everything but fight battles singlehandedly.

The deterioration was particularly evident on the Capellan front. In 3001, the Liao Warrior House regiments Lu Sann and Immara routed the Seventh Crucis Lancers off Lincoln. Then, in 3003, the Ijori regiment pounced upon the Dry River Devils, a Davion mercenary regiment, as they attempted to invade Tsamma. Because of a snarl in logistics, reinforcements meant to back up the Devils were delayed, and the mercenaries were forced to surrender and turn over their 'Mechs to the victorious Ijori Warrior House.

Though there were many reasons for the AFFS's poor performance, much of the blame goes to Duke Hasek and his son.

To counter the growing tide of ill-feeling against the Haseks, Michael orchestrated a subtle propaganda campaign to encourage the people of the Capellan March to question their loyalty to the Davions.

Hasek-Davion had to be very careful, however. The fact that he was not Duke of New Syrtis hindered his efforts, as all of Michael's aspirations depended on his father's good will. George Hasek would never have encouraged or supported outright rebellion as a tool for seizing power. Michael therefore carried out his plans behind closed doors, and ensured that Prince Ian had no cause to investigate what was going on in New Avalon. Supplies and support continued to flow smoothly into the Draconis March, where the Prince and his brother were busy fighting the Kuritans.

Hasek-Davion's efforts to keep his brothers-in-law militarily supplied served to enhance their military reputations and to ensure continued successes for the AFFS. In 3002, Davion troops led by Prince Ian routed a numerically superior Kurita force, which included a Sword of Light regiment, during the Third Battle for Harrow's Sun. Other victories, such as the seizure of Cylene II and the expulsion of Kurita forces off Elidere and Deshler, helped stabilize the border for the Federated Suns. Compared to the marshals in charge on the Capellan front, the two Davions seemed invincible. That would prove useful when the crisis finally came.

Private Nick Thomas is a crewmember of the Federated Suns Ship *Exeter*. He wears a special thin but protective garment under the gold jumpsuit, which can be used as a pressure suit in emergency situations. The crewman can roll the undergarment over his neck and head and attach a face plate to it. He has a ten minute air supply in the box on his belt to give him a chance to get to a pressurized section of the ship or to another supply of air. On his wrist is a communications device. He is carrying a large welder-grappler.

MICHAEL'S TREACHERY

I've no delusions about my brother-in-law. If my brother was known as the Hound, and I'm the Fox, then the Duke of New Syrtis is the Weasel darting though the underbrush, looking to take advantage.

—From the private journals of Hanse Davion, 3013

In 3012, the situation in the Capellan March finally became serious enough to make the First Prince take notice. Though recent Capellan attacks were small, they had been successful against larger and supposedly superior Davion forces. The Maskirovka, the Capellan equivalent of the MIIO, was increasing terrorist activities on border worlds, blowing up munitions plants, hospitals, and other vital facilities.

Thoroughly occupied in the Draconis March, Prince Ian sent Hanse Davion to take command of the Capellan March. Before long, Hanse discovered just how deeply the rot of confusion and corruption had set in there. Enraged, he set out to discover who was responsible for the military's pitiful state. The effort would be as difficult as any military operation he had ever undertaken.

Michael Hasek-Davion began to prepare for the confrontation that must surely come when Hanse's investigation inevitably led to him. He won a slight reprieve, however, when the Capellans launched a new offensive. Hanse had to drop his inquiries in order to rally the dispirited AFFS troops into finding the guts to fight the Capellans.

His efforts soon bore fruit. At the battle of Wright, Hanse's revitalized troops utterly smashed the Liao offensive. One Capellan unit, the LaFarge Hussars, was virtually wiped out during the fight. This victory and Hanse's cleaned-up military bureaucracy helped give the AFFS enough spirit to embark on a front-wide offensive.

Hanse's efforts in the Capellan March enhanced his reputation back at New Avalon, but the whole affair had left a bitter taste in his mouth. The need to weed out the corruption that had threatened the AFFS in the Capellan March became his overriding concern. Though there was little hard evidence, the Prince Imperial had plenty of clues about where he should be looking.

The next year, both Duke George Hasek and Prince Ian Davion died. The Duke's peaceful death at his home on New Syrtis gave the Duchy to his son, who now had full control over his own affairs. The First Prince died on Mallory's World, fighting the Draconis Combine to the last. Prince Ian had neither wife nor children, which left Hanse Davion as the heir-apparent. The new Duke of New Syrtis had his own plans about who should be Prince of the Federated Suns.

It has never been proved that Duke Michael hired the assassin who made an attempt on the Prince Imperial when his entourage stopped at Emerson for rest and resupply before heading to New Avalon. A sniper with a laser rifle shot at Hanse just as he was leaving the ramp of his personal DropShip. The shot missed, and the Prince's bodyguards whisked their charge back inside the ship before the sniper could get off another shot.

The sniper killed himself before he could be captured, and his corpse offered no clues. Though the official investigation remained inconclusive, Hanse Davion had no doubt that Duke Hasek-Davion was connected to the sniper.

This failure left Duke Hasek-Davion with only two choices: open rebellion or the acceptance of defeat. He chose the latter, but began to fortify his position as the Prince's investigation came ever closer to the source of the recent conspiracies.

Pictured here is a member of the Avalon Hussars. The gray jersey and light tan jerkin are traditional because uniforms in these colors were the only ones available to the original Avalon Hussars in the early days of the First Succession War.

THE FOX

Can it be a coincidence that leaders who are able to see the value of diplomacy, and even of alliances, should arise in all five realms simultaneously?

—From *The Resurgence of Diplomacy*, by Professor Randolph Nelson, New Avalon Institute Press, 3024

During the past decade, the Federated Suns has taken the first definite steps toward ending the long stalemate of the Succession Wars. Under the rule of Hanse Davion, the Federation has enjoyed more social, economic, technical, and military recovery than in the two centuries before him. Even enemies of House Davion have been forced to label this period a renaissance of Federated Suns culture. The past decade has also been critical for the whole Inner Sphere, which now stands on the brink of either peace or another round of bloody warfare.

POLITICAL BLOODING

His Highness, the First Prince, is a man to be reckoned with. Under his brother, we saw much of war and were often victorious, but I think we can look forward to far greater success now. The current Davion is not just a ruler but a leader, someone who knows what he wants and how to get it without compromising his position, his popularity, or his honor. That's rare these days, but he's not called "The Fox" for nothing.

—From a letter by Vanessa Sandoval, Countess of the Pearl Islands, to His Grace, the Duke of Robinson, 3013

Hanse Davion became the First Prince of the Federated Suns in 3013. The attempt on his life made the new Prince more determined than ever to tighten his grip on the administration and to safeguard his position. He also pushed forward the investigations into the corruption in the Capellan March, hoping that his people would uncover conclusive proof that Duke Hasek-Davion was behind it. Though Hanse ordered them to spare no expense, the investigations have never uncovered a shred of proof linking Duke Michael to either the assassination plot or to the malfeasance that he practiced for so long in the March. The Duke of New Syrtis is a careful man and a talented schemer, who apparently knows how to cover his trail.

Prince Hanse could not take action against the Duke on the basis of mere suspicion, for the Hasek family is very popular in the Capellan March and also has a number of influential friends at the New Avalon Court. A too-precipitate action against him could have done irreparable harm to the unity of the Federation, which was the one thing that Prince Hanse was not willing to risk. All his reforms would be for naught if he inadvertently set off a civil war.

The one action Hanse could take had to do with the choice of his ministers and advisors. No one was surprised when the Prince removed Hasek-Davion's name from the roll of federation government officials in the new Prince's Honors List. Everyone at Court was aware of the ill-feeling that had characterized the two men's relations from the first day. Outwardly, of course, they treated one another courteously, with many expressions of brotherly devotion. In the celebration of Hanse's investiture, Michael vowed "to loyally serve the realm by personally directing affairs in the Capellan March." Under frozen smiles and friendly gestures, each was testing his opponent's potential as an enemy, probing for signs of weakness.

FORMATION OF THE NAIS

In late 3012, Prince Ian Davion received word that the Kuritans were building huge storage depots on the planet of Halstead Station. If completed, the storage center would vastly speed up resupply for the DCMS, making life tougher for the Federated Suns. Prince Ian had decided to keep that storage center from being built, and had immediately begun to make plans for a strike against that world.

When Prince Ian died, it fell to Prince Hanse to continue the plans for an attack on Halstead Station. In this same period, the MIIO was reporting that Combine construction crews had stumbled upon the ruins of a Star League-era university housing a huge, unopened vault. Realizing how important the contents of that vault might be, Hanse rushed ahead with the raid plans and decide to personally lead the attack.

What the Prince had thought would be an in-and-out affair turned into a month-long battle. Luck was with the young Prince, however. He and his troops managed not only to destroy the Kurita supply center on Halstead Station, but to seize the Halstead Collection, perhaps the most important Star League treasure yet discovered.

When Prince Davion's 'Mechs broke into the vault, he and his men found rows upon rows of books and computer memory cubes. Hanse and his troops carried off as many of the books as time would permit, burning what they could not carry. Unofficial sources say that over three thousand books and other materials were taken from the vault, most of them on highly technical subjects.

The books have proved so significant that the public hailed Prince Hanse as "the Victor of Halstead Station." Capitalizing on this public admiration, he unveiled his plans for the creation of the New Avalon Institute of Science. Today, the Halstead Collection is housed in the NAIS in an environmentally controlled library. Researchers have not yet mined the full wealth of knowledge contained in the Halstead Collection, but much of it is the basis for the NAIS's ability to recover some lost knowledge and technology.

Prince Hanse has made most of the information available for public use. Some of the material is even part of the university dataservice that anyone with a computer can access. This has made Prince Hanse seem to be a peaceful man who believes that everyone—even the lowliest citizen—deserves to reap the benefit of knowledge.

STRATEGIC MOVES

I'm still pretty young, but I've seen enough action to kill any heroic feelings I had toward combat and war. For me, knowing that I have to send men and women into combat is the dirtiest job I can think of.

—Prince Hanse Davion, 3023

Although the war with House Kurita went on, Prince Hanse turned most of his attention to the Capellan situation in the early years of his reign. Things had already gone from bad to worse during his brother's reign, and so Hanse decided to follow up his earlier victory on Wright with a steady stream of new attacks to reestablish the Davion prestige in the region. He made several personal appearances on the front, but he had come mainly to study and observe the situation.

One of Prince Hanse's most strategic moves was to enlarge the role of mercenary units in the AFFS. Both his father and his brother had believed that mercenaries should never be given a chance to show initiative. As a result, they had rarely assigned mercenaries to raids or offensives that would have allowed them independent action. Hanse had his own opinions, however, realizing that mercenaries were the means to keep up pressure on the enemy without exhausting his own military. That meant allowing mercenary units to handle some of the highly mobile actions.

His policy change showed immediate benefits. In 3014, two mercenary units, Kingston's Commandos, a 'Mech regiment, and Tristram's Terrorists, a mixed-arms unit, attacked St. Ives in the Capellan Confederation. Contrary to what some Davion Marshals had predicted, the two mercenary bands showed considerable intelligence and courage, which a pleased Prince rewarded with large bonuses when the unit returned.

These actions provoked a violent and unexpected reaction from Chancellor Maximilian Liao. In 3022, he unleashed McCarron's Armored Cavalry, the Confederation's best mercenary unit, on the Federated Suns. What started out as a raid turned into a three-year odyssey of destruction and pillage deep into the Federated Suns that has only just resolved itself.

Despite the continued strength of the Capellan forces, the AFFS must be well pleased with its standing in the Capellan March. The only cloud on the horizon is the fact of Duke Michael Hasek-Davion's recent talks with Maximilian Liao. Whatever the Duke of New Syrtis has planned, it is unlikely that any plot short of all-out rebellion could shake Prince Davion's hold on the Federated Suns.

As for the struggle against House Kurita, Hanse Davion has made it known that he intends to win back the remaining Federated Suns worlds taken by the Draconis Combine during the First Succession War. This is a surprising boast from a Prince known for his caution. Many feared that the Prince would immediately launch a major campaign to win the worlds as quickly as possible. To the contrary, the Kurita border seems to have settled down into yet another exchange of heavy raids. In the past five years, both sides have carried out a total of 20 raids. Some are worth noting, like the Battle for Harrow's Sun, the Davion raids against Dobson and Yance, and the Battle of New Ivaarsen. Nevertheless, the border has changed little in the past decade, and Prince Hanse seems content to let it remain so.

FUTURE OF THE FEDERATION

An Alliance is a delicate thing, but I think ours will live beyond us. The day will come when our people will view our two states as a single realm with a single desire for peace.

—Toast made by Prince Hanse Davion to Archon Katrina Steiner, 3022

In 3020, Archon Katrina Steiner issued a Peace Proposal to the leaders of the other realms. Though her initiative was really nothing more than a cease-fire, it was the first time in decades that one of the Successor States had made a bid for peace. All but one of the other Successor Lords viewed her offer with scorn, however.

Like Katrina, Hanse Davion realized that the constant fighting was becoming an exercise in futility. Though forced to reject the conditions of the Archon's proposal, he could sense the sincerity behind the offer. Despite the objections raised by some of his closest advisors, the Prince opened secret communications with the Lyran Commonwealth.

His faith did not go unrewarded. The talks between the two realms uncovered considerable common interests and desires. Indeed, the exchange proved so successful that they quickly bloomed from discussions of armistice and economic treaties to earnest conferences on an alliance.

The fruit of these many long hours, weeks, and months of meetings came in 3022, when the two leaders signed the Federated-Commonwealth Alliance Document, or F-C Document, on Terra. The effects of this treaty have already been felt. The most obvious advantage of the alliance is the flow of military intelligence between the AFFS and the LCAF about their common enemy, the Draconis Combine. In 3022, using information given them by the Commonwealth, Prince Hanse unleashed the AFFS in a major offensive against the Draconis Combine. The results were spectacular, as the AFFS recaptured such worlds as Tancredi IV, Galatia III, and Rome. The Commonwealth has also benefited from the complete ouster of Combine forces off Sevren, Carse, and Port Moseby.

Another outcome of the treaty is less visible but potentially more important than the sharing of military information. The stimulation of trade between the two Houses could revitalize the economy of the Federated Suns, which has suffered from an inability to exploit its strong resources. Allied with the Commonwealth, the economic and industrial giant of the Inner Sphere, the Federated Suns can look forward to an explosion of economic activity.

Of course, this would all come to naught if the two ruling families could not get along. The Steiners and the Davions are both proud families, yet practical enough to know the importance of friendship. Members of both Houses have visited one another's realm, and returned home pleased by what they saw and how they were treated.

The most crucial clause of the Davion-Steiner alliance was the betrothal of Melissa Steiner, the Archon-Designate, to Prince Hanse. Though the two have met only a few times and both realize that the marriage is political, there is evidence of real attraction between Hanse and the beautiful young Melissa. What this marriage could mean to the Inner Sphere is the potential creation of a single power strong enough militarily, economically, and emotionally to dominate the rest of the Successor States.

The closer relations between Davion and Steiner have given the Draconis Combine, the Capellan Confederation, and the Free Worlds League plenty to worry about. A united Federated Suns and Lyran Commonwealth will make a fearsome combination. To combat this threat, the three other Houses have signed the Concord of Kapteyn, allying them against Davion and Steiner. It remains to be seen whether the Concord will last, as neither the leaders nor the people of these three realms trust one another.

SOCIOPOLITICAL STRUCTURE

Governments should never be static, inflexible creations of a people anymore than a person should have a static and inflexible personality. The best governments have proven to be quick and decisive in dealing with outside threats, yet cautious and contemplative in dealing with the affairs of its own people.

—From *The Political Philosophies of the Davion Family*, by Countess Yvonne Narwhal, New Avalon Political Press, 3010

Let the Davions worry how many stars fly the Sunburst. I worry about how few coins there are in my pockets and how many rocks sit in my fields.

—A common saying among the people of the Federated Suns

INTRODUCTION

Individual freedom versus the need to follow natural leaders has been the central issue facing the people and the government of the Federated Suns since its founding. Should government bow to the wishes of citizens who believe that the government is a servant of the people's individuality? Or should the people suspend their "rights" to follow men and women with superior leadership abilities?

In the Federated Suns, these questions have been answered by a political system that is, on the surface at least, an absolute monarchy. Though a monarchy might imply that the common citizen has forfeited all his rights, in the Davion realm, the common person can make himself heard in several official and unofficial ways.

The commoners have come to acknowledge the wisdom of the Davion family in most great matters affecting the Federated Suns. The Prince, aided by his family, holds enormous power and generally receives the wholehearted support of most of the population. In return for their support, the common people expect to enjoy certain personal freedoms on the local level. By laws and unwritten agreements, the common folk in the Federated Suns may participate in the decisions of their local governments. This protects the average person against possible excesses of the local nobility, which would affect them more directly than the larger questions occupying the minds of the Davions on far-off New Avalon. With this compromise, the Davions and their people have created a political system of semi-independent planetary governments bound to the actions of the First Prince by a harness of nobles.

No one has ever resolved the debate as to whether citizenry or nobility won the compromise, but neither can anyone deny that the decisive leadership of the Federated Suns central government is what has made it the strongest realm in the Inner Sphere today.

NOBILITY

The idea of a strong noble aristocracy came into the Federated Suns in 2417 when Simon Davion introduced a plan to split the realm into five Principalities. He reasoned that the Federated Suns would never be an effective and responsive government if based totally on so-called "democratic" principles, for it was simply too large. In a time when the Federated Suns needed to become stronger and more decisive, Simon Davion believed that the creation of five domains, each with its own network of nobles, would achieve a government able to react quickly when necessary.

Though the Era of the Five Princes ultimately culminated in a bloody civil war, the concept of nobility did prove effective. No longer was the government bound to the gathering of planetary representatives to get things done. Though most nobles were planetary government leaders assigned a title on a temporary basis, they could still call upon that title and authority to accomplish tasks that would normally have taken years of discussions and light-years' worth of travel.

In the early years, some nobles tried to take advantage of their titles to do as they damn well pleased while others were not shrewd or decisive enough to use their power wisely. Though this could have led to a popular uprising demanding the repeal of the nobility system, it resulted instead in the common people's search for able leaders. Once such a leader had proven himself, the people often consented to reward him by legally changing the title from temporary to hereditary. Their hope was, of course, that the competent leader would produce equally competent offspring.

Many in the Federated Suns believed that the concept of five separate principalities would lead to a clumsy, ill-coordinated government rather than a strong, flexible one. Even before the Civil War, there was a movement for government by a single Prince with authority over the entire Federated Suns. Despite the recent excesses of Princess Edward and Edmund, many also believed that only the Davions had the prestige to uphold so much power. There were those who did not share their view, of course, which led to Civil War between the Davions and the Varnays.

Despite the wars over who would become the ruling family, there was no backlash from the common people against the nobility system itself. Moreover, the nobility system had become fairly well developed by the end of the Civil War, with the people firmly in favor of an aristocratic class as long as they reserved the right to rid themselves of the bad apples. By the end of the Star League Era, most Federated Suns planets had entitled their ruling families, for the nobility system was flexible enough to divest even the worst tyrant of his power. Since that time, the noble titles have not changed much, though their duties have expanded to meet the demands of the times.

BARONS AND BARONESSES

This is the lowest rank of nobility that has certain political duties attached to the government of a section of land. (Though some Knighthoods include land grants with their title, Knights do not automatically have a say in a planet's political system.) Barons usually administer important cities or important sections of continents in the Federated Suns. Though they do not own or control the actions of private firms who use their land, Barons can often use taxes or access to natural resources to make these corporations responsive to their wishes.

In the Lyran Commonwealth, it is quite common for a Baron to control a company and *not* the land, while the people of the Federated Suns have a land-oriented sense of nobility. The rare Baron who owns only a company or companies is often scornfully referred to by other nobles as a "Lyran Lord" or "second-rate Baron".

Because Barons and Baronesses are the smallest of the several classes of nobles, they are the ones subject to dismissal by their subjects.

COUNTS AND COUNTESSES

Count and Countess are the next highest titles used in the Federated Suns today. These nobles are appointed by a Duke to administer whole continents, moons, and sometimes even whole planets. Occasionally, a Count is given control of a single city of extreme importance, such as one containing a BattleMech factory.

Should misfortune befall a Count and his family, the planetary Duke selects a new candidate. This individual must then be approved by the federation government before he can assume his duties. Counts are, by legal definition, more powerful than Barons. A Count does not usually answer to the local civilian government as a Baron must. The only exception occurs on worlds that have passed laws to specifically limit the Count's powers or where a planet's Duke steps in to overrule the action of a Count. Despite his considerable power, experience has taught most Counts to delegate at least some of their powers to a civilian government and civil service.

Counts and Countesses in the Federated Suns can and often do override the actions of civilian governments or of nobles subordinate to them, but not without making some explanation. Counts of the Federated Suns may also use military force to guarantee obedience from the people or from lesser nobles, but few have invoked this right of force.

THE CRUCIS PACT

The Crucis Pact was signed on 26 June 2317 by delegates from 35 planetary governments at a Constitutional Convention held on New Avalon. Though its provisions would be interpreted and reinterpreted many times over the ensuing centuries, this document continues as the basis of the Federated Suns's existence. The Crucis Pact is divided into a Preamble and 42 specific articles, several of which have since been suppressed or amended. Some of the more important articles are summed up here.

Article IV provided for the incorporation of worlds into the Federated Suns. Member worlds were to have full representation in all aspects of federation business. Associate worlds could monitor but not participate in decisions affecting the entire interstellar state. Later provisions would add a bewildering maze of trusteeships, territories, commonalties, and other similar versions of semi-membership that did not include the right to participate in the government at the federation level.

Article VII established the structure of the Federated Suns government. Each member world and associate world would be represented by a delegate in the High Council, though only members could actually vote. Any delegate could voice his opinion in the Council, however, which gave associate members a forum where they could at least air any problems or grievances. Supervising the Council was a President elected from among the Member world delegates; his term was for life or until impeachment by a 75 percent-majority of Council members.

Article XII guaranteed signatory worlds (regardless of actual status) complete internal freedom. Later amendments would make it clear that this applied only to Member and Associate planets; others were subject to whatever arrangements the High Council decided to create. Several other articles spelled out the specific powers and duties of the federation government to maintain trade, settle disputes between members, and to raise and support armed forces for the common defense.

The last ten articles established detailed guidelines for the duties of the High Council, the President, and other elected or appointed officials within the government structure. The President, for instance, was given extensive judicial powers as the ultimate reviewing authority on any dispute involving two or more worlds in the Federated Suns. Charles Davion, however, began the practice of appointing Presidential Representatives to review these cases instead. This later become the rule under Etien, Paul, and Marie Davion in the last quarter of the 23rd century. The High Council was originally given the power to elect other individuals to conduct foreign policy and/or military affairs, but even in Lucien Davion's day, the Presidency tended to function as a Chief Executive office for such purposes. The Davions tended to usurp more and more High Council prerogatives as the years went on, until that practice was entrenched enough to become permanent. Several later amendments actually made these traditions part of the law.

The Crucis Pact might have been considered another Magna Carta or U.S. Constitution, which was certainly the lofty intention of Lucien Davion. In an era of interstellar expansion, however, it was not a Magna Carta that mankind needed, but strong leadership and a determination to survive and grow. The men with the character for such tasks were the same ones most likely to turn the liberal ideals of the Crucis Pact into a personal writ for tyranny on a grand scale. Let it also be said, though, that even the most ruthless Davion had some respect for individual freedom. Otherwise, the Federated Suns might have gone down in history as the most oppressive of dictatorships rather then as a benign, if somewhat rigid, constitutional monarchy.

—From *Idealism vs. Pragmatism: A Comparative Study of Inner Sphere Governments*, by Jean-Francis Davion-Allen, University of Saso Press, New Syrtis, 2999

RIGHT OF APPEAL

The public of the Federated Suns thinks very little of rising up and deposing a noble who has not proven himself worthy of his position. The two most common ways are public appeal or direct action.

In public appeal, the people go before the planetary government to state their case against the noble's administration of a baronage or county. (Counts can be deposed, but with greater difficulty). Most planetary governments have provisions for the public airing of grievances against a noble written into their charter or constitution. On the federation level, the Laws of Noble Conduct and Review, written in 2634 by Prince Alexander Davion, guarantee these public inquiries into the actions of nobles.

Nobles who attempt to censor such criticism face a wide range of actions from the federal government, from a public reprimand to full-scale military action. The most recent military response occurred in 2701 when Count Donvier Rossini of Tancredi destroyed the Justice Building in his city-state and executed his critics. It was not long before a unit of Prince Davion's elite Royal Guards had arrived to deliver him to justice. Rossini was subsequently found guilty, and sentenced by federation judges to the kind of public execution that would discourage other nobles from trying to silence their critics.

The right to public inquiry does not necessarily guarantee immediate removal of the offending noble. Rather, it brings the matter to the attention of the nobleman's superior, who will then decide his fate. If enough complaints are lodged against a noble, he or she is replaced by the next in line of succession. If enough complaints are lodged against an entire noble family, then the superior's task is to strip the entire family of its title. This requires much time and effort, however, as well as the approval of the First Prince. Most recently, Baron Defreders of Coal Valley on the planet Mountcoal was removed from power after three years of legal maneuvering. After the local mine union accused the Baron of ignoring their concerns about the safety of the mines, it still required the action of the Federation Judiciary and the assistance of the Ministry of Administrative Services to depose him.

The two most common methods of direct action are through massive strikes or violent insurrection. Such means tend to be effective because a Baron's or Count's immediate superior is usually more interested in preserving peace within the realm than in supporting some lesser noble. Strikes work particularly well in a baronage that has only one important industry, such as a large coal mine or water purification facility. If the strike is well organized, the offending noble is immediately placed under pressure from his fellow nobles and the people in other realms who trade for what the baronage produces. Such strikes are always tense and often violent, but if properly prepared, the commoners can usually hold out longer than the Baron or Count.

The last time a strike was used to remove a noble was in 3001, when textile workers in the Jasmine Jungle Barony of Marduk stopped the textile factories and went so far as to chain themselves to the Weave Trees which provide their Tree-Cloth raw materials. Though the Baron attempted to wait out the strike, the Duke of Marduk forced him into "early retirement," for he was not about to endanger the local economy because of a political squabble.

Violence is usually a last resort, with assassination attempts most common in extremely rich or poor baronies or counties that have no single product or resource on which to base a strike. In such cases, terrorist organizations with large public support sprout up like weeds. Such extreme methods usually guarantee many deaths on both sides as well as occupation by planetary militia units.

The Davions hate such violence so much that they have created the Planetary War Powers Act. This act gives the Prince the power to place a heavy military presence on any world that becomes a hotbed of internal violence for any length of time. The Prince then names a Military Governor to head a completely new planetary government. The terrorists are hunted down, usually by MIIO agents, and later brought to trial. Similarly, federal judges will scrutinize the activities of the offending noble family, who will also likely be tried for crimes against the people. In these cases, punishments for both terrorists and nobles are severe.

—From *A Street-Smart Guide to Planetary Politics*, by Edward Cummings, Montcoal Revolutionary Press, 3021

DUKES AND DUCHESSES

A Duke is, in some ways, more powerful than even the Prince of the Federated Suns. As ruler over a whole planet or a group of worlds in several star systems, a Duke's decisions have a more immediate and lasting impact on the ordinary citizen than most anything the distant Prince might do. The people, therefore, reserve most of their respect or scorn for their Dukes.

Most Ducal families in the Federated Suns began with one of their members as the elected leader of a planetary government during the Five Princes Era. The various Dukedoms gradually became hereditary whenever the people were particularly pleased with their Duke's performance or when the Prince wished to reward a staunch supporter.

The era of the Five Princes dissolved into Civil War, followed by the Davion victory over all challengers. To consolidate their newly regained power over the Federated Suns, the Davions carried out a dramatic and thorough purge of the nobility. Many Dukes lost their lives, or at least their titles, because they had backed the wrong faction during the wars. In their place, the Davions appointed men and women whose loyalty was unquestioned. Several worlds attempted to overthrow the new foreign Dukes, but the Davions were too strong to let their appointees to be cast aside.

The boundaries of certain Dukedoms conform to the defense leagues and other similar small pacts that existed before the founding of the Federated Suns. For example, the Maharaja of Brantapur is the Duke of several star systems that were originally worlds of the United Hindu Collective. The number of Dukes and Duchesses in the Federated Suns has always fluctuated with deaths and marriages. Currently, there are between 100 and 120 Dukes and Duchesses.

MARQUESSES AND MARQUESSAS

The title of Marquess has recently regained prominence in the Federated Suns because of the Davions' effort to promote interest in the worlds near the Periphery.

A Marquess is, technically, higher than a Count, but below a Duke in importance. In the Federated Suns, the title has been used as a means for a Duke to bestow control of one of his worlds on a son or a daughter. Some Dukes use a Marquesship to compensate children not in line for the Dukedom. How the public perceives a Marquess and the power he wields is not much different than their perception of a Duke's.

A Marquess of the Federated Suns was originally the ruler of a frontier or a section of border territory that had yet to be exploited. To promote economic activity among the many poor worlds of the Periphery, Prince Ian and, after him, Prince Hanse Davion have revived this earlier use of the title.

Prince Hanse has taken to awarding Marquesships of worlds near the Periphery to young nobles. These endowments are temporary, with usually a five- or ten-year contract. During those years, the young noble must show an ability to exploit the resources of his world or else lose the title of Marquess. If the young noble proves competent to the task of developing the land, educating the people, and generally improving the conditions of the world, his title becomes permanent, with the possibility of rising to a Dukedom in the future.

PRINCIPALITY

The highest power in the Federated Suns is, of course, the First Prince. Indeed, the First Prince of the Federated Suns may be the most powerful person in the Inner Sphere.

The first modern Prince of the Federated Suns was Alexander Davion, who swept aside the last vestiges of the awkward Five Princes Era and fought down any claimants to the throne that he intended to hold for himself and the Davion family. It was in his time that most of the Princely powers known today were forged. The most important of these were the complete control that Alexander won over the AFFS, the realm's revenue, and the Federated Suns bureaucracy.

Prince Alexander Davion was the first to assume the title and position of Marshal of the Armed Forces of the Federated Suns. In later years, that title would be upgraded to Supreme Marshal of the AFFS. This means that every First Prince is the supreme commander of the vast military organization of the Federated Suns. No one can overrule his orders or decisions. Fortunately for the Federated Suns, the Davions' consistently high level of military experience has resulted in leaders who, at worst, have been only fair strategists.

Through the Ministry of Ways and Means, the Prince controls the flow of money through the realm. In the long and complex path taken by federal tax money, there comes a point when the vast wealth actually becomes the property of the Davion family. This has given them the unprecedented power to restrict or even stop the flow of money and services to any section of the realm that might somehow displease the Prince. The Davions have used this economic cudgel only sparingly, however. Its first use, during the Freedom Rebellion on Colorado in 2609, was so effective that the threat of economic isolation is usually more than enough to bring a balky planet into line.

The Ministry of Administrative Services is the branch of the Davion government that actually moves the government along on its sea of paper and computer memories. As the bureaucracy of the Federated Suns, it has a place on every planet and in almost every possible interaction between peoples and worlds. The Minister of Administrative Services has always been the traditional job of the Prince or Princess Imperial (the title given to the Heir-Apparent). Indeed, the Davions believe that training in administration is as important as the five years the Heir-Apparent must also spend in the military. When there is no Prince Imperial, the Ministry usually goes to one of the elder members of the Davion family so that this vital branch of the government remains under Davion control.

In return for these and other powers, the Prince agreed that the nobility and the citizenry would retain a certain amount of freedom to conduct their lives as they see fit. To guarantee that freedom, a long list of articles and acts have been drawn up to allow citizens and minor nobles the right to petition the Prince for aid against a government or a Duke that has usurped individual freedoms.

The Prince spends most of his time dealing with interstellar affairs with the other Successor States. So much of his reputation and support depend on the effectiveness of his foreign policy that Davion Princes tend to approach every diplomatic incident as though their lives depended on it. To carry out his foreign policy, a Prince must have a strong diplomatic and information network through which he can judge what actions to take toward other realms. In recent reigns, the Davions have succeeded in creating a large network of spies throughout the Inner Sphere, which has paid off with several spectacular diplomatic coups and other political skullduggery.

What many people do not realize is that the power of the Prince and the Davion family is based largely on their control of a huge empire of lands and corporations. Though their economic clout is not as great as that of the Steiners, their counterparts in the Lyran Commonwealth, what the Davions do own gives them quite a bit of unofficial power with many of the realm's most powerful companies. Though records on the wealth of the Davions are not for public consumption, our agents have managed to uncover the fact that they have major landholdings on twelve worlds, including land on each of the Golden Five worlds. They also have major holdings in twenty companies: six in the military industry, five in the information services, and four in mining and metal works. With that kind of economic influence, the Prince can sometimes use his power silently through the denial of lands to developers, for example, or in votes on the boards of various companies.

RULING FAMILY

Some mention must be made of how deeply the rest of the Davion family is involved in the affairs of the Federated Suns. All Davions, unless obviously mentally or physically unable, are expected to serve at least five years in the AFFS. Most enter service as MechWarrior cadets, though quite a few have trained and fought as AeroSpace Pilots, tank crew commanders, and even as members of the infantry. Rules regarding how the other soldiers are to react to a Davion in their midst are clear: It is strictly forbidden to give any preferential treatment whatsoever to a Davion, especially during training and combat situations. Though this is the standing order, history is filled with stories of men and women risking or giving their lives to save a Davion. Indeed, many made such a sacrifice without hope of later reknown or even knowledge of their deed. It is a testament to the high esteem, even awe, that most Davions have inspired in their people.

The Davion family expects every member to participate in politics with as much skill and devotion as they give to their military careers. Most Davions will spend at least a few years in active political service, either in the military or civilian bureaucracy or in the diplomatic service. Those who do not have the finesse or patience for a life in politics often just retire from public life and serve the Davion family and the Federated Suns as noble landowners.

BATTLEFIELD ROYALTY

In 2762, the big brass back at New Avalon decided to start yet another push against the lice-infested Liaos. The offense, codenamed "Lace Curtain" for some godawful reason, was to be aimed at Tsanna, Wel, and Redfield.

So we packed up and off we went. Once aboard the *FSS Exeter,* we were briefed on our mission. Our target was a small industrial city near the capital of Tsanna. Being jump troops, we weren't expected to do much more than mop up the Drop Zone once the 'Mechs had had their fun and pushed on for the capital. Having participated in enough missions to get a feeling on how things should go, I didn't like the briefing. It was too short on facts and too long on the pretty boys from the computer rooms and their "educated guesses". I looked over at my CO, Captain Michael Edders, and could see that he didn't like what he was hearing, either.

But orders were orders and soon my company was suited up and sitting ready in our DropShip, the *Kick Ass Special*. I've always loved the *Kick Ass*. It's one of those rare *Vampire* Class infantry DropShips that looks and flies almost like a large fighter. What really made it special was that it was one of the rare designs that allowed us jump troops to drop from it while it was still in the air. That way we could be in the fight much faster and didn't have to wait for the damned DropShip to land to let us out. 'Course you had to have a real ace for a pilot to make best use of such a special ship and we had that in Sergeant Fredricks.

The run into Tsanna had been fairly uneventful, meaning that our flyboys had managed to clear a corridor down past nightside Tsanna into dawn Tsanna, where our drop zone was just seeing the sunrise. The *Kick Ass Special* was given a Wave Two Priority by the Drop Manager, who was sitting above the planet and coordinating the actions of the 20-plus Davion DropShips. That suited me fine. I've always hated waiting on the ground while somewhere above my head were 30 'Mechs, each weighing at least 20 tons, plummeting to land nearby. I've heard too many stories about whole squads being lost to a landing 'Mech who didn't have sense enough to avoid falling on them. Gives me shivers.

Anyway, we were in the second wave. Listening in to the chatter over the communications channels, I could hear that the Jasper company of the 12th Avalon Hussars had made the drop unopposed and were just waiting for us to show up to secure the city's center so that they could start advancing on the capital.

The *Kick Ass* was just about on the drop zone, when Sergeant Fredericks hit the green ready light in our drop bay. Captain Edders and I gave the warm-up gesture to the troops. Everyone stood up, stowing the chairs and getting the kinks out of their legs. I suppose I should have felt the DropShip give a few jerks from the laser blasts, but I was too busy watching the oxygen and fuel pressure gauges on my jump pack to notice.

Next thing I know, the *Kick Ass* has flipped onto its back. Bodies were flying everywhere. Someone must have hit their jump button, or else something hit the *Kick Ass,* 'cause out of the corner of my eye I could see a burst of light. The scream lasted only a second before it dwindled to a choking gurgle.

Finally, the DropShip righted and the company began to pick itself up. I dragged myself from between the legs of Trooper York, who was apologizing profusely because he had given my breast a squeeze in the confusion. He always was a gentleman. Anyway, in my ear, I could hear Fredericks yelling at his flight crew as he tried to control the ship.

Suddenly, he's yelling at me that he was going to blow the Drop-doors open. I had only a few seconds to shout and give the Hot Drop signal when sections of the floor split open. The wind was hot and the ship was almost immediately filled with smoke. The *Kick Ass* had apparently been hit in the engines. I heard the captain yell the "Go" signal. My company heard the order and began dropping through the Drop-doors. Four at a time, they disappeared through the smoke.

Edders and I were the last to jump out of the *Kick Ass*. As I jumped, the noise of the dying DropShip and the stinking smell of smoke was instantly replaced with the sound of wind and the cool, sweet smell of the morning air. Looking around at the quiet landscape as I fell, I could almost imagine that this was just another peaceful practice jump. Then I caught sight of the *Kick Ass*, trailing smoke and fire. With a sick feeling in my stomach, I watched as the ship attempted to climb. I found myself yelling for Sergeant Fredericks to bail out, knowing that he wouldn't. The *Kick Ass* hit some sort of factory and exploded.

I couldn't see anyone around me. I suspected that the *Kick Ass* must have pitched me and Captain Edders away from the rest of the company when we jumped. Trouble was that I couldn't find Captain Edders around me either. Worse still was the fact that I was very far from our assigned drop zone.

I was pretty near the end of my free fall. The 15-second buzzer warned me to prepare to fire my jump pack so I could land safely and not end up driving myself into the ground. Looking down at my gauges, I felt my mouth go completely dry. Both my fuel and oxygen levels were dropping faster than I was. I must have landed wrong on my jump pack and cracked open a feed line when the *Kick Ass* had rolled over. By the time I figured that out, the five-second buzzer rang and the two levels on the gauges had reached double-0.

I had only enough time to say "damn" and close my eyes. The first impact wasn't what I had expected because it hit me from behind. Suddenly, I could feel two arms reach around and grab at me. Next thing I know, I could hear the scream of a jump pack with its throttle pushed past the red line. My legs got damnably hot as they dangled in the exhaust of the jump pack. Then we hit the ground with a thud. I blacked out.

When I woke up, I saw the brand-new knee the impact had created in my left leg. Though the pain was manageable, for the moment, I knew there was no way I was going to walk around in this fight. I was thinking about how I'd look with crutches when I heard someone moaning behind me.

Following the sound, I crawled into the bushes, where I found Captain Edders. He was the one that had saved my bacon. He had become separated from me after the jump, but then spotted me again and saw the oxygen and fuel spewing away from my jump pack. While he tried to keep me from tending to the cut on his head, I learned that it took everything he had to make his way to me and grab me. That he'd risked his own life didn't seem to enter his head.

—From *Memoirs of a Fighting Davion,* by Duchess Alexandra Davion, New Avalon Press, 2780

Over the years, the Davions have drawn up several documents to lay out concrete rules on the order of succession. The latest is the DeBurke Treatise on the Davion Lineage, which was published in 2846. Not only is it a detailed description of the entire Davion family tree from its origins in Napoleonic France to the present day, but it offers clear-cut rules and terms to calculate the order of succession. The DeBurke System of Succession, as it has come to be called, has been added to the Acts of Succession written by Prince Alexander to resolve questions of succession for most of the noble families in the Federated Suns.

Currently, Michael Hasek-Davion is the Heir-Presumptive, meaning that he would assume the Princeship should Prince Hanse Davion die. If the proposed marriage between Prince Davion and Melissa Steiner, the Lyran Archon Designate, does occur, then Melissa would become the Heir-by-Marriage, a weaker but still strong claim to rule the Federated Suns. Should the marriage result in a son or daughter, then that child would become the Heir-Apparent, also known as the Prince or Princess Imperial, an almost unassailable claim to the leadership of the Federated Suns. Should that child reach his majority and serve the proscribed years in the AFFS, he would become the undisputed successor to Hanse Davion. The child would also have direct claim to the Archonship of the Lyran Commonwealth. Should the day come when a child of the Davions and Steiners sits as ruler of both Houses, this would surely be the first step toward reunifying the Inner Sphere under one rule.

ROYAL COURT

The term Royal Court actually has two separate meanings. First is the collection of buildings sitting just below the Royal Palace, which are known as the Royal Court. Within their medieval facade are the offices of the Ministries and the countless bureaucrats that form the backbone of the government implementing the Prince's decisions.

In stark contrast to this governmental drudgery is what most people think of when they hear the words "Royal Court". Every spring, the Davion family opens sections of the Royal Palace, that huge castle overlooking Avalon City, to the nobility of the Federated Suns. For three months, the Davions entertain the nobles with grand dinners, elegant balls, and hunts on the many game preserves around the castle. Usually, about 40 dukes and as many lesser nobles attend a Royal Court. Important nobles, such as the Dukes of New Syrtis and Robinson, attend the Royal Court regularly. In some ways, the Royal Court is a replacement for the High Council, which still meets but has lost most of its powers.

The Davions host this expensive series of entertainments for a number of reasons. One of the most important is simply to learn the news of what is happening on other worlds in the Federated Suns. Even though ComStar is supremely efficient and fast in relaying messages between New Avalon and the rest of the realm, few nobles can really afford to say all they want over the ComStar lines. The Royal Court is their chance to talk personally to the First Prince and for the First Prince to get a real feel for the mood of the realm.

Another function of the Royal Court's importance is as a forum for the Davions to announce and discuss important new policies. In the three months of rubbing elbows with the nobles, the Davions have ample time to soothe the ruffled feathers of any who disagree with a particular decision. Needless to say, the political wheeling and dealing that goes on during the three months of Court result in the trading of vast amounts of money and resources.

COUNCILS AND MINISTRIES

The High Council, originally a body equal in power to the Prince, has lost most of its power and is now little more than a rubber stamp of the Davion government. The Council was originally a body of representatives from every member world in the Federated Suns. As the realm grew, the distances between worlds made it impractical to wait for the 100 Council members to agree before a decision could be implemented. The Council was gradually deprived of its authority because it sometimes stood in the path of effective government.

In its place, the Privy Council, which Alexander Davion created as an alternative to the High Council, has grown in power. Composed of the leaders of the various Ministries and the commanding generals of the AFFS, the Privy Council is small enough to react quickly and effectively to changing conditions. The Privy Council has become the Prince's closest group of official advisors, bringing considerable power to the individual members.

Ministries are sections of the Federated Suns government devoted to either one region or one particular aspect of government. They are responsible for carrying out the policies of the Prince and of reporting back on their effectiveness. As a result, Ministries wield a great amount of power to accomplish their tasks. A Minister, or head of a Ministry, has powers surpassed only by the Prince he serves, which can make a Minister more powerful than the most powerful Duke. More than once, a Duke has challenged that power, only to be run over by a virtual steamroller of legalities and bureaucratic red tape.

There are currently eight Ministries in the government of the Federated Suns.

MINISTRY OF WAYS AND MEANS

The Ministry of Ways and Means is responsible for collecting taxes and carrying out economic policies. (See **Socioeconomics.**)

MINISTRY OF EDUCATION

The Ministry of Education is responsible for the level of education of the citizens of the Federated Suns. (See **Culture and Arts**.)

MINISTRY OF THE CRUCIS MARCH

This is one of the three regional March Ministries largely responsible for the operations of a major division of the Federated Suns. The policies of this Ministry are considered law unless overridden by one of the other service-oriented ministries such as the Ministry of Ways and Means. Each of the three Marches is further divided up into Administrative Areas (see sidebar) to help ensure that all worlds receive their fair share of services and participation.

Prince Hanse Davion, as Duke of New Avalon, is the titular Minister of the Crucis March. By tradition, however, the Prince now bestows the title on the Heir-Apparent. As Prince Hanse has no offspring as yet, and has not named an Heir-Apparent, the Ministry of the Crucis March is currently held by Melinda Davion, one of the Victorian Davions and a close friend.

ADMINISTRATIVE AREAS

There are currently 26 separate Administrative Areas, divided among the three Marches, in the Federated Suns today. Each Administrative Area has anywhere from 10 to 30 separate worlds and colonies under its control. Most areas are named after the planet where are located the various head ministry offices for the area.

The Draconis March

There are currently eight Administrative Areas in the Draconis March.

Administrative Area	Chief Administrator
Ozawa	Sub-Minister D. Henders
Kentares	Sub-Minister K. Dogall
Robinson	Minister Jacob Sandoval
Sakhara	Sub-Minister U. Frethers
Fallon	Sub-Minister L. Petre
Tancredi	Sub-Minister E. Drison
Delos	Sub-Minister T. Ferris
Boondock	Sub-Minister W. Edders

The Capellan March

There are currently eight Administrative Areas in the Capellan March.

Administrative Area	Chief Administrator
Almach	Sub-Minister Y. Hasek-Davion
Acala	Sub-Minister U. Kilos
Emerson	Sub-Minister R. Loper
Daniels	Sub-Minister W. Tenglat
New Syrtis	Minister R. Hasek-Davion
Hobbs	Sub-Minister Q. Aserton
Sirdar	Sub-Minister S. Simoth
Pleiades	Sub-Minister H. Capah

The Crucis March

There are ten Administrative Areas in the Crucis March.

Administrative Area	Chief Administrator
Marlette	Sub-Minister Y. Tewwe
Kestrel	Sub-Minister F. Xerxes
New Avalon	Minister Melinda Davion
Tsamma	Sub-Minister Q. Dontent
Point Barrow	Sub-Minister J. Poert
Remagen	Sub-Minister H. Sweret
Anjin Muerto	Sub-Minister O. Kaxu
Broken Wheel	Sub-Minister R. Edgrer
Killarney	Sub-Minister X. Frether
Islamabad	Sub-Minister A. Jinah

MINISTRY OF THE CAPELLAN MARCH

Michael Hasek-Davion, as the Duke of New Syrtis, is the Minister of the Capellan March. It is a title that he has entrusted to his younger sister, Rebecca Hasek-Davion.

Because it is a war zone, the Capellan March exists under a group of rules and laws slightly different than those in the peaceful Crucis March. Matters such as military governorships, counterinsurgency efforts, and the movement of thousands of troops are commonplace in the Capellan March. The set of laws governing wartime conditions in the Federated Suns are called the War Zone Actions of 2787. Most of these are modifications of certain peacetime laws or new laws that deal with the military's power when interacting with a planet's government.

MINISTRY OF THE DRACONIS MARCH

Duke Aaron Sandoval is the Minister of the Draconis March. Unlike Prince Hanse, who has too many other duties, and Duke Michael, who is uninterested, Duke Sandoval takes his role as Minister seriously. He actively rules over the bureaucracy of the Draconis March, and can usually be counted on to show up for serious High Council meetings. When he cannot attend to his duties as Minister, his eldest son, Jacob Sandoval, the Count of Olkman, fills in for him.

The Draconis March has been under the War Zone Actions ever since they became law in 2787.

MINISTRY OF ADMINISTRATIVE SERVICES

The function of the Ministry of Administrative Services is to be the oil that lubricates the activities of the other Ministries. Any bit of law or policy that falls outside the jurisdiction of the other ministries is handled by Administrative Service. Also, the Ministry is responsible for the flow of communications between all the Ministries as well as for coordinating laws and activities that require participation by more than just one or two ministries.

Administrative Services is also responsible for the operation of Federal Missions on each member world. These Federation Missions handle all the work created as a result of relations between a planet and the Federated Suns government. They also act as unofficial ambassadors. For the vast majority of people in the Federated Suns, the local Federal Mission is the only contact they have with the outside world.

In reality, the Ministry of Administrative Services has become the bane of the Davion government and the butt of countless jokes. (How many sub-ministers from Ad-Ser does it take to....?) Some blame the Ministry's almost perpetual poor performance on the fact that the three Regional Ministries tend to take the best bureaucrats away from Administrative Services, leaving behind a vast number of poor performers. Others blame the Ministry's woes on a perpetual lack of funds.

The current Minister of Administrative Services is Nelitha Green-Davion, Countess of the Gold Isles on Galax. She is a recent appointee by Prince Hanse, who has lost patience with the incompetence of the Administrative Services. He has given Countess Nelitha five years to improve the workings of the Ministry. To help her, the Prince has expanded her budget and ordered a stop to the raids on personnel by the Regional Ministries.

JUDICIARY

The Department of the Judiciary is a part of the Ministry of Administrative Services. Because the Davion government allows planetary governments a certain amount of freedom, planetary laws usually meet most of a world's judicial needs. There are relatively few federal government laws that override planetary laws in the matter of personal freedoms. Areas such as interplanetary relations, inter-realm and interplanetary trade, human rights, and loyalty fall within the jurisdiction of the Federated Suns Judiciary.

The Court of the Old Bailey on New Avalon is the highest court of law in the Federated Suns. The buildings, patterned after pictures of the old courts of England, house the Judicial Review Court, which hears cases involving major conflicts between the laws of the Federated Suns and those of the planetary governments. There are nine judges, usually chosen from the 200 Regional Review Courts in the realm.

Though most of the cases heard by the Judicial Review Court are of an economic nature, such as a dispute over whether a load of Victorian Truffles should require a Federal Interplanetary Commerce tax if shipped to the moon of Victoria, the court is occasionally asked to hear a case involving Human rights, which every Federated Suns citizen enjoys. These rights, called the Six Liberties, are the right to personal liberty, to own property, to own weapons, to expect fair treatment, to privacy, and to participate in planetary government.

PLANETARY GOVERNMENTS

Planetary governments are of many types and styles. Some are patterned after governments from the history of Terra, such as Russian Communism or French Socialism. Other planetary governments are bold experiments in governmental theory such as the electronic democracy of Torrence or the representative commune system on Xenia. The average planet in the Federated Suns is ruled in a manner similar to the old British parliamentary system.

No matter what the type of government, it is closely joined with and often depends on the noble aristocracy. How tightly the nobles control the government tends to depend on the world and on the noble families. Many noble families are content to deal only with interplanetary affairs or with the management of their lands, which allows the planetary government a great deal of latitude. Other noble families believe that their title authorizes them to completely dominate a world's government. Needless to say, these noble families are seldom popular with the citizens or the federation government, which needs a well-liked nobility to help it govern properly.

The only exceptions to the stated Federated Suns policy of allowing the most freedom possible to its people and its planets are those worlds under Military Governorship. A Military Governor is usually assigned to planets recently captured from another realm and whose population probably still harbors a great many citizens loyal to their former rulers. The Military Governor is an officer with experience in military matters, particularly in areas of terrorism and small-unit tactics, as well as political matters such as the use of propaganda to sway ambivalent populations.

How long a Military Governor rules a world depends largely on the people and the military situation around the planet. Some captured worlds readily embrace the freedoms granted them by the Federated Suns, while others take generations to change their allegiances. When a Military Governor feels that the world he has been administering is ready to show loyalty to the Federated Suns, he will help it form a new, more representative government. He is also empowered to choose candidates for nobility from the world's population. Once done, the Military Governor leaves the planet to its own future.

Military Governors are also sometimes used far from the frontlines to help calm rebellious worlds or those that have suffered severe disruptions. Sometimes, a Military Governor is sent in to administer a world that has undergone a catastrophic natural disaster, such as a series of earthquakes or volcanos. More often, a Military Governor is called in to take control of a world whose government has violently collapsed. On such a planet, the Military Governor often has as much trouble with rebellious populations as on worlds just captured from the enemy.

OVERT POLITICAL GOALS

The Federated Suns is the largest and most powerful of the five Successor States. All the other states watch what it does and react to it. Even we of ComStar, with our superior intelligence and technology, find ourselves watching and waiting to see what the Davions will do next.

The Federated Suns and its Davion rulers understand their position as the creators of change in the Inner Sphere. Ever since the collapse of the Star League, the realm has considered itself to be the last bastion of the Star League legacy. Despite the many years and the devastation of three horrible wars, the Federated Suns claims to be attempting to preserve the honor and traditions of the Star League in its own government and military.

Central to the political goals of the Federated Suns is the founding of a new Star League with a Davion as the First Lord. Though this is true for the other four Houses, too, the Federated Suns believes its "legal" claim is strongest (if legality still counts after all these years of fighting and misery). This is based on Richard Cameron's alleged desire that John Davion be his successor.

Katrina Steiner

The Davions' claim to the First Lordship has not blinded them to political and military realities. There have been several instances in the Succession Wars when the Federated Suns have renounced the achievement of certain goals, such as the capture of Terra, that would have meant a great deal to their claim of the First Lordship. The Davions did this because they felt it more important to seize other opportunities for more pragmatic military or political goals.

This political levelheadedness is why the current Prince was so eager to respond to the Peace Proposal put forth in 3020 by Katrina Steiner, the Archon of the Lyran Commonwealth. Only he seemed to realize the sincerity of the Archon's initiative, and was able to keep the lines of communication open despite the necessity of rejecting her original proposal. When others scoffed at the idea of peace in the Inner Sphere, Hanse Davion realized that the next best thing to peace would be a strong alliance.

MINISTRY OF FOREIGN RELATIONS AND ITS POLICIES

It is the duty of the Ministry of Foreign Relations to advance the aims of the Federated Suns in foreign relations. To achieve this, the Ministry receives a large percentage of the realm's yearly budget, much of which goes to maintaining diplomatic missions on the capital worlds of the other four Successor States.

The Ministry also maintains several large administrative and research centers in the Federated Suns. In these centers, information about the four realms is gathered from numerous sources such as foreign news programs. Ministry experts then summarize the main points in a daily report on interstellar events. These reports are sent daily to New Avalon, where Prince Davion reads them religiously.

The planet Robinson is the Ministry's center on the affairs of the Draconis Combine. New Syrtis is the location of the Capellan Confederation center, while Caph is where news and opinions concerning the Free Worlds League is gathered and developed. Since the formation of the Lyran Commonwealth-Federated Suns alliance, the Ministry of Foreign Relations has moved its center of Lyran information from Caph to the Lyran world of Thorin.

Olivia Fenlon

The Minister of Foreign Relations is Olivia Fenlon, Duchess of Chesterton. Earlier in her career as a diplomat, Duchess Fenlon was Ambassador to the Free Worlds League. Before that, she was an agent in the MIIO. Since assuming command of the Ministry when Hanse Davion took the throne in 3013, she has proved most capable. Her policies are far-seeing, which is exactly why Prince Hanse appointed her to the post. It is said that she was the original architect of the alliance between the Federated Suns and the Lyran Commonwealth and apparently enjoys an amiable friendship with her counterpart in the Commonwealth, Cynthia Eddies, Chancellor of Foreign Affairs.

INTERSTELLAR RELATIONS

FREE WORLDS LEAGUE

Before the turn of the century, the Federated Suns and the Free Worlds League were on relatively good terms. Both had a common neighbor and enemy, the Capellan Confederation, and both realms needed to promote trade. Because the Free Worlds League could not hope to compete with the variety of goods offered by the economically superior Lyran Commonwealth, their enemy, they chose instead to offer a few items for export at much cheaper prices. This proved tempting to the Federated Suns, whose traders were often cash-poor. For many years, the flow of trade between the two realms was strong and consistent.

This economic relationship helped ease political tensions between the realms so that many observers were speculating that a pact, if not a full alliance, between the two realms might occur toward the end of the century. This was despite the fact that agents from the Federated Suns had been responsible for sparking the civil war between Captain-General Janos Marik and Anton Marik—and everyone knew it. Somehow the Federated Suns was able to later convince the victorious Janos Marik that the Federated Suns had actually done him a favor in exposing Anton's treasonous ways! Indeed, when rumors of an alliance between the Federated Suns and another Successor State leaked out, most assumed that the Free Worlds League was the other realm.

The alliance between the Federated Suns and the Lyran Commonwealth has placed the Federated Suns relations with the Free Worlds League in jeopardy. Already, economic contracts between the two Houses are being allowed to expire without renewal, and diplomatic relations between the two States are definitely cooler. Apparently, this chill in relations is something that the Davions would like to remedy. There are rumors that Prince Hanse might even be contemplating offering House Marik a chance to join the Federated Suns-Lyran Commonwealth alliance.

Though anything is possible, such a triple alliance is unlikely. For one thing, the Free Worlds League joined the Concord of Kapteyn, formed in reaction to the Federated-Commonwealth alliance. For the moment at least, this counter-alliance has ended hostilities between the Free Worlds League and the Capellan Confederation. It is also known that the Free Worlds League has sent sensitive information about the Federated Suns to the dreaded Draconis Combine.

The second obstacle to bettering relations between the two realms is that the Lyran Commonwealth stands firmly against it. The Free Worlds League is a long-time enemy of the Lyrans, with considerable mutual hate having built up over the years. Even though relations between the two have warmed recently, the Lyran government fears that House Marik might somehow poison the Davion-Steiner alliance if they became too friendly with House Davion.

Grethers Fredric, Count of the New England continent on Salem, currently serves the Federated Suns as Ambassador to the Free Worlds League. He is quite at home in the maelstrom of fragmented political regions and sub-realms that is the nature of Free Worlds League politics.

CAPELLAN CONFEDERATION

Relations between the Federated Suns and the Capellan Confederation have waxed and waned over the years. At times, there seemed to be a real chance of peace between the two realms. At other times, a deep well of hate has existed between the House Liao and House Davion.

Current relations are definitely at a low point. With typical deviousness, Maximilian Liao, the ruler of the Capellan Confederation, has attempted several plots in recent years, all aimed at weakening the Federated Suns. These plots have ranged from attacks on the key Federated Suns world of Redfield to attempts to bribe the power-hungry Michael Hasek-Davion to an attempt to substitute a double for Hanse Davion, whom they had imprisoned far from New Avalon. Though the Federated Suns has thus far managed to remain one step ahead of Liao's efforts, the success of any of these plots would have had interstellar implications.

Ambassador Edwina Sanders, Countess of the Yuler Mountains on Galax, is the current Federated Suns representative in the Capellan Confederation.

DRACONIS COMBINE

Relations between the Draconis Combine and the Federated Suns have been consistent—consistently bad and consistently tense. There have been too many atrocities on both sides, too many vicious battles, too many people with long memories to think that any attempt at peace between these two most powerful realms stands any chance of success.

At the moment, the border between the two is quiet. The MIIO has been paying special attention to its intelligence network in the area and to establishing a web of spies in the Draconis Combine. Their actions seem to be paying off. Recent attempts by Kamikaze Units (small squads of elite Kurita Special Forces units) to raid targets along the border have all been notable failures.

The terms for peaceful coexistence between the two realms, according to the Federated Suns, have been the same for the past hundred years. The Davions say that there will be no peace unless the Draconis Combine withdraws behind its Star League borders, which would return some 20 worlds taken from the Federated Suns in the first years of the Succession Wars. Of course, the chances of this happening are nil, and so these two Houses are likely to remain the bitterest enemies in the Inner Sphere today.

There has been no Federated Suns ambassador to the Draconis Combine in over a hundred years.

LYRAN COMMONWEALTH

Relations between the Federated Suns and the Lyran Commonwealth are currently the subject of constant speculation. Not since the collapse of the Star League has any single action had such Sphere-shattering implications as the alliance between these two Houses. Any attempt to describe even the most likely consequences of the Federated-Commonwealth alliance would take far more space than we have here.

When Hanse Davion politely but firmly rejected Archon Katrina Steiner's Peace Proposal of 3020, Katrina saw that he did so with genuine regret. With the lines of communication open, the two began tentatively to discuss ways to reduce the tension and barriers between their two realms. It was not long before the talks advanced from merely improving relations to discussion of a full alliance, the first of its kind since the fall of the Star League.

In 3022, when the treaty known as the Federated-Commonwealth Alliance Document, or the F-C Document for short, was signed on Terra, its effects were immediate. Within mere hours after the signing, communiques between the military High Commands of the two realms began to flow. Information traded between the two states concerning their common neighbor and foe, the Draconis Combine, has resulted in both the Lyran Commonwealth Armed Forces and the Armed Forces of the Federated Suns winning back worlds from House Kurita. The Commonwealth military is especially grateful because it has tended to come up short when confronted by the better-trained Kurita troops.

WHY NOT HANSE AND KATRINA DAVION?

Why did Katrina Steiner and Hanse Davion not cement their treaty with their own marriage instead of one between the Prince and the Archon's daughter? They certainly seemed attracted enough when they met on Terra for the treaty. Some scandalous rumors even accused them of "private" meetings, during which they acted out their attraction.

Of all the proposed explanations to this riddle, the one that stands up best is that they did not want any clash between them to threaten the alliance. By establishing a marriage between Hanse and Melissa, the two rulers were giving their alliance at least eight years to prove itself. If it should hold, then the marriage to a decidedly young Melissa in 3028, when she has reached her majority, would cement the alliance. Obviously, a marriage between Katrina and Hanse would not be likely to produce an heir to a united state, an heir Melissa would be capable of bearing. Also Prince Davion may have felt that marriage to a young woman inexperienced in the political arena would give him the upper hand in any policy arguments. Of course, this assumes that Melissa Steiner would meekly back down in an argument, an assumption definitely *not* borne out by the evidence we have—A.M.

Also vitally important are the improved economic relations between the two Houses. As described in the **Socioeconomics** section of this book, the introduction of Lyran high-tech items into the Federated Suns is most welcome, especially in the less affluent sections of the realm. Also, the presence of the renowned Lyran traders will go a long way to help get the so-called Skid Row worlds of the Federated Suns back on their feet.

There is no doubt that, personally, Prince Hanse Davion and Archon Katrina Steiner get along quite well. When they met on Terra to sign the treaty, the two spent hours trading old war stories. Archon Katrina even dropped her usually reserved attitude and attempted to teach a charmed Prince Davion one of the native dances of her Tharkad homeworld.

Moreover, if the first meeting between the Prince and Melissa Steiner is any indication of the future, their proposed marriage may turn out to be more than a political marriage of convenience. Though only twelve years old at the time the treaty was signed, Melissa seemed thoroughly enamored of the dashing Prince Davion, while he appeared to be entranced by the young child-woman.

Currently, there is no Ambassador to the Lyran Commonwealth. Last year, Ambassador Daniel DeKirthus was recalled to his homeworld in the Federated Suns because of illness in the family. When most of his family perished in a tragic plague, it became doubtful that he would return to his post. Until Prince Davion finds a new ambassador, all diplomatic duties in the Lyran Commonwealth are being handled by Jessica Fanna, Countess of Rewas on Numenor.

PERIPHERY

For most of the past hundred years, the Davions have had little interest in what lay beyond their most distant worlds. For many years, they considered the Periphery to be merely a hideout for brigands and pirates. The prevailing logic decided that House Davion was already too large and unwieldy.

All this changed with Hanse Davion's accession to the throne. He encourages an active interest in reexploring the Periphery to rediscover lost worlds and reestablish closer ties with the remnants of the Periphery republics. He wants to expand the Federated Suns by absorbing worlds from the Periphery, "pushing back the curtain of ignorant night and waking worlds long thought dead with sleep." This sudden interest in the dark Periphery is a puzzle even to those who are close to Davion.

One possible explanation is that Prince Hanse's desire is based on hope of encountering potential major worlds rich in either resources or Star League artifacts. Yet, it seems unlikely that the crafty Hanse Davion would put so much effort into a wishful hope. He is too clever and intelligent to gamble purely on the luck of the draw.

Another theory is that he wishes to expand his realm's borders in order to outflank and encircle one of the neighboring realms. Certain facts also seem to make this theory unlikely. First of all, it would require generations to accomplish such a military strategy. Another fact overlooked is that the current explorations are expanding around the Outworlds Alliance, not toward the Capellan Confederation. Surely, Prince Davion is not attempting to encircle the Draconis Combine!

The third, and most plausible, theory is that Prince Hanse wishes to generate some economic activity in the blighted Outback of his realm. Even though he first made his plea for exploration a mere eight years ago, the effect of corporations and colonists moving into the Periphery border regions are already being felt, as old markets closed because of the lack of consumers are now opening up again. These positive effects have even spurred some government officials to propose the creation of a Periphery March to better handle growth of the region. It is uncertain whether the Federated Suns will redraw its map anytime soon, however.

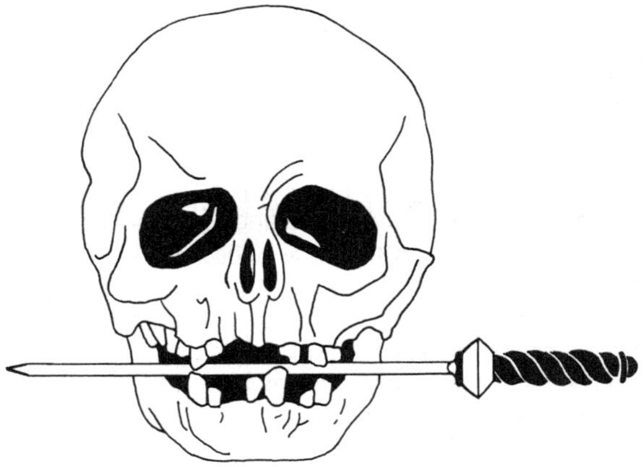

COMSTAR

ComStar relations with the Federated Suns tend to be a picture of extremes. As noted elsewhere in this book, our representatives are treated either as demi-gods on the poorer worlds, or as second-rate-magicians by inhabitants of the sophisticated richer planets. This has made for some unusual and difficult relations between our Blessed Order and the Federated Suns.

The greatest bone of contention between ComStar and the Federated Suns is the question of how much local participation to allow when it is necessary to repair or upgrade our facilities. The Davion government has taken a very hard line, saying that local labor should carry out at least half of the work done on ComStar facilities, with the rest handled by our own labor units. Of course, we find this unacceptable and Precentor Huthrin Vandvel has so informed the Davion government. Ignoring our decision, the Davion people stubbornly continue to press their labor demands.

Among several possible reasons why the Davions demand local participation in the servicing and maintenance of our communications centers, all bode ill for future relations between ComStar and the Federated Suns. Of course, the public reason behind Prince Davion's demand is that he "wishes to stimulate the economic growth of his realm and to further the bonds that exist between House Davion and ComStar."

It is entirely possible that he merely wants to spread some work to local construction companies so they can improve their corporate profiles. This would be most unlike Prince Hanse Davion, however. Seldom, if ever, do his public wishes match his hidden desires, and most everyone here on Terra feels that the Fox is up to no good. It is the general opinion that Hanse Davion wishes to observe and steal our most sacred technology. No doubt his much-vaunted local construction crew would be staffed heavily by members of Military Intelligence, the MIIO, and scientists from the NAIS masquerading as laborers. They would scrutinize every centimeter of our facilities to discover any clue to how our communications equipment works. Despite the efficiency of our ROM agents in protecting our facilities, the confusion of repairing a facility would make it easy for some precious bit of technology to fall into the hands of any outsiders present.

Though this is the most serious and most public of our disputes with the Federated Suns, it is certainly not the only one. Privately, our officials have been exposed to an ever-increasing wall of bureaucracy apparently meant to separate us from the Davion family. It used to be that our precentors had almost unlimited access to members of the Davion family. Now we have to wait for days for an appointment, during which time we must fill out reams of forms and inquiries concerning the purpose of the meeting. Whether intentional or not, this has strained our relations with the Davions and the government.

COVERT POLITICAL GOALS

The Ministry of Intelligence Investigations and Operations (MIIO) is the name given to various departments and secret agencies that the Davion government has created over the centuries to gather information and to act covertly against their enemies. Though perhaps not the most aggressive secret agency in the Inner Sphere, the MIIO has a reputation for being the cleverest and most effective group. Their agents provocateur are undoubtedly the best at slipping into positions of importance in a foreign realm, then inciting so much unrest that the agent can easily slip back to the Federated Suns in the resulting confusion. Certainly, the MIIO's most spectacular success to date was in so heightening the hatred between Janos Marik, the Captain-General of the Free Worlds League, and Anton Marik that it led to all-out civil war.

The Ministry has, of course, been shrouded in mystery, but our own skilled agents have uncovered a few facts about its various departments.

INFORMATION-GATHERING SERVICES DIVISION

The IGS is the section of the MIIO responsible for passively gathering information from the public media of all the other Successor States as well as attempting to intercept their private communications. They have listening posts on most border worlds, tuned into nearby worlds. They also regularly attempt to bribe our officials for transcripts of foreign messages.

The current leader of the IGS is Coordinator Davis Rterry, Count of Lake Washington, Ozawa.

COVERT OPERATIONS DIVISION

The Covert Operations Division is the department of the MIIO responsible for actively gathering information about other Successor States. Agents trained at the division's ten secret training facilities are sent into foreign realms to actively, and at times aggressively, gather secret information. Though some agents, the Free-Floaters, are sent into foreign worlds with nothing more specific than orders to gather whatever they can find, others have specific missions and are called Diggers.

The current leader of the COD is Coordinator Edwina Silvan, Countess of Dry Thomas, Lackland.

MILITARY INTELLIGENCE LIAISON

The Military Intelligence Liaison is the department within the MIIO that coordinates the Ministry's actions with that of Military Intelligence to prevent unnecessary, and sometimes dangerous, duplication of missions. The department also handles the borrowing of Special Force units from Military Intelligence, for special search-and-destroy missions.

The current liaison officer of the MIIO is Coordinator Jacoff DeVries, retired colonel of the AFFS.

BUREAU OF INTERNAL INVESTIGATIONS

The Bureau of Internal Investigations is the department within the MIIO that scrutinizes the Federated Suns itself, monitoring the public media as well as gathering secret information about the political climate of the Federated Suns. The BII is estimated to be the largest department within the Ministry, with agents supposedly reporting in from every planet and colony in the Federated Suns. Though spying on the actions of planetary governments and the private lives of nobility is illegal, according to all the public declarations of the Davion government, it must be noted that the other Successor States also practice internal investigations of their own people. There have been only three documented cases of BII agents interfering to stop a perceived threat to the Federated Suns. On two occasions, its agents assassinated the leaders of planetary governments that posed a direct threat to the Federated Suns. The third incident was the blackmail of an unknown Duke to guarantee his support of Prince Ian Davion's policies.

The current Coordinator of the BII is Count Trimons Retweth of Numenor.

COUNTER-INTELLIGENCE DIVISION

The Counter-Intelligence Division originally began its service as a sub-division of the BII, but its role as spy-hunters and anti-terrorists made it so important that it became a full-fledged division of the MIIO during the Second Succession War. Since then, the CID (nicknamed the Rat Catchers) has so grown in power and importance that it now vies with the Covert Operations Division for the biggest hunk of the Ministry's budget. Without the CID, the Federated Suns could easily fall prey to crippling attacks from its neighbors as well as lose valuable information to spies.

Quintus Allard, Count of Bristol, is the current Coordinator of the CID.

The Ministry of Intelligence Investigations and Operations is currently without a minister. The last minister, Count Nicholas Truston, was killed five years ago in a DropShip accident. Since then, Quintus Allard, Coordinator of the CID, has been interim minister of the MIIO. Though Prince Hanse Davion intends to find a new minister, he has not yet appointed anyone.

Prince Davion must be pleased with Coordinator Allard's leadership to have permitted him to run the Ministry this long. As the temporary leader of the MIIO, Coordinator Allard has been responsible for setting up relations with the Lyran Intelligence Corps, the MIIO's counterpart in the Commonwealth. The successful sharing of mutual information in recent months is due largely to the efforts of Coordinator Quintus Allard and Landgrave Simon Johnson, Chancellor of the LIC. Speculation around New Avalon is that Davion is considering promoting Allard to Minister and letting him run the MIIO permanently.

PRINCE HANSE DAVION

Prince Hanse has certainly lived up to his nickname of "the Fox". Attempts to guess at his true motives and intentions are chancey, at best. The only thing certain about him is that he has proven to be a consummate politician and an excellent strategist during his relatively short reign as Prince.

Publicly, he has gone to a great deal of trouble to portray himself as a peaceful man. His alliance with the Lyran Commonwealth has been presented to the public and to ComStar as the next best thing to a total peace. He says that it will guarantee that his realm must never suffer the kind devastating attack that it did during the opening days of the First Succession War. His propagandists go out of their way to deemphasize the Prince's prowess as a 'Mech commander, choosing instead to focus on his skills as a statesman and humanitarian.

Privately, his actions are far from what might be expected of a peace-loving ruler. Rumor has it that he has been contemplating a major military action ever since finalizing his alliance with the Lyrans. True or not, the threat alone might cause the Draconis Combine and the Capellan Confederation to tone down their aggressiveness for fear of angering Prince Davion.

Many in ComStar believe that the key to understanding Hanse Davion's private goals is to decide how strongly he believes in the Davion family's right to sit upon the throne of a reconstructed Star League. Certainly, the coming marriage between Davion and Melissa Steiner will give the Prince's claim to the throne new vigor. Whether Hanse Davion entered the marriage because he harbors some dream of Davion destiny or because of more pragmatic goals is still unknown.

The MIIO, after its surprising success in the Free Worlds Civil Wars of a decade ago, has been remarkably quiet. There have been few MIIO covert operations against its neighbors in recent years, while the Ministry appears content with just gathering information. They can only be operating on direct orders from the Prince, but we still do not know why it is so.

Considering the vast resources at the beck and call of Hanse Davion, our inability to accurately assess his motives makes him potentially the most dangerous of all the Successor Lords.

FEDERATED SUNS NOBILITY

The nobility of the Federated Suns has no real forum for expressing itself beyond the casual discussions at the Royal Court. Interstellar distances and the structure of the federation government both keep the nobility from extending their influence outside the affairs of their world and its local region of space. Even those who have ambitions to rise beyond their current domain do not have access to sophisticated information-gathering services as the Prince does.

This tends to leave the Dukes of the Federated Suns with limited ambitions. Nobles aspire either to be richer or more important in their world's government. Young nobles dream of being appointed as Marquess on the Periphery so that one day they can become Duke of an entire world. Dukes dream of owning more worlds. Finally, all nobles want to protect what they have against the greed of others, whether it be a foreign power or another noble.

The only exceptions are those nobles active in the federation government and those who have any claim to the throne of the Federated Suns (the Davion family of the planet Victoria, for example). These nobles attempt to fulfill their greater desires by creating private spy networks and by scheming for the loyalty of other nobles. Though many of these nobles are at least semi-loyal to Hanse Davion, it does not prevent them from attempting to improve their position in the government.

Michael Hasek-Davion, the Duke of New Syrtis is the most obvious and most powerful example of a noble who will go to great lengths to fulfill his ambition. To accomplish his desire to be the Prince of the Federated Suns, the Duke has built up an extensive network of spies both inside and outside the federation government. Drawing upon the information supplied to him, the Duke has become a shrewd political wheeler-dealer who has won the loyalty of many nobles in the Capellan March.

His ambition is so large that he has received offers from Chancellor Liao through secret emissaries from the Capellan Confederation. Whether his ambition will drive him to actually enlist the aid of the enemy realm remains to be seen, but the fact that he has not rejected Liao's emissaries out of hand is evidence of how much Duke Hasek-Davion wants the throne of the Federated Suns.

THE COMMONERS

The common person is often more involved in the actions of the vast Federated Suns than are the nobles of his world. Planetary governments are notorious for complaining and demonstrating either for or against federation policy. Though these local actions have no official power, they are noticed by the Prince and the Davion family, and often result in a change of some especially unpopular law. But the actions of the public are never coordinated to the point of actually threatening the government of the Federated Suns.

The current interest of the general public has been toward the recent alliance between their realm and the Lyran Commonwealth. Many are impatiently looking forward to a better economy because of the alliance, while some fear that the realm will be overwhelmed by the vastly superior manufacturing ability of the Lyran Commonwealth. Whatever the belief, the public of the Federated Suns has been eyeing the neighboring realm's economy and debating over every change for the better or worse.

STRENGTHS AND WEAKNESSES

The current political system of the Federated Suns has stood for at least a century without a major change. Though the public is sharply divided between the rich and poor worlds, all seem generally content with the performance of the Davion family.

The nobility system in the society of the Federated Suns has proven to be capable of attracting good planetary leaders, and it gives those leaders enough power to do their job. This nobility system, while strong and strictly defined, is flexible enough that the average citizen can hope to become a noble through hard work and skill. In a kind of reverse corollary, the people can have an incompetent noble removed from power. Though the process is almost unforgivably slow, it does exist. The people's right to call down a noble actually protects the nobility as a whole because it prevents people from being so frustrated that they overturn the whole applecart.

The one major weakness of the political system of the Federated Suns is the sharp contrast between poor worlds and rich ones. The Davion government cannot afford to let the less-developed planets become too dissatisfied because so many of them possess untapped resources. Recent official policy such as Prince Davion's call for Periphery exploration and the proposed creation of a Periphery March, show the political significance that these worlds have taken on in recent years.

A more imminent danger to the political system is the approaching showdown between Prince Hanse Davion and Duke Michael Hasek-Davion. Hanse Davion is aware of most, if not all, of Duke Hasek-Davion's ploys to gain power and win control of the throne, so it is something of a mystery that he has not confronted the Duke directly. Perhaps the Prince realizes that open conflict within the Royal Family would do the Federated Suns more harm than good.

RELIGION

RELIGION

I deeply respect what the many beliefs have brought mankind over the millennia, but I often feel that major religions are a bit like flowers in a bed of manure: the stench from all the evil acts committed in their name tends to overpower the good that comes from them.

—From *The Collected Sayings of Lucien Davion*, edited by Charles Davion, New Avalon Press, 2336

I, Hanse Davion, rightful heir and successor to my brother, do accept from His Most Honored Chairman Elder, Cardinal Diego Lavera, the Unfinished Book and the Scepter of State as symbols of my rights and responsibilities toward all the peoples of the Federated Suns. God bless the freedom-loving people of the Federated Suns.

—From the official Coronation Speech of Hanse Davion, Prince of the Federated Suns, 3013

MAJOR RELIGIONS

Many observers of the Federated Suns are puzzled at why a state that proudly proclaims its love of freedom and independent thought should have such a large and active religious community. Though no pat answers exist, this report offers at least a few paths of further exploration. It also attempts to show how the religious community has at times affected the Davion realm's history and how it continues to color the daily life of the average citizen.

There are five major religions in the Federated Suns today, plus the Unfinished Book phenomenon, which has shown rapid growth since it first emerged.

BUDDHISM

The many sects based on the teachings of the Buddha represent the smallest number of believers among the major faiths in the Federated Suns. Despite this, the Buddhist faith has recently become one of the fastest-growing religions in the realm.

A few hundred years ago, the people of the Federated Suns had a much less tolerant attitude toward Buddhism. The "Oriental Purge" is seldom mentioned in the histories; whether from guilt or disbelief, no one can say. The facts are there in the archives, however, and so must be examined.

During House Kurita's invasion of the Federated Suns during the First Succession War, a wave of paranoia and prejudice aimed at all things Oriental swept the realm. Every institution, business, or service run by Orientals was under immediate suspicion, including those that chanced to have vaguely Oriental-sounding names. Many of the businesses were burned to the ground, but the madness did not stop there. Countless thousands of citizens were arrested as Kurita and/or Liao spies, and thrown into makeshift prisons. Many who attempted to stand up for their rights were killed.

Buddhism took the brunt of this hatred during the Purge, though neither the Kuritas nor the Liaos were Buddhists at that time. It is estimated that 90 percent of all Buddhist temples and monasteries in the Crucis and Capellan Marches, including the beautiful Byodo-in Temple on New Avalon, were destroyed. The Buddhist priests and monks at those temples often suffered the same fate.

The Purge did not end until the Wesley Yoshido Incident and the Davion Appeal. The latter occurred on 15 September 2796, when members of the Davion family and the Pope of the New Avalon Catholic Church appeared on the steps of the Royal Court with the surviving priests from the Byodo-in Temple. The Davions made an impassioned plea that the public stop the senseless violence and focus their attention on the real enemy, the Draconis Combine. They went on to say that the madness was a stain on the realm's honor, and was also costing the war effort men and supplies. The Appeal concluded when the Buddhist priests, some still in bandages and casts, led everyone in a prayer for the victims of the Purge.

Since then, Buddhism has been making a slow but steady comeback. In 2850, Michael Davion gave a major boost to Buddhism. When worlds with large Asian populations were liberated from the Draconis Combine and the Capellan Confederation, Davion sent Buddhist priests and monks, at his own expense, to reestablish the religion on those planets. The people of these worlds, who had suffered religious persecution under their former rulers, were so deeply moved that they became among the most loyal and devoted Davion subjects.

There is still a bit of lingering prejudice against Buddhists in the interior of the Federated Suns, though the ruling family frowns on it. Local governments and nobles are encouraged to stamp out such paranoia. Esoteric Buddhism, with its religious art of brightly painted mandalas, is in such vogue with the young aristocracy that priests from the sect are a common sight in the Royal Court.

The most recent advance made by the Buddhist faith was its admittance into the Inter-Faith Congress (i.e., the Unfinished Book) in 2957.

THE WESLEY YOSHIDO INCIDENT

Captain Wesley Yoshido was a lance commander in the 16th Avalon Hussars and a former member of the Davion Special Forces. Though still very young, his military career had already been noteworthy. It was rumored around the regiment that he would soon be promoted into the regimental commander's staff, which was a popular decision because Yoshido was so well-liked.

In late 2795, the Hussars were posted to Sodus to defend that industrialized world against the possibility of a Kurita attack. Sodus was Captain Yoshido's homeworld, and he was overjoyed at the news.

A few days before the Hussars arrived, however, a crowd of hooligans bent on "ridding the planet of Asian scum" brutally murdered Captain Yoshido's parents, owners of a small electronic firm. The rest of the Yoshido family was in the base hospital, some of them barely clinging to life. Witnesses later described Captain Yoshido's expression as contorted with hatred and anger after he visited them in the hospital.

By a cruel coincidence, another anti-Asian riot occurred that day in the city's main square. The usual band of ruffians had seized ten Oriental-looking men and women and were about to lynch them on a crudely constructed gallows.

As the thugs were ready to hang their first victim, a bolt from a laser pistol caught one of them full in the face. Confused, the mob looked about but could see nothing. Suddenly, a figure dressed in a Special Forces uniform rushed the gallows from among some nearby trees. With a laser pistol in one hand and a cutlass in the other, the figure shot and slashed its way up to the gallows, leaving behind him a wake of dead and wounded. The man cut down the noose and freed the hostages before the mob could react.

As he was about to have his revenge upon the leader of the anti-Asian gang, however, the rescuer was stabbed from behind. At that moment, three Davion 'Mechs lumbered into the city square.

The mob was about to kill the wounded Captain Yoshido and recapture the freed hostages, when the three 'Mechs, all from Yoshido's lance, moved forward and swatted them away from the dying commander. One gently picked up Captain Yoshido in his 'Mech's hand and hurried to the hospital, while the other two MechWarriors stood over the hostages until help arrived.

Captain Yoshido died of his wounds. The story of his plight was picked up by the major Davion news services and broadcast realm-wide. Even Prince Davion was moved by the courage of Captain Yoshido. Yoshido was posthumously awarded the Medal Excalibur, the award given only to citizens who have sacrificed their lives in the name of freedom.

—From *Race Relations in the Federated Suns*, by Countess Fabrina DuMont, New Avalon Independent Press, 3011

CHRISTIANITY

Christianity is undoubtedly the largest religion in the Federated Suns, with over 50 separate denominations. The most obscure of these is probably the Neo-Mennonites of Alsek, who eschew all technology created after the 1800s. There are also Commonwealth Lutherans near Terra, Presbyterians around the Pleiades Cluster, and Fundamentalists on Memphis. Indeed, all manner of Christian beliefs are tolerated in the Federated Suns.

By far the largest, and perhaps most controversial, of these Christian groups is the New Avalon Catholic Church. The existence of this vast church is based entirely on a single, garbled communique from Terra to New Avalon made some 250 years ago. Since then, the New Avalon Catholic Church (NACC) has created considerable tension between its believers and those of the Roman Catholic Church.

During the twelve-year reign of terror at the hands of Stefan Amaris, the Vatican, heart of the Catholic Church on Terra, suffered grievously. Rome had been seized and occupied by the infamous Greenhaven Gestapo when Stefan Amaris took control of the planet. Mercenaries of the worst type, the Greenhaven Gestapo's lack of morality was exceeded only by their lust for money. They ransacked the churches for priceless art objects and historical artifacts, which they either carried off or shot up just for the fun of it.

In 2770, the Greenhaven Gestapo decided to put the entire College of Cardinals on trial, in a cynical attempt to extort more riches from the Church. With the lives of everyone in the Vatican hanging in the balance, Pope Clement XXVII was chosen to represent his Church. The choice put to him by the mercenary commander was simple: pay a huge ransom or watch his staff put to death. The Pope had nothing left of value with which to pay, but then, he had no assurance that the mercenaries would not kill everyone anyway.

In desperation, the Vatican managed to broadcast a short message to the ranking church members throughout the Inner Sphere. The message read that his Holy Father was setting aside the mantle of the Church for the duration of the emergency in an attempt to avoid further violence. He gave the cardinals in the capitals of the five member-states certain powers needed to carry on the workings of the Church until such time when he could reassume his position.

Cardinal Kinsey de Medici of New Avalon never received key parts of the message because a burst of static briefly interrupted transmission. In one of those fateful quirks of history, the Cardinal interpreted what he did understand as a command that he alone was to assume complete control of the Catholic Church.

Cardinal Medici was a pious but egocentric man. Despite the urgent messages from church officials in the other states about the true nature of the Pope's message, he stood firm. When news reached Cardinal Medici that Pope Clement and almost everyone else in the Vatican had been murdered by the Greenhaven Gestapo, Medici saw it as proof of his holy mission to revive the wounded Catholic Church. The Cardinal proclaimed himself Pope Thomas X on 15 December 2772, demanding the immediate obedience of all Catholics in the Inner Sphere.

Five months after Terra was recovered in 2779, Cardinal Vladimir Kosiv of the People's Union of Kazkakh was elected Pope John Paul V. For the second time in history, there were two Catholic Churches, with two Popes competing for allegiance.

Though many efforts have been made to reconcile the rift, all have failed. It is an interesting footnote of history that the Davions were responsible for scuttling the early reconciliation talks. The Roman Catholic Church, centered squarely in the Vatican on Terra, had always been a staunch supporter of the Star League. The Davions saw the New Avalon Catholic Church as a chance to draw support away from Terra and the League. After the Star League collapsed, the Davions saw the NACC as a way to promote patriotism in the Federated Suns.

On individual planets, the conflict between the two churches tended to be anything but a sane and civil debate. On many worlds, the Catholic community was split down the middle and the question was settled, more often than not, with blood. On worlds such as Barstow and Goderich, clashes between the two factions grew so intense that the planetary governments were forced to call upon the Davion military for help.

Today, an uneasy truce exists between the two Churches. The Roman Catholic Church recognizes the New Avalon Catholic Church as a wayward flock, and every few years sends a Cardinal to New Avalon to try to negotiate reconciliation. The New Avalon Catholic Church, on the other hand, sees the Roman Catholic Church as the lost and confused one. Its leaders occasionally send one of their Cardinals to Terra in an attempt to win its rivals to the NACC.

The followers of the NACC currently outnumber followers of the Roman Catholic Church in the Federated Suns five to one. Worlds close to Terra tend to follow the Roman Catholic Church, while most of the others follow the NACC. There are still many worlds whose populations live under the cloud of the schism. As long as the Davions continue to acknowledge the NACC as a legitimate religion (out of political, if not theological, necessity), the question is unlikely to be resolved any time in the near future.

ISLAM

Followers of the Prophet Mohammad form one of the smaller religious groups in the Federated Suns. Due to religious uncertainty during the first centuries of space travel, Moslems did not participate fully in the mass exodus into space. Only 400 years after the first mass migration from Terra did the Islamic faith allow its believers to travel offworld.

The majority sect of Islam, the Sunnis, have settled on two marginally habitable planets near the Federated Suns borders with the Draconis Combine and the Capellan Confederation. Ankaa, a small world beneath a hot white star, was long considered worthless, but the Sunni settlers have discovered deposits of rare metals. On Raman, which the Sunnis share with Christians and Buddhists, the Sunnis were responsible for exploring the large and harsh Arab Continent, where they discovered hidden pockets of fertile land. There are also sizeable populations of Sunni Moslems on Islamabad, Necedah, Greeley, and Inman.

The Azami, a stricter, messianic sect of Islam, are also found in the Federated Suns, though scattered and in very small numbers. The Azami appear in numbers large enough to be a political force only on Inman. On Islamabad, the Azami are in direct conflict with the smaller Sunni Moslem population. The feud between the two has roots extending back at least a millennium. The two religious groups have carried on a guerrilla war on the planet's desert continent for the past 50 years, despite the heavy Davion military presence garrisoned there to stop it. Recent discoveries of oil and radioactives on the planet have suddenly given a new urgency to the religious feud.

Politically, the Sunnis are staunch supporters of the Davions, one of the few major powers that allows freedom of worship.

The Azami in the Federated Suns are in a far different position. A band of Azami tribesmen in the Draconis Combine long ago came to an agreement with House Kurita, who promised them freedom of worship in exchange for service from the feared Azami warriors, who have struck many times at Davion border worlds. Though some Azami in the Federated Suns actively support the Davions, the actions of their kinsmen in the Draconis Combine have placed all Azami under a cloud of suspicion. Because of the sometimes violent prejudice the Azami encounter, especially on Kurita border worlds, the Azami have adopted a low profile in the Federated Suns, preferring to go unnoticed rather than chance being attacked.

JUDAISM

The center of Judaism in the Federated Suns is the planet Robinson. There, the famed Yeshiva of New Tel Aviv has continued its centuries-long study of the Talmud. Study at the Yeshiva is funded in large part by the nobility of Robinson and by the donations of the realm's many believers.

The Federated Suns Jewish community is currently locked in a great debate over the Unfinished Book and its Council of Elders. The more conservative Jewish thinkers want nothing to do with what they consider an attempt to synthesize a single belief from those of all the major religions. The more liberal Jews welcome anything that promotes greater understanding between the religions.

Jews have traditionally settled the planets around Robinson, though there has been recent interest in heeding Hanse Davion's call for explorers to the Periphery. Because of the location of their worlds, the Jewish community suffered greatly during the First Succession War, when the Draconis Combine drove deep into the Federated Suns. The horrors they experienced at the hands of the Kurita forces has left the Jewish community almost unanimously pro-Davion. Since that war, many Jews have been eager to join the Davion military. Continuing a centuries-old tradition, Jewish officers are noted for their ability as AeroSpace Pilots and tank commanders, as well as tacticians.

HINDUISM

The center of Hinduism in the Federated Suns was once a small, but powerful realm. The United Hindu Collective was a group of star systems populated almost entirely by Hindus whose ancestors had left India during the first Exodus. Not satisfied to settle on worlds already colonized by others, the Indians bought several Jump-Ships and pushed deep into unexplored space. They settled on Panpour, As Samik, and Basantapur, three worlds near the Periphery and Crucis-Capellan March border.

By 2534, the United Hindu Collective was 17 star systems strong, with thriving industries and a well-trained military. The pride of that military were the Messengers of Shiva, a large AeroSpace-BattleMech regiment. The Federated Suns was in the grip of a civil war, with Cassandra Varnay fighting Alexander Davion for control of the realm. Because the United Hindu Collective and its military would represent a strong ally for either side, the leaders of both factions attempted to persuade the Maharaja of Basantapur to join them. The Maharaja warned both sides to respect the sovereignty of the Collective.

Alexander Davion did. Cassandra Varnay and her overconfident generals did not. In 2535, forces from Varnay's rebel army attempted to pass through the Collective in an effort to outflank Prince Davion's forces. They stole what they needed from the Collective worlds and burned the rest.

They were intercepted by the Messengers of Shiva on the Collective world of Jodipur, now known as Neukirchen. The highly trained and motivated Messengers crushed the Varnay forces. Humiliated and disgraced, the remaining trespassers were dragged back to Basantapur, where they were held for the duration of the war.

Immediately after Alexander Davion had defeated Varnay and secured his throne, he launched diplomatic missions to bring the United Hindu Collective into the Federated Suns. His efforts paid off in 2540, when he and the Maharaja signed the Basantapur Treaty, which made the Collective a full, semi-autonomous member of the Federated Suns. Many attribute the economic boom that followed soon after to the Collective's wealthy and highly industrialized worlds.

Today, Hindus make up a strong and vigorous portion of the population. Almost five centuries after the Basantapur Treaty, Hindus have spread out from the domains of the old United Hindu Collective into almost all parts of the Federated Suns and they are present among all classes and occupations.

Though Hindus generally live peacefully with other citizens of the realm, a bitter religious war between Hindus and conservative Christians threatens to destroy the loser on Kantwood, a world near the Periphery. As Kantwood is so far from New Avalon, the government is having difficulty finding resources to calm the unrest. As hostilities continue, the prospect of the spreading hatred becomes more likely. Our representatives have already reported riots on the nearby worlds of Hivanee and Abbeville, apparently sparked by the war on Kantwood.

UNFINISHED BOOK MOVEMENT

In 2955, Father Jasper Ovidon of Delphos, a noted religious scholar, petitioned Prince Peter Davion to create a council of religious leaders to promote good will between the various faiths. It was to be patterned after the Commonwealth Religious Council, which for centuries has been successfully promoting interfaith understanding in the Lyran Commonwealth. This idea seemed so obvious and so simple that the priest's appeal was championed by the New Avalon Catholic Church and several influential members of the nobility. In 2956, Prince Davion announced the creation of the Federated Inter-Faith Congress and appointed Father Ovidon as its first Chairman Elder.

Under the leadership of Father Ovidon, the first Unfinished Book (unfinished in the sense that there was so much spiritual wisdom yet to be discovered) was published. This "book" actually consisted of ten large volumes containing the holy words of many religions, and it was the culmination of Father Ovidon's overriding desire for peace. He felt that if mankind could write down all the scriptures of all the religions and bind them into one series of books, true harmony and peace for mankind would follow. How this would come about, whether through divine intervention or some act of man, the priest never said.

Nevertheless, his enthusiasm was infectious. Editions of the Unfinished Book soon became unexpected bestsellers throughout the Federated Suns. Immediately after publication of the first edition, people began to offer their aid to the Inter-Faith Congress, not out of any desire to join a specific faith, but to see the Unfinished Book completed. This phenomenon did not stop with the death of Father Ovidon in 2966. Under Lama Potec, the second Chairman Elder, this extensive research project to collect the writings of countless holy men and women continued.

Today, the influence of the Unfinished Book—as the Inter-Faith Congress has come to be known—is phenomenal. Those who helped with the Unfinished Book seem to have created a quasi-religious order. They even have a symbol, an open book with blank pages set against the Federated Sunburst, which many dedicated young men and women wear like a badge.

The people who work on the Unfinished Book baffle any attempt to characterize them as a group. No single faith predominates among those who donate their energies. Indeed, a fair number of agnostics exist among the workers. There is not a common political belief, nor is there a common class affiliation. About the only thing these people have in common is their interest in humanity's many religions and philosophies.

This interest in spirituality gives ComStar some difficulty. In recent years, the appeal of working on the Unfinished Book has drawn away many people whom we might have recruited for our Blessed Order. Even more disturbing is that the Unfinished Book is becoming a political power. Priests trained at the Unfinished Book's learning center on New Avalon and who are accepted in two or more faiths, have become the official chaplains for the Davion military. The true measure of their growing power and our distress is that some in the Unfinished Book cadre have recently been questioning our Order. Recent articles published by members of the Inter-Faith Congress have called us "a pack of voodoo priests and mumbo jumbo chiefs." Their slanderous attacks have, as yet, gone unpunished by the Davion government, though we protested mightily.

The most convincing example of the prestige of this quasi-religion occurred at the coronation of Hanse Davion. Instead of swearing his oath on a traditional Bible, the Prince swore upon a copy of the new fifth edition of the Unfinished Book.

STRENGTHS AND WEAKNESSES

The Federated Suns shares with the Lyran Commonwealth, and to a lesser extent, the Free Worlds League, the boon and bane of religious freedom. On the positive side, the freedom to worship as one believes is a crucial factor in helping the citizenry feel "free" and unoppressed by the government. Freedom of worship also promotes an open culture able to accept and discuss new ideas and changes without fear.

On the dark side, there is always the specter of religious conflict. In times of stress and uncertainty, the common people, especially the uneducated, often feel impotent. At such times, they often lash out in frustration at the nearest minority. Once a religious conflict occurs, it often takes years to heal the wounds.

The other disadvantage of religious freedom is that it becomes impossible to gain support for a cause by claiming it is God's will. Diversity of religion guarantees a diversity of opinions on almost every subject. It is futile to attempt to stir up the emotions of the realm with scripture, as so many past leaders have done. Indeed, a few Davion princes have tried and failed miserably.

The Federated Suns tolerance and interest in religions—as seen in the Unfinished Book phenomenon—often makes things difficult for ComStar. We cannot simply awe the people of the Suns with stories of the golden years of the Star League and the lives of our Blessed Founders. Diplomacy and intelligence must be exercised when dealing with the more sophisticated citizens of the Federated Suns.

The Armed Forces of the Federated Suns is the largest military force in the Inner Sphere, with the capability of fielding over 110 BattleMech regiments and many times that number of armored, aerospace, infantry, and support units. Regiments in the AFFS are reasonably well-equipped with a wide variety of weapon systems, and every soldier's training emphasizes wits and intuition to make even the lowliest trooper a thinking, scheming, dangerous opponent. Considering the vastness of the Federated Suns, the AFFS military bureaucracy is amazingly swift in responding to the needs of its forces, while the officers trained in the Federated Suns academies are perhaps the best educated and certainly the most daring in the Inner Sphere. These factors make the AFFS not only the largest, but the best fighting force among the five Successor States.

Despite these advantages, the AFFS is still too small to do much more than keep the forces of the Draconis Combine and the Capellan Confederation at bay. The occasional effort to seize a world, the periodic raid, or the rare offensive have been the extent of the AFFS offensive capacity in the last few decades. This inability to take the fight to the enemy is as frustrating to enlisted men as to their officers, for both groups are highly motivated to prove their worth.

The biggest disadvantage the AFFS must overcome is the vastness of the Federated Suns. The extent of the realm, with its incredibly long borders, means that units must defend huge volumes of space. To mount an offensive means that the High Command must take troops away from the defense, leaving those who remain behind to watch over an even larger volume of space. The chances for disaster are many. Many a Field Marshal has lost sleep over the fear of finding his offensive outflanked and the enemy pouring into the Federated Suns from behind.

The recent alliance between the Federated Suns and the Lyran Commonwealth may help to alleviate this problem of distance versus available manpower. Now that the AFFS has a strong, though sometimes inept, ally in the Lyran Commonwealth Armed Forces, they can share intelligence and other kinds of assistance that will make a new Federated Suns offensive more feasible. One way or the other, any action by this highly trained and motivated military will affect the destiny of the other four Great Houses of the Inner Sphere.

STRUCTURE OF THE AFFS

The AFFS has developed through the centuries from a loose collection of fighting forces contributed by member worlds into a highly organized military that is the most powerful and one of the most efficiently run modern fighting machines.

HIGH COMMAND

The High Command of the AFFS is composed of all 14 Field Marshals, as well as the three principal strategists from Military Intelligence. The High Command usually meets in the Strategy Room (aka The Fox's Den) somewhere in the AFFS Headquarters.

There, the members of the High Command use advanced computers to project maps of the Federated Suns and any other realm in the Inner Sphere. The computers contain details of the terrain of every world, down to the smallest brook or hill, and so holographic representations can be called up at a moment's notice. If fed the proper data, the computers can also suggest what strategy and tactics offer the best chances for success. The computers have also analyzed data on the military actions of all the previous Davions, which allows them to suggest tactics in the style of any of the former leaders of the Federated Suns.

Presiding over the High Command is the Marshal of the AFFS, the Prince of the Federated Suns. Responsibility for everything the AFFS does is on his shoulders, and so the more conscientious Davion Princes have religiously attended every High Command meeting. Anytime the Prince cannot attend, the Prince's Champion, an ancient title bestowed on the most senior Field Marshal, presides over the meeting. The Prince's current Champion is Field Marshal Yvonne Davion, one of the famed Victoria Davions and a fifty-year veteran of the AFFS. As she is a staunch supporter of Prince Hanse, he is confident that she will handle command of the military just as he would wish it.

DRACONIS AND CAPELLAN MARCH REGIONAL COMMANDS (DMRC & CMRC)

Because the Draconis and Capellan Marches bear the brunt of all the fighting, the AFFS has done everything possible to streamline the command structures of these two regions. Anything not directly involved with the effort to attack or defend against the Capellan Confederation and the Draconis Combine was turned over to the New Avalon Regional Command for handling (e.g., transportation and supply). As a result, the DMRC and CMRC are focused almost entirely on military matters.

Field Marshal Michael Hasek-Davion, Duke of New Syrtis, currently commands the CMRC. Field Marshal Aaron Sandoval, Duke of Robinson, commands the DMRC. Though the Dukes of New Syrtis and Robinson are automatically political leaders of the two marches, they do not automatically assume command of all military forces in the region. Even a Duke must prove his military prowess if he expects to command the armed forces of a whole March. If the Regional Commander does not prove himself worthy, command falls to the senior Field Marshal in the March.

COMBAT THEATER COMMANDS

All three Marches of the Federated Suns are divided into Combat Theaters to allow for quicker and more responsive leadership in each section of the March. The Draconis and Capellan Marches are each divided into a Coreward and an Edgeward Combat Theater. Field Marshals command the Combat Theaters. The commanders of the March Militia and Frontline RCTs make up his command staff.

The current Commanders of the Capellan March Combat Theaters are Field Marshal Damien DeKirl, who commands the Coreward CT, and Field Marshal Jennifer Lawson, who commands the Edgeward CT.

In the Draconis March, the two Combat Theater Commanders are Field Marshal James Sandoval, who commands the Coreward CT, and Field Marshal Kinsey Tronley, who commands the Edgeward CT.

March Militias and PDZ Commands

Dividing up each Combat Theater are the Polymorphous Defense Zones. Within each of these is a March Militia unit. Unless a PDZ hosts a Frontline Regimental Combat Team, the commander of the March Militia is also the commander of the PDZ. He must coordinate the actions of all the units in his PDZ as well as seeing to it that all are properly supplied.

March Resources

This section of the Regional Command acts as a distributor of the precious flow of supplies and replacements to the forces within the March. It is also responsible for requesting the necessary supplies and replacements from the NARC as well as transportation to move the supplies about the March. It is sometimes necessary for the March Resources to ask the NARC for extra transportation to move the troops of the March from world to world during offensives or defenses.

Marshal Timothy Nilson is the current March Resources Commander for the Draconis March and Marshal Geraldine Revson is his counterpart in the Capellan March.

NEW AVALON REGIONAL COMMAND (NARC)

The defense of the Crucis March is handled in much the same way as in the other Marches. The only real difference is that the March is not divided up into PDZs but uses the older Combat Regions system, which are defensive regions corresponding to Administrative Areas. Otherwise, the military handles defense of the region in ways that closely parallel those of the Capellan and Draconis Marches.

The Crucis March is divided into three Combat Theaters: one Coreward Combat Theater and two Edgeward Combat Theaters. The current commanders are Field Marshal Nicholas Geheart, who commands the Coreward CT, Field Marshal Aziz Khan, who commands the Edgeward Alpha CT, and Field Marshal Barbara Topaz, who commands the Edgeward Bravo CT. The commander of the entire March is Prince Hanse Davion, though the Prince's Champion handles many of those duties for him.

In addition to responsibility for defense of the Crucis March, the NARC also handles the AFFS bureaucracy. This is what truly separates the NARC from its sister commands and makes it the core and the power of the entire AFFS. It has been said that if the bureaucracy of the AFFS were evenly distributed among the three Marches, two possible results would occur. One would be that the Federated Suns would fall prey to its neighbors because the Field Marshals would be far too busy managing their supplies to pay close enough attention to the fighting. The other outcome would be the complete shattering of the Federated Suns as the three equally powerful March Commanders fought among themselves for control of the realm.

Transportation and Resupply Command

The sense that the Prince of the Federated Suns controls the realm because he controls the workings of the AFFS is felt nowhere as strongly as here in the T & R Command. Every request for a spare part, every ton of grain, every roll of bathroom tissue needed by the AFFS must pass this way to be approved and fulfilled. This means that Duke Michael Hasek-Davion, Prince Hanse's wiliest enemy, must come hat-in-hand to the T & R Command to have his troops supplied. Thus, Duke Hasek-Davion cannot try to build up his forces by requesting large orders of supplies because Prince Hanse would immediately know about it.

RANKS IN THE AFFS

The AFFS ranking system is heavily influenced by the Code of Ranks formerly used by the Star League Regular Army and the British military before that. Though the AFFS system does not approach the number of ranks in the Lyran military, it does have more ranks than the House Marik, Liao, or Kurita militaries. This system, which has been in use for the past hundred years, suits the AFFS well. It allows for promotions, an important key to morale, yet the number of ranks allows plenty of time to judge a soldier's ultimate abilities. This helps to prevent incompetent soldiers from being promoted too far too fast, which can sometimes happen in militaries with too small a number of ranks.

The enlisted ranks are designated by a system of stripes usually worn on the right shoulder. Because of the small black representation of the Federated Sword, the soldiers call these stripes "Scars of the Sword," or Scars for short. The color of the stripes shows to which branch of the service the soldier belongs. Red is reserved for MechWarriors. Blue is reserved for the armor, infantry, artillery, and other similar branches of the military. Gold is for AeroSpace Fighters. White denotes members of the navy, security, and various miscellaneous sections of the military.

Private

Recruits and volunteers who have reached their 16th birthday enter the AFFS with the rank of private. Every member world of the Federated Suns must provide for the initial training of its soldiers and therefore fund and operate their own boot camps. The drill instructors at these camps are not active-duty members of the AFFS, but retirees and reservists who work as instructors to free up active sergeants and lieutenants for more important tasks.

Having completed boot camp training, the young soldier attends one of the hundreds of Specialty Schools scattered across the Federated Suns. These schools teach recruits particular skills such as Battlefield Communications or Tracked Vehicle Operation. After successful speciality training, the private is assigned to a regiment. Most soldiers are assigned to a regiment within their native March, if not one close to their homeworld. Once assigned to a regiment, the soldier is likely to remain with it for his entire six-year tour with the AFFS.

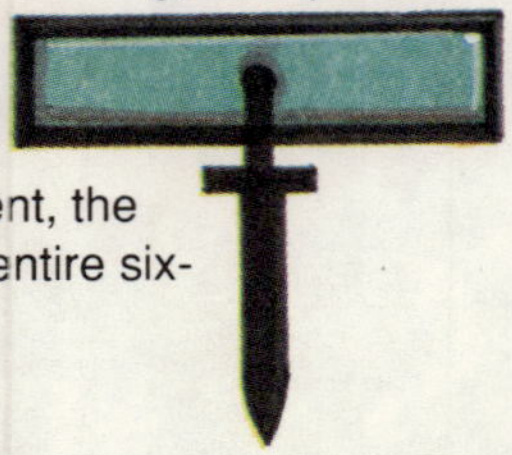

Privates wear one Scar.

Though this system may sound unwieldy, it is not. It is true that all requests must travel to New Avalon, but Regional T & R Warehouses are located near the Headquarters of every March Militia. These regional storehouses will eventually fill the March Commander's requests, assuming that the T & R Command on New Avalon has approved and confirmed the order.

The unspoken political power of this department makes the individual in command of it an important figure. Prince Hanse and his brother Ian before him have both paid special attention to ensuring that this department's personnel are particularly loyal.

Marshal Jisosa Miller is the current commander of the T & R Department and has been for the last 20 years. Though it is rumored that she wishes to retire, Prince Hanse has asked her to stay on for the next five years. No one knows why.

DEPARTMENT OF MILITARY EDUCATION

Equally important to the running of a capable and loyal military is an effective program to provide recruits and cadets with the best possible education, not only in military matters but in historical and scientific fields as well. In the Federated Suns, the Department of Military Education is responsible to provide such an education to its many thousands of soldiers.

One section of the Department of Military Education oversees the training methods at boot camps scattered across the realm to be sure they conform with the rules and objectives laid out by the AFFS. Observers, usually retired officers, travel the realm to inspect boot camps, checking on proper skill training and that the discipline and physical requirements are neither too difficult nor too soft. In some cases, the observers must take over a boot camp that fails to meet the standards.

The rest of the Department is concerned with the operation of the realm's military academies. The Department of Military Education must approve anything taught in the military academies, which stress a broad range of subjects for its cadets. The subject of history gets special emphasis, for "those who forget the past are doomed to repeat it." Sciences are also emphasized.

As for political education, the AFFS academies preach a pro-Prince, pro-Federated Suns doctrine. This line tries to give each cadet the sense that he is akin to the ancient knights of Europe, that he is the embodiment of the best that the Federated Suns has to offer. They are indoctrinated with a sense of duty that is not directed toward a particular political aim, such as elevating a Davion to First Lordship. Rather, the academics impress the idea in the minds of their cadets that they must act for the betterment of the people of the Federated Suns. Officers are supposed to live and die for the citizenry rather than for some "Davion destiny". That the needs of the people and the needs of the Prince are one and the same is viewed as just a happy coincidence.

The Department of Military Education is currently headed by Marshal Grentha DeBurton, who also heads the section concerned with academy curricula. Her immediate aide is Master-Sergeant Dirk Kentrick, who is responsible for overseeing the realm's boot camps.

DEPARTMENT OF THE ARMY AND THE NAVY

This Department handles the posting, the advanced training, and the eventual retirement of the soldiers of the AFFS. Its members must consider the homeworld, the skills, and the apparent trustworthiness of the soldiers fresh from their military training before assigning them to posts. If all goes well, the soldier will become a valuable asset to the Federated Suns. If incorrectly assigned, he may become a liability or even a traitor to the realm.

The Department is also something of an unofficial spokesman for the common soldier. Complaints about unfair treatment, poor officers, poor supplies, or general ineffectiveness are funneled to the Department. Its officials then act on them, either through unofficial channels, or by referring the complaint to the Department of Military Justice.

The Department of the Army and the Navy is currently commanded by Fleet Admiral Della Trine, who is also the commanding officer of the entire AFFS Navy.

Corporal

After a short time, usually less than a year, the soldier is promoted to Corporal. Though there is little distinction between the duties of a Corporal and a Private, Corporals receive higher pay and begin to collect Death Compensation (a fund for the next-of-kin in case of the trooper's death). Corporals who have shown leadership ability will be sent to a Small Units School for training in tactical skills.

Corporals wear two Scars.

Sergeant

Sergeants play an important role in the AFFS. Most graduates from the various military academies begin their careers in the AFFS at the rank of Sergeant. In infantry units, Sergeants must play the dual role of tactician and protector of the men and women in their squads. In other branches of the AFFS, Sergeants are assigned to be vehicle commanders, gunnery chiefs, or other positions of importance. The many demands facing Sergeants are why few common soldiers in the AFFS ever attain this rank. For this reason, the AFFS considers the Sergeant to be a non-commissioned officer with pay and privileges far above the lower ranks.

In most service branches, Sergeants wear three scars. Graduates from the military academies enter active duty wearing three curved Scars as a sign of both their branch of service and their inexperience.

DEPARTMENT OF MILITARY JUSTICE

The Department of Military Justice is responsible for enforcing military law in the AFFS. To carry out this work, it has its own judges, lawyers, and police force. Though the AFFS is not an overly repressive military, it will not hesitate to seize and prosecute any soldier who does not conform to the rules and regulations.

The Department of Military Justice is currently under the command of Marshal DuPrichard Mechalam, a former member of Team Banzai.

DEPARTMENT OF MERCENARY RELATIONS

The AFFS is the largest employer of mercenary units in the entire Inner Sphere, which makes it vital for the Regular Army to maintain friendly relations with these professional soldiers. There have been enough serious "incidents" between the AFFS and disgruntled mercenaries to make even the most prejudiced AFFS officer accept the need to keep the mercenary units content.

Most mercenary units consider the Federated Suns a good employer. The standard terms are generous enough that only the most prestigious mercenary units would dare quibble. One of the most attractive clauses of a Federated Suns contract is Hanse Davion's willingness to allow a mercenary unit the courtesy of semi-independence. That means the AFFS issues objectives to a mercenary unit in a combat zone; how the mercenaries fulfil the objective is totally up to them. In return for this rare show of trust, the AFFS demands that a mercenary unit fight to the best of its ability. There should be no unexpected retreats, and outright cowardice is reason to reassess the mercenary unit's contract. The AFFS would consider any hint of treachery reason enough to fall on the merc unit and to confiscate all its property and weapons.

Each mercenary unit has a Liaison Officer from the Department of Mercenary Relations. It is through him that the merc unit's requests are funneled. Liaison officers are almost always chosen for their links with the Davion family because the presence of a Davion, no matter how distantly related to the Prince, is a sign of the AFFS's eagerness to listen to the merc unit's problems.

Marshal Tobias Green-Davion currently heads the Department of Mercenary Relations.

DEPARTMENT OF MILITARY ADMINISTRATION

All tasks that do not comfortably fit into the jurisdiction of one of the above departments is assigned to the Department of Military Administration. These matters include record-keeping and payroll for the millions of Federated Suns soldiers. The Department also acts as a communications center between the departments, a task of major significance that is constantly under scrutiny for improvement.

Another vital part of the Department of Military Administration is the Medical Corps. Up until late in the First Succession War, the Navy, the BattleMech regiments, and the other fighting services were each using their own separate medical organizations. Because of the waste of valuable equipment, ill-management, and duplication of effort among the three medical branches, the Davions decided to strip the fighting units of their doctors and create a Medical Corps independent of all but the Department of Military Administration. Today, the Department of Military Administration runs the AFFS's countless medic stations, MASH units, Evac hospitals, and Hospital JumpShips.

The Department of Military Administration is commanded by Marshal Clarissa Maynall. The Marshal of the Medical Corps is Dr. Jason VanTallson.

Sergeant-Major

Sergeant-Major is the highest rank an enlisted soldier can expect to attain. In addition to his combat duties, the Sergeant-Major's responsibilities include the role of an unofficial liaison between the officers and the enlisted soldiers. Though this may seem trivial, the ability of a Sergeant-Major to discuss the men's problems with the unit's officers, who have a tendency to be more preoccupied with plans then with people, has improved the abilities of many a unit and prevented many a mutiny. There are usually no more than three or four Sergeant-Majors in a typical AFFS regiment. In 'Mech units, Sergeant-Majors often command lances.

The AFFS has no separate ranks for Technicians and Engineers, for which the LCAF has created the Warrant Officer rank. If a Technician or Engineer is sufficiently skillful, he will reach the rank of Sergeant-Major. The authority he wields with this rank is restricted to matters concerning the repair and maintenance of equipment.

Sergeant-Majors wear what have become known as V-Scars. This rank insignia consists of two downward-pointing chevrons with a disc and Federated Sword. The insignia for a MechWarrior Sergeant-Major uses two curved chevrons.

DEPARTMENT OF MILITARY INTELLIGENCE

The Department of Military Intelligence's duty is to acquire knowledge about the enemies and suspected enemies of the Federated Suns through any means necessary and at no matter what cost. The MI's second, though no less important mission, is to strike out at enemies with acts of misinformation, sabotage, and even terrorism.

The Department of Military Intelligence is divided into seven separate divisions.

MI1 – Command

MI1 is the pinnacle of the department, for it commands the whole Military Intelligence effort. MI1 is composed of the six Coordinators from the other divisions, who answer to the Intelligence Marshal. The six Coordinators and the Intelligence Marshal must approve all missions. In the case of a particularly risky mission, they must have at least oral permission from Prince Hanse before the operation can go forward.

Marshal Ivan Tevanol is currently the Intelligence Marshal of MI1 Command.

MI2 – Analysis and Speculation

All the information gathered by the other sections eventually ends up in the MI2 Department. Deep inside the Fox's Lair, the Analysis and Speculation computers puzzle over all these various bits of intelligence. Guided by the members of the section, nicknamed "The Brains of MI2", the computer eventually spits out predictions based upon its analysis of the information gathered. Though it cannot be specific, the computer's broad hunches on what the other realms will do have been correct enough often enough to prove its worth.

MI2 is currently under the command of Marshal Rebecca Derwith.

MI3 – Electronic Information Gathering

The Department of Military Intelligence maintains a large number of electronic listening posts on and around foreign worlds near its borders. Such a listening post is usually nothing more than a crude receiver listening in on radio communications and channeling the results to a ComStar station, which will beam it to a Davion world. MI3, whose agents are nicknamed "the Snoops," is responsible for these listening posts.

Marshal Simone DeCrimmon is currently in command of MI3.

MI4 – Covert Operations Division

"The Stealthy Foxes" is the nickname given the agents of MI4. Over the years, this section of Military Intelligence, aided by MI6 and the MIIO, has proven itself equal to any covert agency in the Inner Sphere.

All recruits and cadets entering the AFFS are scrutinized for traits or potential that might make them candidates for MI4. These skills range from a facility for languages and a flair for costuming to martial arts abilities or a photographic memory. Candidates for MI4, MI5, and MI6 are trained at one of two locations so secret that even ComStar has not a clue to where they are.

Marshal Ulinov Debreber is said to be the commanding officer of MI4 and the head instructor at one of the agent training schools.

MI5 – Counter-Insurgency

Responsibility for the detection and capture of enemy agents within the AFFS is the responsibility of MI5. Those entering MI5 train alongside the future agents of MI4. Where the MI4 agents are called the Stealthy Foxes, the members of MI5 are known as the "Relentless Hounds" for their determination to keep the AFFS clean of alien spies. The Relentless Hounds have an impressive history of accomplishing their duty. They have even managed to capture five of our own ROM agents over the years.

Marshal Tatyana Debreber is said to be the commanding officer of MI5 and a counter-insurgency instructor at one of the agent training schools.

MI6 – Special Forces

Special Forces is the name given to teams of elite saboteurs and terrorists who conduct missions into enemy realms. These Hit Teams carry out missions ranging from abductions and assassinations to demolitions and random murders to terrorize civilian populations. Any use of the Hit Teams requires Prince Davion's approval.

Nicknamed "the Rabid Foxes," these Hit Teams have been known to show as much ferocity as any Draconis Elite Strike Team (DEST). Because the Federated Suns peace-loving propaganda is so successful, few outside the MI or MIIO realize to what extent the Davions condone terrorist activities.

MI6 is commanded by Marshal Clint Walker.

MI7 – MIIO Liaison

The Ministry of Information, Intelligence, and Operations is the civilian counterpart of the Department of Military Intelligence. Because the actions of the two departments often overlap, it is essential that each know what the other is doing so that their agents will not stumble over one another or duplicate efforts.

Coordinator Jacoff DeVries, a retired member of MI4, has been the Liaison Officer between the two intelligence-gathering services for the past five years. Under his guidance, coordination between the two agencies improved and a program of sharing information and personnel has begun.

OFFICERS

Officers in the AFFS and other militaries are not primarily charged with the safety of those they command. That responsibility falls to the Sergeants and Sergeant-Majors. An officer's primary task is to make and carry out plans. They must rid their minds of the fact that the blips of light on a computer screen represent people if they want to do their job efficiently, because only when the officers are efficient do the majority of soldiers survive.

Cadet/Subaltern

Subalterns are the first of the commissioned officer ranks. To designate this, all Subalterns wear white shoulder boards, or epaulettes, no matter what their branch of service. This announces to others that the man is at the beginning of his military career and is therefore inexperienced. Graduates from the academies enter their chosen service within the AFFS at this rank and serve a six-month probationary period until advancing up to Leftenant. The exceptions are MechWarriors who become Sergeants immediately upon graduation. MechWarriors who win promotion from Sergeant-Major will skip the Subaltern rank and advance directly to Leftenant.

TYPICAL AFFS FORCES

FRONTLINE REGIMENTAL COMBAT TEAMS

Ever since the Second Succession War, the AFFS has combined most of its major 'Mech regiments with regiments from other service branches to support them. Together, these groups form Regimental Combat Teams, or RCTs. The usual makeup of an RCT is as follows: one BattleMech Regiment, three Armored Regiments (1 Heavy, 2 Medium), 5 Infantry Regiments (1 Jump, 2 Mechanized, 2 Regular), 1 Artillery Battalion, and two AeroSpace Fighter Wings (1 Regimental, 1 Independent). Of course, in this universe of diminishing supplies, the actual composition of an RCT may vary widely from this model.

The RCTs practice constantly to coordinate their various skills into a single potent fighting force. The command structure of the RCT is also unique because of the AFFS philosophy that only the best man or woman should be chosen for the job, no matter what was his former posting. Thus, an RCT might well be commanded by a Marshal who rose up from the ranks commanding infantry units. [This does not mean that there is not a hierarchy of services. An RCT has yet to be commanded by an ex-artillery officer, for example—Ed.] For most major RCTs, the commanding officer is usually a Marshal.

An RCT is trained to enter a combat zone as a single unit and to use its superior inter-service cooperation to overwhelm its opponents. Modern combat being what it is, however, an RCT must often split up into its component units to face multiple threats on more than one world. An RCT is deployed in its area of responsibility with about half its forces stationed around its headquarters, and the rest fanned out on other nearby worlds.

The RCT is known by the BattleMech regiment in it. Thus, the RCT containing the 11th Avalon Hussars is known as the 11th Avalon RCT. An RCT is considered a single unit. Once a conventional regiment is assigned to one, that unit can expect to stay attached to the RCT for some time. While conventional units retain their names and traditions, they often become known as the "lesser" units of the RCT. Units become identified as "11th Avalon Armor A" or the "11th Avalon Infantry D" and not by their proper names. This often irks conventional units, for they may be as highly individual and tradition-proud as any BattleMech regiment.

Leftenant

Leftenants are usually responsible for command of 'Mech lances, platoons, tank squads, and artillery lances in combat units.

Leftenants wear an epaulette colorcoded to their branch of service.

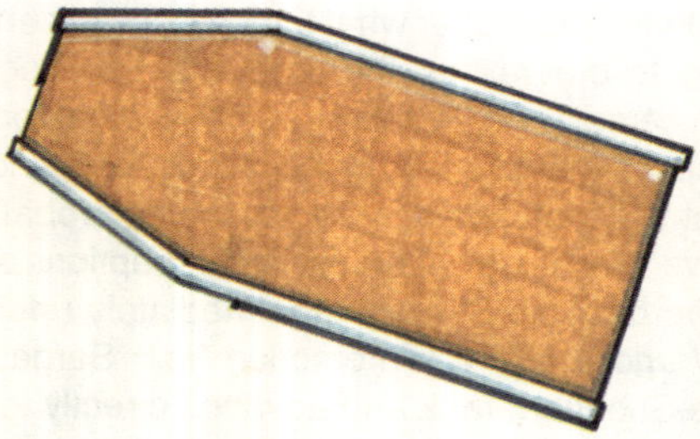

Captain

A Captain commands a company of soldiers. Common troopers and the Sergeants treat Captains with considerable respect because this officer faces as much action as they do. Captains are responsible for carrying out plans given to them from above and for improvising new tactics if the plan begins to fail. This ability to think on the run is a trait highly prized in the AFFS, and is something no amount of training can teach, though the academies try.

Captains wear epaulettes with one narrow white band across the base.

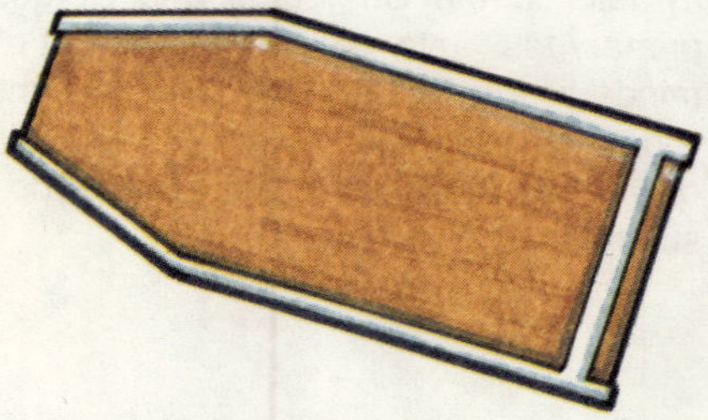

MARCH MILITIA AND PLANETARY GUARD UNITS

March Militias function as both a reserve army and as fighting forces in their own right. Each PDZ in the Capellan and the Draconis Marches raises the money and provides the manpower to field a single RCT, consisting of one BattleMech regiment (of ancient but still combatworthy 'Mechs), two armored regiments, five infantry regiments, a company of artillery, and AeroSpace Fighter support).

The skill and abilities of these Militia units, though not as great as the elite units, do make them a formidable defensive force. Many of the troops serving in Militia units are regular military soldiers and officers who either requested or were transferred to a Militia unit near their homeworld. The remainder of Militia units are composed of soldiers trained at the PDZ's military training school. Command of an MM usually goes to a Major General familiar with the PDZ.

Planetary Guards, nicknamed the Ugly PUGs, are regiments—usually infantry—raised from a world's population to serve as a last-ditch line of defense. Their personnel are usually retirees from the regular forces and men and women who have been excused from regular service. Ugly PUG regiments seldom receive the equipment they need, and so must often rely on local industries to churn out the weapons and vehicles to field a proper fighting force. A certain portion of each PUG unit is made up of Scout units whose sole purpose is to become intimately acquainted with the planet's terrain. If war should come to their world, these Scout units would then be able to guide the regular AFFS defenders in ways that even the best maps could not.

BATTLEMECH REGIMENTS

The AFFS tends to be loose in breaking down its BattleMech regiments into categories such as "heavy" or "light". The AFFS tends to favor assigning the best possible MechWarriors to a regiment, no matter what class 'Mechs they might pilot. Once the assignments are made, the regiment will be arranged into heavy and light battalions, companies, and lances. As a result, the average Davion regiment tends to be lighter than its counterparts in other Houses.

Within the 78 BattleMech regiments of the AFFS are a vast number of different 'Mech designs and variations. It was once estimated that every variation of every type of 'Mech, from the *Locust* Model 1E to the *Atlas* Model RS, can be found among House Davion's regiments. Nevertheless, most Davion BattleMech regiments consist mainly of the handful of 'Mech designs that are currently produced in the Federated Suns.

Light lances in the AFFS are built around the *Valkyrie* BattleMech. A least one of these 'Mechs unique to the Federated Suns can be found in any light unit, usually piloted by the lance's commander. Grouped with it are the usual collection of *Locust*s, *Stinger*s, and *Wasp*s. There are also many *Javelin* light 'Mechs in the AFFS, but parts for this type are in such chronically short supply that the Davion *Javelin*s seem to spend more time strung up in hoists or repair cranes than on the field.

Medium lances in the Davion military usually have at least one *Enforcer* medium 'Mech, also a design unique to the Federated Suns. Accompanying the *Enforcer* may be a wide variety of medium 'Mech designs: *Phoenix Hawk*s, *Hunchback*s, *Trebuchet*s, *Dervish*es, *Griffin*s, or *Shadow Hawk*s. Rarer designs, such as the highly experimental *Hatchetman*, with its wicked hatchet, also appear in frontline 'Mech regiments for battlefield testing.

Heavy lances in the AFFS show the widest diversity. Though *Jagermech*s and *Marauder*s are the most commonly found design, there are all manner of other types. In heavy and assault lances, the AFFS makes no real effort to balance the firepower of one 'Mech design with that of another. Again, the Davion military is more interested in combining four MechWarriors with good skills and the ability to work as a team.

Assault lances in the AFFS usually include the famed *Battlemaster* 'Mech as one of their members. Other Assault 'Mech designs include the *Stalker*, *Victor*, *Banshee*, and the occasional *Atlas*. Rumor has it that the AFFS may soon purchase some *Zeus* Assault 'Mechs from the Lyran Commonwealth military. Some reports indicate that the Lyrans have already lent *Zeus* 'Mechs to Davion units serving near Terra.

Major

A Major usually commands a battalion in the AFFS, though an elite company of 'Mechs or tanks might also be led by a Major. At the battalion level, the Major is expected to integrate strategic plans that his Colonel has given him with the tactical realities reported by his Captains and Leftenants. If this proves impossible, the Major must develop a plan to save the majority of his forces without placing the rest of the regiment in danger.

Majors wear an epaulette with one broad band across its base.

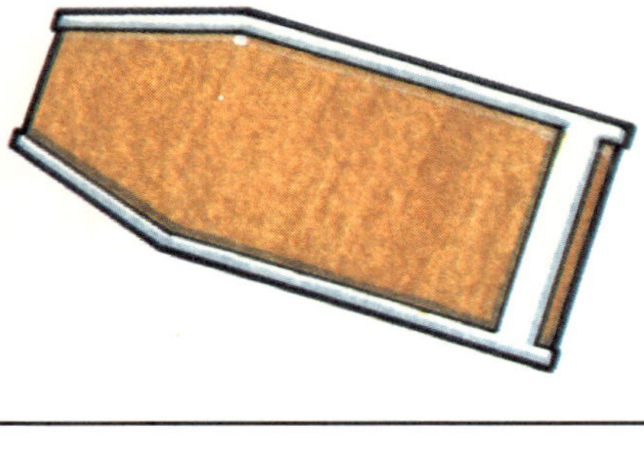

Leftenant Colonel

A Leftenant, or Light, Colonel is usually the Executive Officer of a regiment. As the unit's second-in-command, it falls to the Leftenant Colonel to see that the orders of the Regimental Commander are carried out. It is also the duty of the Light Colonel to handle most of the unit's administrative duties to free the Colonel for more planning time on the regiment's future. In battle, the Light Colonel is often placed in temporary command of a special group, such as a battalion being held in reserve or a group of combat engineers. The Naval equivalent to the rank is Light Commodore.

Leftenant Colonels wear an epaulette with one broad band and one narrow band.

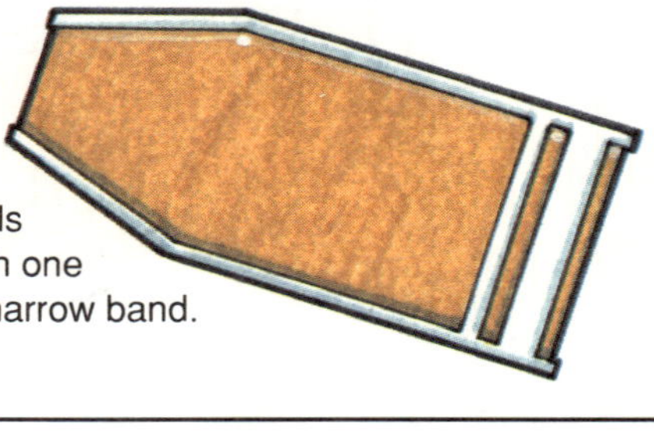

AEROSPACE UNITS

Aerospace units in the AFFS are usually divided into two classes. Pilots attached to service in a BattleMech regiment are called Regimental Wings. One of their main duties is to protect the regiment's DropShips while voyaging to and from a target world. Regimental Wings must also be prepared to perform surface attacks to fly freelance combat air patrols in an effort to knock opposition fighters from the sky. There are usually 20 AeroSpace Pilots to a regiment.

Most RCTs carry equipment that allows them to quickly construct fighter bases near the frontlines. Once the unit has located a large, open but defensible field near the frontlines, huge earthmovers flatten the ground so that other machinery can lay down a thick carpet of poly-turf strengthened by strips of steel to take the weight and impact of landing fighters. A Command Post, usually a van filled with communications equipment and sensing devices, sits in some secluded spot overlooking the field. Shelters for the fighters, ammunition, and machine shops are also built and camouflaged to match the scenery so as not to be immediately visible from the air. VTOLs, missile batteries, and conventional craft are also stationed at these fighter bases, carrying out a variety of missions as well as defending the base from attack. As protection, most of these fighter bases are assigned at least one 'Mech lance from the 'Mech regiment. From this base of operations, fighters listen in to the battlefield communications and respond to regimental requests for their help.

The other class of aerospace unit is the Independent Wing. About the same size as a Regimental Wing, the Independent Wing is totally free to pursue its own missions, such as aerial bombardments or engaging enemy aerospace fighters. These Wings respond to only the most urgent of requests made by ground troops.

In the Armed Forces of the Federated Suns, all Wings are composed of three basic fighter designs unique to the realm. The *Sparrowhawk* is the mainstay of light aerospace units. The *Corsair* is the core of medium aerospace units, while the *Stuka* is the heavy-hitter of the AFFS. Around these three fighter designs the AFFS gathers more common designs such as the *Sabre*, *Lightning*, *Hellcat*, and *Thunderbird*.

ARMORED UNITS

The Armored regiments are second only to the BattleMech regiments in firepower and speed. In the AFFS, most armored regiments are a balance of highly mobile hovertanks with slower but far more deadly heavy tanks.

Heavy Tanks such as the rare Von Luckner, the Demolisher, or the new Rommel and Patton designs are usually reserved for special duties such as crushing an enemy stronghold or dealing with dirty city fighting. The lighter Hunter, Vedette, and other similar Davion designs usually make up the majority of tanks in a Davion armor units. Hovertanks such as the J. Edgar, the Pegasus, and the Condor, along with wheeled vehicles like the Striker, are usually assigned the task of exploiting breaks in an enemy's line of defense.

INFANTRY UNITS

Infantry units are treated slightly better in the AFFS than in other realms, where they are usually considered little more than guards sent to police territory won by the more glamorous units. Enough members of the Davion family have spent their military careers as officers in infantry units to make them realize how much of the dirty work of war is left to the poor sod carrying a rifle.

Infantry regiments in the AFFS are not tailored to fit certain tasks, though special units such as the 3rd Ozawa Light Infantry, the 4th Numenor Mountain Troops, or the 9th Tecumseh Marines do exist. The average infantry unit is created with an eye toward developing camaraderie and esprit de corps. Once a soldier is assigned to a regiment, he is not likely to be transferred to another. Most regiments are manned by soldiers and officers from a single world or a group of neighboring worlds. The only exceptions are units from worlds of questionable loyalty, whose commanding officers and sergeants are usually brought in from other, more dependable regions.

Colonel

A Colonel commands a regiment or a minor department within the AFFS bureaucracy. Battle Colonels are expected to be both tacticians and strategists. Supplying his regiment is also the responsibility of the Colonel, which often means that he spends as much time fighting the bureaucracy of the AFFS as he does the enemy. This is the last rank where the officer is expected to show loyalty to one particular regiment. The rank of Commodore is the Naval equivalent.

Colonels wear epaulettes with two broad bands.

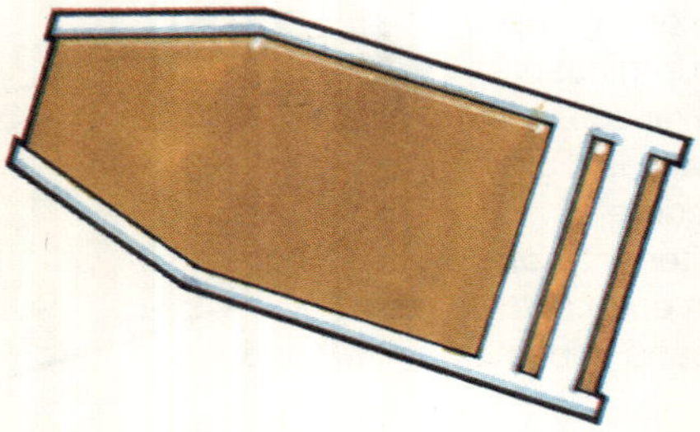

Leftenant General

The rank of Leftenant General, sometimes known as Lieutenant General, is a relatively new rank in the AFFS. It was created in 2953 when the military leaders discovered that Major Generals and Marshals were fine on the battlefield but were often severely lacking in administrative skills and totally unprepared to handle multi-regimental logistics. They therefore decided to create a new rank that would require officers to train their administrative skills as well to become acquainted with grand strategy. Upon leaving the academy, the Leftenant General is usually assigned to the Command Staff of an RCT. During a crucial battle, the Leftenant General is often called upon to command key regiments in the RCT.

Leftenant Generals wear epaulettes with two broad bands and one narrow band.

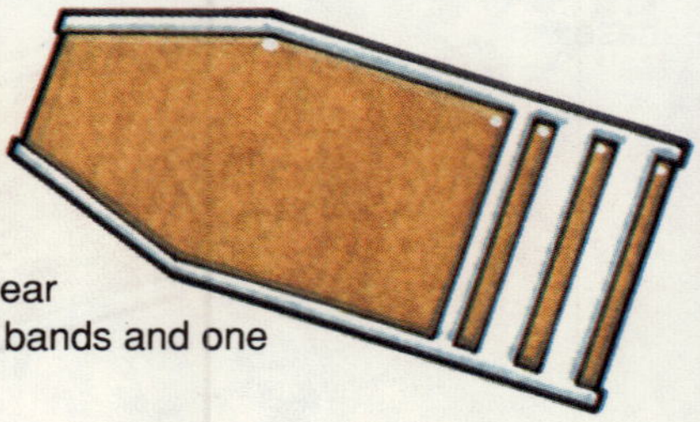

UNIFORMS IN THE AFFS

In any military organization, pride in the uniform is an important factor in promoting camaraderie and esprit de corps. Though the AFFS does want to honor the traditions built up by the various regiments over the centuries, it seeks to avoid gaudy embellishment. To find this happy medium, the AFFS approved a uniform style that can be altered to show distinctions between groups of regiments but that is still functional and free from useless adornment.

OFFICER UNIFORMS

Officers in the AFFS do not have a specific uniform when in the field as do officers in some other realms. Instead, Federated Suns officers wear the same standard uniform as the soldiers they command. Only the shoulder epaulets and the small rank insignia on the front of the officer's helmet or cap show the wearer's status. In particularly dangerous situations, AFFS officers often shed the epaulettes and rely on the small insignia, which are painted a flat black, to show their rank.

The reason for blurring the distinction between officers and common soldiers serves two purposes. First, it tightens the bond between the officer and his men. The leaders of the Davion military prefer not to emphasize an officer's superiority over his men, but rather they want the officer to concentrate on the responsibility of his position. It is rather difficult for an officer to put on airs when he or she looks essentially like everyone else.

The simple issue of survival is the other reason to keep the two groups similarly dressed. An officer represents a considerable investment of AFFS time and money, and so it just good sense to keep him from being even more of a target because of some special clothing. It is a favorite tactic of Kurita snipers to pick out the officers in a troop and shoot at them first. After all, what better way to upset the performance of a combat unit than to kill of its officer.

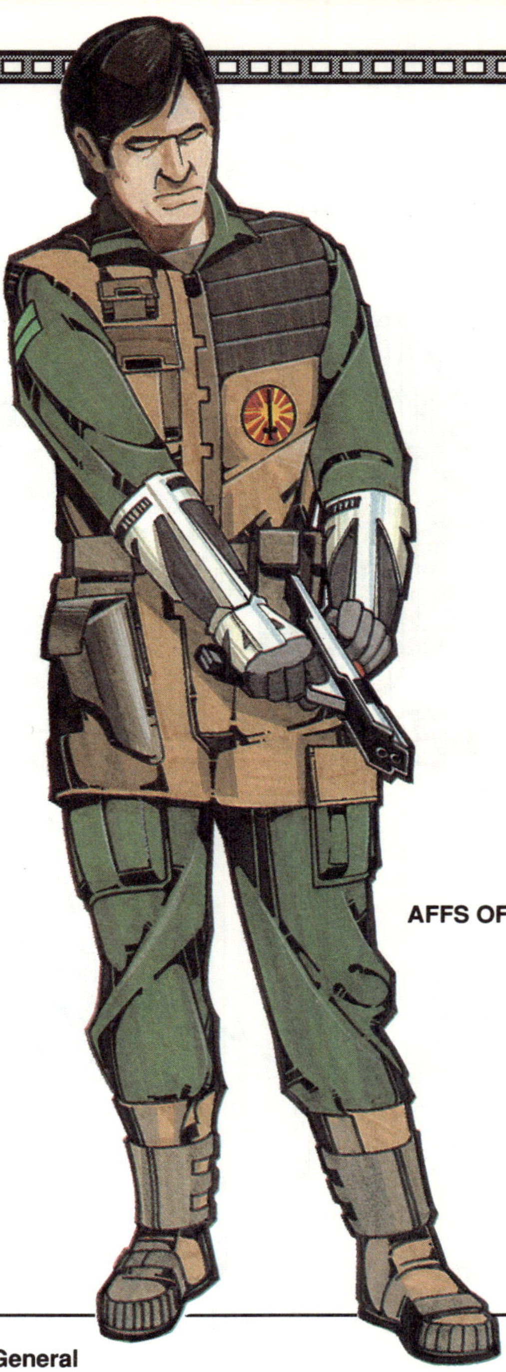

AFFS OFFICER

Major General

The Major General is usually the Executive Officer in a March Militia. Because Militia units are composed of many regiments that are often spread out over several worlds, the officer must learn both tact and politics to manage such a large group of soldiers. In battle, the Major General often assumes command of a regiment vital to the commanding officer's plans. The Naval equivalent to this rank is Vice Admiral.

Major Generals wear epaulettes with three broad bands.

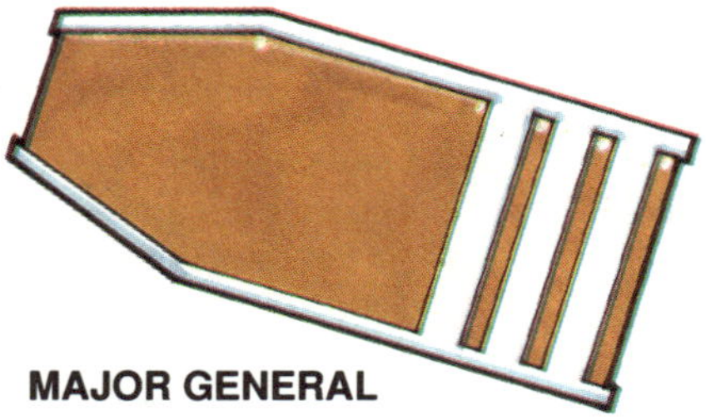

MAJOR GENERAL

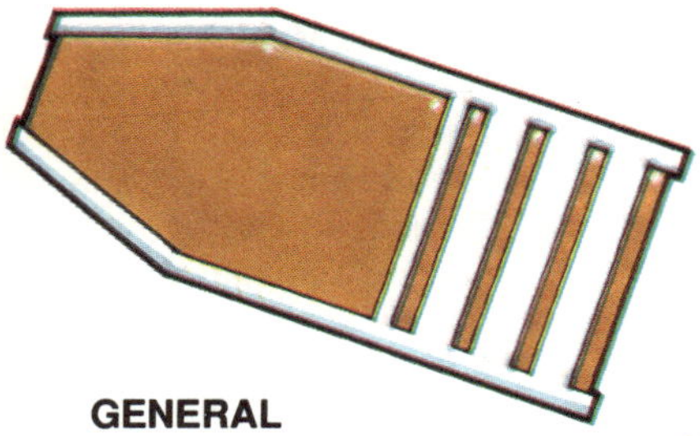

GENERAL

General

Once a Major General has proven his abilities to lead widely scattered regiments, he is promoted to command of a March Militia. When commanding a March Militia, a General must be intimately acquainted with the planets entrusted to his care. He must also somehow manage the mountains of supplies needed to keep his regiments capable of fighting.

If the officer proves to be a strong strategist, he is often given the post of Executive Officer in one of the major RCTs. When acting as the Executive Officer in a so-called Frontline RCT, the General must act as his superior's eyes and ears, often traveling to the heart of the battle to assume command of key regiments while constantly reporting back to his commander.

The Naval equivalent to this rank is Admiral.

Generals wear epaulettes with three broad bands and one narrow band.

FIGHTER PILOT

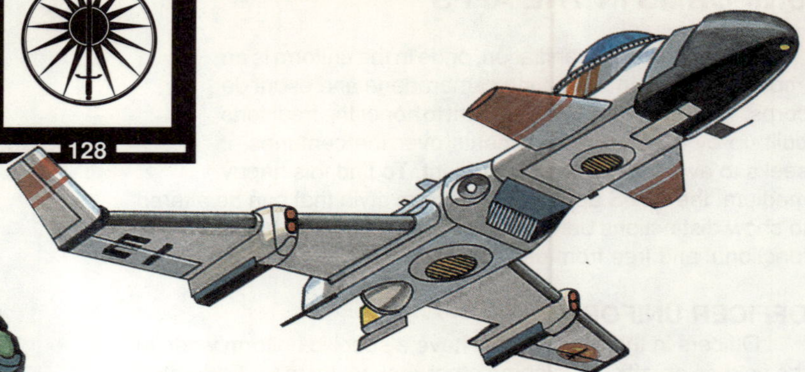

FIGHTER PILOT UNIFORMS

AeroSpace Pilots are the other elite group of warriors in the AFFS. Long known in the Federated Suns as "The Knights of the Void," the pilots of sophisticated modern spacecraft prefer to imitate the traditions of World War I and World War II fighter pilots, those free-spirited men and women who lived by their own code of honor and loyalty.

As part of their tradition, the AFFS fighter pilots show an almost total lack of respect for showy traditions or finery. Indeed, Federated Suns fighter pilots wear perhaps one of the least pretentious military uniforms in the entire Inner Sphere. They wear a simple g-suit, with its tubes and bladders to prevent the pilot's blood from pooling in his legs during high-speed turns, covered by a simple green flight jacket. The pilots further honor the tradition of unpretentiousness by eliminating symbols of rank from their dress. After all, the pilots tend to know everyone else in their unit and what their rank is. Besides, what everyone really considers significant is how many fighters and 'Mechs a pilot has destroyed. The pilots keep this tally on their duck-billed caps, using the symbols of stars and skulls. The stars represent destroyed 'Mechs and the skulls stand for enemy fighters shot down.

MECHWARRIOR UNIFORMS

As the elite of the AFFS, MechWarriors represent perhaps the most fiercely independent group of soldiers in the military. Yet their traditions and loyalties are often divided between the cult of the MechWarrior, attachment to a regiment, and to regional loyalties. MechWarrior uniforms usually reflect all these varying loyalties as well as the warrior's independence.

Marshal

A Marshal is given command of a Frontline Regimental Combat Team, a group of highly trained and motivated regiments, often up to 15 in number, spearheaded by a 'Mech regiment of proven skills. A Marshal is appointed by a Field Marshal after he has proven his ability to plan strategy and handle the immense communication problems involved in the command of so many regiments scattered over several worlds. In the AFFS Navy, the equivalent rank is Fleet Admiral.

Marshals wear epaulettes with four wide bands.

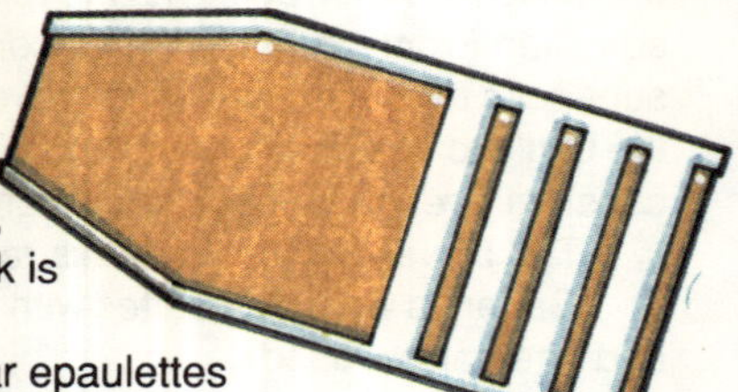

Field Marshal

There are currently 14 Field Marshals commanding combat troops in the AFFS today. Seven Field Marshals command the AFFS forces in the seven Combat Theaters that divide up the Federated Suns (two each in the Capellan and Draconis Marches and three in the Crucis March). Two Field Marshals act as overall Commanders of the Capellan and Draconis Marches. Five Field Marshals are advisors to Prince Hanse Davion, with no specific command.

Field Marshals wear epaulettes with a silver Federated Sunburst to show their rank.

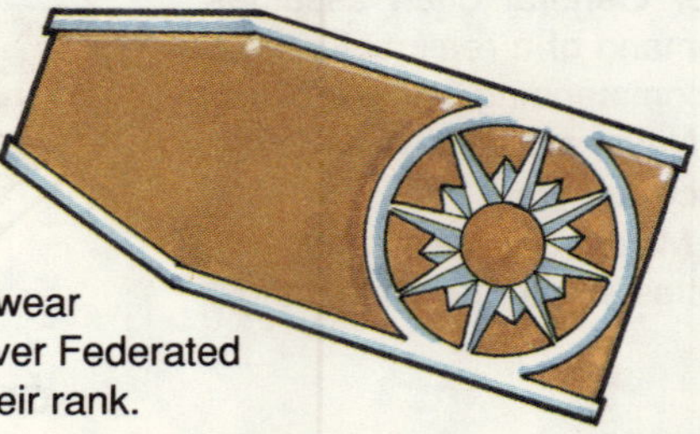

Keeping the 'Mech pilot cool enough is the basic function of the MechWarrior's uniform. In a constant search to help its warriors be more comfortable in the hell of a 'Mech in combat, the AFFS introduced the Cooling Undershirt about ten years ago. Developed by the scientists of Team Banzai and the NAIS, this short-sleeved red undershirt is basically a miniature version of the MechWarrior's cooling vest. Each strand of the undershirt is actually an extremely small but tough tube that carries water or similar fluid around the wearer's torso. Though it cannot cool as much as a regular cooling vest, the cumulative effect of the shirt and the vest allows the 'Mech pilot to withstand even the hottest cockpit in relative comfort.

Spurs, on the other hand, are more a symbol of the pride MechWarriors take in their profession than a utilitarian item of clothing.

During the era of knighthood in medieval Europe, the knight dressed in his coat of chainmail or plate armor and seated upon a horse was the ultimate warrior. A whole set of traditions and myths developed around what the knight did and what he stood for. The Federated Suns, with its French and English heritage, has been especially attached to the concept of knights and chivalry, as evidenced by the image of the broadsword on the Federated Sunburst. To show their kinship to those knights of the long-dead past, the Federated Suns MechWarriors wear spurs on their boots just as the ancient knights, though in the modern era the spurs are pure anachronism. Most MechWarriors continue to wear the spurs long after their BattleMech days are over.

Another example of this pride in tradition is the colorcoding of the MechWarrior's cooling vest and shorts to indicate to which group of regiments he belongs. If dressed in red, white, and blue vest and shorts, the warrior is a member of the elite Davion Brigade of Guards. If he wears gray vest and shorts, he is a member of one of the Ceti Hussar regiments. Two-tone green shorts and vest denote a Crucis Lancer. MechWarriors in light green shorts and vest, wearing a regimental patch instead of the Federated Sunburst, are Syrtis Fusiliers. MechWarriors in mercenary units are issued cooling vests and shorts in basic green.

Because the bulky cooling vest covers the shoulders, the epaulettes denoting the wearer's rank are worn over the upper arm instead.

MECHWARRIOR

Marshal of the AFFS

Marshal of the AFFS is the position held by the First Prince as commander of the AFFS. Epaulettes with a golden Federated Sunburst are the insignia of the Prince's rank and position in the AFFS. Throughout history, however, the leaders of the Federated Suns have tended to prefer to wear the insignia and uniform of a Field Marshal rather than the epaulettes.

ARMORED VEHICLE CREW UNIFORMS

The development of the Cooling Undershirt considerably changed the uniform of AFFS armored vehicle crews. No longer must they wear bulky uniforms like the vehicle crews of the other militaries. Instead, they wear a blue cooling undershirt underneath a regular infantry uniform. The color of the uniform depends on what group of regiments the wearer is from, just as the uniforms of MechWarriors reflect the regiments they serve. Armored vehicle crewmembers wear a tan harness over the uniform. The harness straps and belts the crewman securely into his seat as his vehicle bounces and lurches about in combat.

The unusually shaped helmet worn by crews was also designed with safety in mind. The seats of all AFFS armored vehicles have a large and complex headrest system into which the oddly shaped helmet fits like a key into a lock. Once secured by the headrest-helmet system, the crewmember is protected from the sudden jerks and shudders that have snapped the necks of unfortunate crewman whose heads were not restrained. The helmets also allow crewmembers to talk to one another through the mike and speaker system built into the helmet's interior.

INFANTRY UNIFORMS

The differences between groups of regiments is most readily seen in the uniforms worn by AFFS infantry.

The standard-issue uniform consists of dark-green shirt and pants covered by a step-through brown padded jerkin with the Federated Sunburst stitched on the left-hand side. Most regiments wear a brown helmet similar to those of Lyran Commonwealth soldiers, though some troopers prefer to wear fabric caps—sometimes because of tradition and sometimes out of pure foolishness. Officers wear what they call Comm-Gauntlets. These are thick steel-and-leather gloves that, besides being wicked weapons in hand-to-hand combat, have communications equipment built into them. This unusual system has given rise to the AFFS expression, "There's no use swearing at the Captain's hand," which refers to the futility of fighting orders from above.

Regional and regimental variations in the basic uniform abound. Syrtis Fusiliers wear greenish-tan shirts and boots, as well as a jerkin that matches the standard-color green pants. The Fusiliers, always a prideful group, consider it cowardly to wear a helmet and so favor green fabric caps instead. As a last bit of regimental individuality, Fusiliers wear their regimental patch instead of the Federated Sunburst on their cap and jerkin.

Crucis Lancers wear a uniform similar to that of the Fusiliers except that their jerkin is tan and green. They have no quarrel with wearing a helmet, though. Avalon Hussars, on the other hand, wear standard-issue jerkin, pants, and boots, but with gray jerseys underneath.

To further add to the confusion, the AFFS issues different uniforms for various climates and locales. The colors of these range from white for cold-weather troops to night-black. There are also a variety of camouflage uniforms to match settings such jungle vegetation, forest, rocks, and urban environments.

SYRTIS FUSILIERS

SUPPORT PERSONNEL UNIFORMS

The technicians and engineers whose job it is to repair the machinery of the AFFS wear standard infantry jersey and pants. Over those, they wear a loose-fitting pullover Tech's Jacket made of a synthetic fabric resistant to heat and electricity. The jacket's color indicates in which branch of service the wearer is most proficient (red for a 'Mech Tech, gold for an aerospace engineer, blue for a mechanic).

The Tech's jacket has one large chest pocket on the chest, where the Tech can keep his important tools as well as a small, hand-held computer-communications device. The rank of the Tech is shown on this pocket.

DRESS UNIFORMS

Dress uniforms in the AFFS are simple yet elegant, particularly when compared to the cluttered extravagance of the Lyran Commonwealth dress uniforms. The dress jacket and trousers are in the standard Davion green. MechWarriors can attach spurs to the cuffs of the trousers.

The construction of the dress jacket allows the Federated Sunburst design to extend from the left shoulder, with sun's rays radiating down and across the chest and back of the jacket. Officers wear what is known as the Sunburst Vest, a shiny golden semi-breastplate over the left shoulder. It is attached by a series of buttons on the chest and on the back.

Ribbons representing certain medals are worn on the rays of the vest, though some of the medals are prestigious enough to be worn on the right-hand side of the uniform instead. Because the Sunburst Vest takes up space on one shoulder, AFFS officers wear only one epaulette. Enlisted personnel wear their rank on the right shoulder, as usual.

DEPLOYMENT OF AFFS FORCES

During the early years of the First Succession War, the AFFS High Command paid a tragically high price to learn how utterly flawed was their deployment of forces along the Draconis border. The Marshals of the AFFS had divided the front lines into six Combat Regions, each with a set number of troops to deal with the enemy. The command structure hindered communications, let alone cooperation, between neighboring Combat Regions.

When the Combine mounted its offensive, the Kurita commanders chose to focus their opening attacks within the Robinson Combat Region. These attacks overwhelmed the Federated Suns forces there, and the lack of rapid communications kept the Davion commanders from being able to obtain reinforcements in time. When the Robinson Combat Region collapsed, it split open the entire Davion front, leaving very little standing between the Draconis Combine and New Avalon.

When the Marshals of the AFFS later reviewed the situation, they devised a new way to divide up the Draconis and the Capellan Marches. Under the name of Polymorphous Defensive Zones, each March was sectioned into a variety of oddly shaped regions. Now, instead of large defensive regions that cut clear across a March, giving an enemy a clear path into the heart of the Federated Suns, there were smaller zones that could back one another up. Large-scale annual military exercises were held to stress cooperation among the PDZs. Though it took the federation bureaucracy some time to become accustomed to the newer, smaller military regions, the PDZs have proven their worth in decades of combat.

Each PDZ is named after its command and logistics center. These centers are the headquarters of the PDZ's March Militia and often serve as the headquarters of any major unit in the area. From these centers, the commanding staff of the zone dispatches orders and receives information about the fighting in their area. They are also in direct communication with the commanders of bordering PDZs, who stand ready to send reinforcements or launch counterattacks if necessary.

To ensure that each PDZ has at least some defensive forces, the leaders of the AFFS created the March Militias. These are regimental 'Mech units raised from local and regular soldiers and officers. Though Militia units do not have the best equipment and their training does not match that of frontline units, they are a definite deterrent to any enemy force that intends to invade a PDZ.

There are no PDZs within the Crucis March, however. Here, the old system of Combat Regions is still in place, with each region corresponding to an Administrative Area. Because the Crucis March is not directly threatened by either the Kuritans or the Capellans, the Davions decided that it was not necessary to redraw the March's military zones.

'MECH UNIT DEPLOYMENT TABLE
(as of 3025)

THE DRACONIS MARCH
CO: Field Marshal Aaron Sandoval

Addicks PDZ

Unit Name	Experience Level	Loyalty	Homeworld
Addicks DMM (CO: Major General Dether DeBurke)	Regular	Reliable	Addicks
2nd Davion Guards RCT (CO: Marshal Jessica Davion)	Veteran	Fanatical	Fomalhaut
22nd Avalon Hussars RCT (CO: Marshal Timothy Ulso)	Veteran	Reliable	Quentin
33rd Avalon Hussars RCT (CO: Lieutenant General Wilson Mandella)	Veteran	Fanatical	Nopah
5th Deneb Light Cavalry RCT (CO: Marshal Lovis Yolis)	Regular	Reliable	Northwind
3rd Crucis Lancers RCT (CO: Marshal Jessie Pedroza)	Veteran	Reliable	Caph
Blue Star Irregulars (CO: General Mona Troy) (3 regiments)	Regular	Reliable	Ozawa
The Crater-Cobras (CO: Colonel Richard Westrick) (2 regiments)	Regular	Reliable	Deneb Kaitos

11 regiments
CO: Marshal Jessica Davion
Second-in-Command: Marshal Lovis Yolis

Clovis PDZ

Unit Name	Experience Level	Loyalty	Homeworld
Clovis DMM (CO: Major General Nicholas Clement VIII)	Green	Reliable	Cartago
12th Deneb Light Cavalry RCT (CO: Marshal Petroff Haveson)	Regular	Reliable	Galatia III
8th Deneb Light Cavalry RCT (CO: Marshal Laverne Denton)	Regular	Reliable	Angol
8th Crucis Lancers RCT (CO: Marshal Neale Wargo)	Regular	Reliable	New Rhodes III
1st NAIS Cadet Cadre (CO: Major General Donna Iona)	Green	Fanatical	Exeter

5 Regiments
CO: Marshal Neale Wargo
Second-in-Command: Major General Donna Iona

Raman PDZ

Unit Name	Experience Level	Loyalty	Homeworld
Raman DMM (CO: Major General Conroy Baden-Powell)	Green	Reliable	Raman
4th Davion Guards RCT (CO: Marshal Eugene Drivers)	Elite	Fanatical	Deshler
41st Avalon Hussars RCT (CO: Marshal Kenney Graham)	Regular	Reliable	David
4th Deneb Light Cavalry RCT (CO: Marshal Vernon Fredricks)	Veteran	Reliable	Harrow's Sun
4th Crucis Lancers RCT (CO: Major General Charles Duncan)	Veteran	Reliable	Klathandu IV
1st Chisholm's Raiders RCT (CO: Marshal Melford Dennis)	Veteran	Reliable	Royal
2nd Robinson Rangers (CO: Colonel Eric Vehrson)	Regular	Reliable	Breed

7 regiments
CO: Marshal Eugene Drivers
Second-in-Command: Marshal Melford Dennis

Robinson PDZ

Unit Name	Experience Level	Loyalty	Homeworld
Robinson DMM (CO: Major General Andrew Cunningham)	Green	Reliable	Le Blanc
1st Robinson Rangers (CO: Field Marshal Aaron Sandoval)	Veteran	Fanatical	Robinson
1st New Ivaarsen Chasseurs (CO: Marshal Nicholas Stephenson)	Elite	Fanatical	New Ivaarsen
3rd NAIS Cadet Cadre (CO: Major General Jeremiah Green)	Green	Fanatical	Tarkio
Kell Hounds (CO: Lieutenant Colonel Patrick M. Kell) (1 battalion)	Veteran	Reliable	Robinson

4 regiments, 1 battalion
CO: Marshal Nicholas Stephenson
Second-in-Command: Major General Andrew Cunningham

Dahar PDZ

Unit Name	Experience Level	Loyalty	Homeworld
Dahar DMM (CO: General William Dobson)	Regular	Reliable	Tishomingo
17th Avalon Hussars RCT (CO: Marshal Addley Verner)	Regular	Reliable	Cassias
10th Deneb Light Cavalry RCT (CO: Marshal Carl Sanders)	Green	Reliable	Rowe
1st Ceti Hussars RCT (CO: Marshal Patrick Weldon)	Veteran	Reliable	Glenmora

4 regiments
CO: Marshal Addley Verner
Second-in-Command: Marshal Patrick Weldon

Bremond PDZ

Unit Name	Experience Level	Loyalty	Headquarters
Bremond DMM (CO: Lieutenant General Mary Tallman)	Regular	Reliable	Bremond
42nd Avalon Hussars RCT (CO: Marshal Henry Offendahl)	Green	Reliable	Fallon II

2 regiments
CO: Marshal Henry Offendahl
Second-in-Command: Major General Mary Tallman

Bryceland PDZ

Unit Name	Experience Level	Loyalty	Homeworld
Bryceland DMM (CO: Major General Lao Tsu XI)	Regular	Reliable	Bryceland
Davion Light Guards RCT (CO: Marshal Jonathan Riffenberg)	Veteran	Reliable	Groveld III
11th Avalon Hussars RCT (CO: Marshal Walther Gothard)	Regular	Reliable	Kesai IV
1st Crucis Lancers RCT (CO: Marshal George Powell)	Veteran	Fanatical	Tancredi IV
1st Conroe Training Battalion (CO: Lieutenant Colonel Harry Vasos)	Green	Reliable	Conroe
Lindon's Company (CO: Acting Captain Sarah Lindon) (1 battalion)	Veteran	Reliable	Kennard

4 regiments, 2 battalions
CO: Marshal Walther Gothard
Second in Command: Marshal George Powell

'MECH UNIT DEPLOYMENT TABLE (as of 3025)

Mayetta PDZ

Unit Name	Experience Level	Loyalty	Homeworld
Mayetta DMM (CO: Major General Vonda DeGreer)	Green	Reliable	Mayetta
2nd Ceti Hussars RCT (CO: Marshal Alan Cline)	Veteran	Reliable	Verde

2 regiments
CO: Marshal Alan Cline
Second-in-Command: Major General Vonda DeGreer

Milligan PDZ

Unit Name	Experience Level	Loyalty	Homeworld
Milligan DMM (CO: General Mitch Xexell)	Green	Reliable	Milligan

1 regiment
CO: General Mitch Xexell
Second-in-Command: Major General Dominic Travers

Kilbourne PDZ

Unit Name	Experience Level	Loyalty	Homeworld
Kilbourne DMM (CO: Major General Wesely French)	Green	Reliable	Kilbourne
Dismal Disinherited (CO: General Ivan Fontral) (3 regiments)	Regular	Reliable	Boondock
Greenburg's Godzillas (CO: Colonel Sarah Dosimion)	Veteran	Reliable	Sterlington

5 regiments
CO: Major General Wesely French
Second-in-Command: General Dan Pockvitich

THE CAPELLAN MARCH
CO: Field Marshal Michael Hasek-Davion

Valexa PDZ

Unit Name	Experience Level	Loyalty	Homeworld
Valexa CMM (CO: Major General Killy Fren)	Regular	Reliable	Valexa
20th Avalon Hussars RCT (CO: Marshal Vivian Chou)	Veteran	Fanatical	New Aragon
1st Aragon Borderers (CO: Colonel Vincent Kooper)	Veteran	Reliable	New Aragon
2nd Crucis Lancers RCT (CO: Marshal Dino Riggs)	Regular	Reliable	Mira
6th Crucis Lancers RCT (CO: Marshal Orval Gossiage)	Veteran	Reliable	Moravian
2nd NAIS Cadet Cadre (CO: Major General Quentin Drathers)	Green	Fanatical	Beten Kaitos
1st Bell Training Battalion (CO: Light Colonel Daniel Brewer)	Green	Reliable	Bell
Redfield Renegades (CO: Colonel Tyrell Redfield)	Regular	Reliable	Demeter

7 regiments, 1 battalion
CO: Marshal Vivian Chou
Second-in-Command: Major General Quentin Drathers

Kathil PDZ

Unit Name	Experience Level	Loyalty	Homeworld
Kathil CMM (CO: Major General Edward Gashton)	Green	Questionable	Kathil
5th Crucis Lancers RCT (CO: Marshal Mathew Agerstand)	Veteran	Fanatical	Kathil
2nd Chisholm's Raiders (CO: Colonel Henry Bauer)	Veteran	Fanatical	Kathil

3 regiments
CO: Marshal Mathew Agerstand
Second-in-Command: Major General Edward Gashton

Alcyone PDZ

Unit Name	Experience Level	Loyalty	Homeworld
Alcyone CMM (CO: Major General Wesley Demurke)	Regular	Reliable	Alcyone
5th Davion Guards RCT (CO: Marshal Kinsely Crossburns)	Veteran	Fanatical	Lee
7th Crucis Lancers RCT (CO: Marshal Robert Steadman)	Elite	Fanatical	Weekapaug
5th Syrtis Fusiliers RCT (CO: Marshal Tamara Hasek)	Regular	Questionable	Kittery
1st Kittery Borderers (CO: Colonel Vivian Dulers)	Regular	Reliable	Kittery
Waco Rangers (CO: Colonel Wayne Waco)	Regular	Reliable	Mentasta

6 Regiments
CO: Marshal Robert Steadman
Second-in-Command: Marshal Kinsely Crossburns

New Syrtis PDZ

Unit Name	Experience Level	Loyalty	Headquarters
New Syrtis CMM (CO: Major General Jara Dahar)	Green	Questionable	New Syrtis
6th Syrtis Fusiliers RCT (CO: Field Marshal Michael Hasek-Davion)	Elite	Questionable	New Syrtis
1st Capellan Dragoons (CO: Colonel Robert Johnstadt)	Regular	Questionable	Wernke
Lexington Combat Group (CO: General Isabella Rahm) (3 regiments)	Regular	Reliable	Wappingers

6 regiments
CO: Major General Jara Dahar
Second-in-Command: Isabella Rahm

Sirdar PDZ

Unit Name	Experience Level	Loyalty	Homeworld
Sirdar CMM (CO: Major General Latrell Greigg)	Regular	Questionable	Sirdar
Davion Assault Guards RCT (CO: Marshal Winston Ashley)	Veteran	Reliable	Frazer
39th Avalon Hussars RCT (CO: Marshal Saunders DuCrimmon)	Regular	Reliable	Aucara
15th Deneb Light Cavalry RCT (CO: Marshal Henry Quentin)	Regular	Reliable	Verlo
3rd Ceti Hussars RCT (CO: Marshal Joseph Goff)	Regular	Reliable	Immenstadt
1st F.S. Armored Cavalry (CO: Colonel Todd Revons)	Elite	Fanatical	Ziliang
Illician Lancers (CO: General Ellaine Steward) (4 regiments)	Regular	Reliable	Jonzac
Wylie's Coyotes (CO: Colonel Enzo Wylie)	Regular	Reliable	Bromhead

11 regiments
CO: Marshall Winston Ashley
Second-in-Command: Marshal Joseph Goff

Ridgebrook PDZ

Unit Name	Experience Level	Loyalty	Homeworld
Ridgebrook CMM (CO: Major General Pete Gullam)	Green	Reliable	Ridgebrook
8th Syrtis Fusiliers RCT (CO: Marshal Iona Hasek)	Regular	Questionable	Avigait

2 regiments
CO: Marshal Iona Hasek
Second-in-Command: Major General Pete Gullam

'MECH UNIT DEPLOYMENT TABLE
(as of 3025)

Warren PDZ

Unit Name	Experience Level	Loyalty	Homeworld
Warren CMM (CO: Major General Ashley Kurtis)	Green	Questionable	Warren
2nd Albion Training Cadre (CO: Major General Orlando Frenc)	Green	Fanatical	Enchi

2 regiments
CO: Major General Ashley Kurtis
Second-in-Command: Major General Orlando Frenc

THE CRUCIS MARCH
CO: Supreme Marshal Hanse Davion, First Prince of the Federated Suns

Unit Name	Experience Level	Loyalty	Headquarters
Marlette CMM (CO: Major General Dennis Wantraub)	Regular	Reliable	Marlette
Miller's Marauders (CO: Major Grissom Miller) (2 battalions)	Veteran	Reliable	Layover
12th Vegan Rangers (CO: Margrave Sheridan Douglass) (4 regiments)	Elite	Reliable	Edwards
Grim Determination (CO: Colonel Fiona Trent)	Regular	Reliable	Muskegon

7 regiments
CO: Major General Dennis Wantraub
Second-in-Command: General Carl VanTan

Kestrel Combat Region

Unit Name	Experience Level	Loyalty	Homeworld
Kestrel CMM (CO: Major General Hilda Delf)	Regular	Reliable	Kestrel
2nd New Ivaarsen Chausseurs (CO: Colonel Reginald Stephenson)	Regular	Fanatical	New Valencia
1st Kestrel Grenadiers (CO: Colonel Rudy Pinada)	Elite	Fanatical	Kestrel
Eridani Light Horse (CO: Brevet General Nathan Armstrong) (3 regiments)	Veteran	Reliable	Colchester

6 regiments
CO: Major General Hilda Delf
Second-in-Command: Lieutenant General Do Seng Ki

New Avalon Combat Region

Unit Name	Experience Level	Loyalty	Homeworld
New Avalon CMM (CO: Major General Edward Valos)	Regular	Reliable	New Avalon
1st Davion Guards RCT (CO: Marshal Stephen Davion)	Elite	Fanatical	New Avalon
Davion Heavy Guards RCT (CO: Field Marshal Ran Felsner)	Elite	Fanatical	New Avalon
Team Banzai (CO: Dr. B. Banzai)	Elite	Fanatical	New Avalon
Markson Marauders (CO: Colonel Richard Markson)	Regular	Reliable	New Avalon

5 regiments
CO: Field Marshal Ran Felsner
Second-in-Command: Marshal Stephen Davion

Point Barrow Combat Region

Unit Name	Experience Level	Loyalty	Homeworld
No 'Mech Regiments			

CO: Colonel Calvin Froshitl (Commander, Point Barlow Military Academy)
Second-in-Command: Lieutenant General Doshyo Kinchawa

Remagen Combat Region

Unit Name	Experience Level	Loyalty	Homeworld
Remagen CMM (CO: Major General Xerxes Davion)	Green	Reliable	Remagen
Harlock's Warriors (CO: Colonel Rodney Viele)	Regular	Reliable	Victoria

2 regiments
CO: Major General Xerxes Davion
Second-in-Command: Colonel Rodney Viele

Islamabad Combat Region

Unit Name	Experience Level	Loyalty	Homeworld
Islamabad CMM (CO: Major General Lucille Carlton)	Green	Reliable	Islamabad
The Fighting Urakhai (CO: General Harold Greenspan) (3 regiments)	Regular	Reliable	Pierce

4 regiments
CO: Major General Lucille Carton
Second-in-Command: General Harold Greenspan

Kearny Combat Region

Unit Name	Experience Level	Loyalty	Homeworld
The Dioscuri (CO: Colonel Timothy Nels) (2 regiments)	Regular	Reliable	Vackisujfal

2 regiments
CO: Major General Elizabeth Henders
Second-in-Command: Colonel Paul Gevin

Tsamma Combat Region

Unit Name	Experience Level	Loyalty	Homeworld
Tsamma CMM (CO: Major General Saffer Mantas)	Green	Reliable	Tsamma

1 regiment
CO: Major General Saffer Mantas
Second-in-Command: Colonel Shaul DuBan

Anjin Muerto Combat Region

Unit Name	Experience Level	Loyalty	Homeworld
Anjin Muerto CMM (CO: Major General David Trime)	Green	Reliable	Anjin Muerto
1st Brockton Training Battalion (CO: Lieutenant Colonel Solas Greerson) (1 battalion)	Green	Reliable	Brockton

1 regiment, 1 battalion
CO: Major General David Trime
Second-in-Command: Colonel Lori Oscar

Broken Wheel Combat Region

Unit Name	Experience Level	Loyalty	Homeworld
1st Argyle Lancers (CO: Colonel Matt Preston)	Regular	Reliable	Broken Wheel
1st Albion Training Cadre (CO: Major General Paula Winters)	Green	Fanatical	Morven
Screaming Eagles (CO: Colonel Walther Hokala) (2 regiments)	Regular	Reliable	Skepptana

4 regiments
CO: Major General Paula Winters
Second-in-Command: Colonel Wlather Hokala

UNIT DESCRIPTIONS

DAVION BRIGADE OF GUARDS (ROYAL BRIGADE)

The eight 'Mech regiments and fifty lesser regiments that make up the Davion Brigade Guards represent the core of the Davion family's strength. The units are also portrayed in the popular media as the pride of the common citizen, because the AFFS makes a point of recruiting the best soldiers from every member planet in the Federated Suns for these regiments.

The Brigade of Guards was originally a small group of bodyguards employed to protect the Davion family when they first took political power on New Avalon. As the Davions' power expanded so did the responsibilities and strength of the Davion Guards. The number of fighting units swelled so quickly that by the 2600s the Davion Guards were renamed the Davion Brigade of Guards. The Davion Brigade of Guards were among the first regiments to use BattleMechs in the Federated Suns, and among the first to adopt the Star League style of organizing its forces into mixed-weapon Regimental Combat Teams.

Though the Brigade long ago took on far greater responsibilities than just protecting the Davion family, the connection between the regiments and the royal family is still quite strong. All the regiments of the Brigade of Guards wear a version of the Federated Suns crest as their unit insignia. Also, most Davions serve part of their military career in a Guards regiment, with many rising to command one of its regiments. Any Brigade unit in which a Prince once served is called the Royal Guards. The Brigade still fulfills its original duty to the ruling family by providing all personal bodyguards and soldiers used to protect the Royal Castle and the Fox's Lair.

Because of the Brigade of Guards' close connection with the royal family, the unit is assigned the best equipment and supplies and the best graduates from boot camps and academies are sent to Guards units. The esprit de corps is so strong that no Guards unit in history has commited a dishonorable act. This camaraderie has also made the infantry regiments famous for their fighting abilities. Some regiments in the Brigade of Guards even sat out the Civil War to preserve their honor when Laura Davion fought Alexander Davion for control of the Federated Suns. Today, there is a growing unease in the Brigade because of the tensions between Prince Hanse and Duke Michael Hasek-Davion.

For ceremonial occasions, the 'Mechs and tanks of the Brigade are painted with blue, white, and red striping, the colors of the Davion family. Otherwise, the 'Mechs and vehicles of the Brigade adopt standard color schemes that depend on the locale of their combat mission. Only the RCT patch marks a Brigade member.

First Guards RCT: The Strength of Alexander

Elements in this RCT were among Alexander Davion's favorite troops. Currently, the First Guards RCT is short an infantry regiment because of the disbanding of the 567th Torrence Light Infantry. The 'Mech regiment is composed of medium to heavy 'Mechs. The RCT patch shows a Greek helmet.

Second Guards RCT: Wrath of the Free

Though the rest of the Inner Sphere can only afford to group its few artillery pieces into lance-sized units, the Second Guards RCT boasts one of the few artillery regiments in the Inner Sphere today. As a result, this RCT of light to medium 'Mechs is a feared opponent. The RCT patch shows a pair of broken shackles.

Third Guards RCT: Don't Tread on Me

Perhaps the weakest RCT of the entire Brigade, the Third Guards RCT is currently short two infantry regiments and one armored regiment. Morale remains high, however, in this medium 'Mech regiment. The RCT patch shows a coiled golden rattlesnake.

Fourth Guards RCT: Dragon's Bane

Dragon's Bane is a highly mobile RCT. Its 'Mechs are all light and fast, as are the tanks and hovercraft used in the four armored regiments. This unit is always stationed in or near the Draconis March. The RCT patch shows a hand holding up a Dragon's head.

Fifth Guards RCT: Bent Sword Brigade

The Bent Sword is a multi-purpose RCT with a wide variety of 'Mechs, tanks, and infantry types. The unit is stationed near or in the Capellan March. The RCT patch parodies the Liao insignia with a limp sword.

Assault Guards RCT: The Crushers

The Crushers are composed of the largest 'Mechs, the largest tanks, the largest fighters, and soldiers carrying the heaviest weaponry available. The purpose of this RCT is to take heavily defended worlds that lighter forces could not hope to handle. The RCT patch shows the head of an *Atlas* BattleMech.

Heavy Guards RCT: Strength of the Davions

This RCT is composed of heavy 'Mechs, fighters, and vehicles, while its support is mostly jump infantry. The combination provides for unique strategies, a supposed Davion trait. The RCT patch shows a grinning fox.

Light Guards RCT: Swift Foxes

The Swift Foxes RCT is composed of light 'Mechs and vehicles, with mostly jump infantry as infantry support. The unit's aerospace support uses heavy fighters. The RCT patch shows a fox's tail.

AVALON HUSSARS

One of the oldest organized fighting forces in the Federated Suns, the Avalon Hussars can trace their origins to the Terran Alliance Marines of the 25th century, which had a tank regiment called the Avalon Hussars stationed on New Avalon. When the Federated Suns was created, the Avalon Hussars was one of the first units activated within the newborn AFFS. When BattleMechs began to dominate the battlefield, the Hussars, which had grown to more than 30 separate regiments, made the change easily. During the First Succession War, the Hussars played a crucial role in stalling the Draconis Combine's invasion of the Federated Suns.

Today, the Hussars represent a large, and mostly loyal, fighting force. Though not as well supplied as the Brigade of Guards, the Hussars never have to wait long for the arrival of their supplies. The Hussars tend to favor a light grey paint scheme for their 'Mechs and vehicles. The unit symbol is a white sword against a gold shield. Individual RCTs do not have unit symbols of their own, but paint their unit number underneath the Hussars's insignia instead.

11th Avalon Hussars RCT

The 11th is a collection of lightly equipped regiments that form a highly mobile fighting force. Recently, there has been talk of adding an extra battalion of Assault 'Mechs to the unit.

17th Avalon Hussars RCT

The 17th is famed for their ability to attack and deploy under enemy fire. Though the unit keeps this ability honed through constant practice, the RCT has occasionally suffered from attempting an invasion too close to the enemy. This is why the unit is currently minus an infantry regiment.

20th Avalon Hussars RCT

The 20th is a collection of regiments groomed for defense. In addition to its regular makeup of fighting units, the 20th also includes a regiment of combat engineers who specialize in building fortifications and installations.

22nd Avalon Hussars RCT

The 22nd is one of the best Hussar RCTs. Its regiments, particularly the armored and infantry, have won impressive victories in recent months. The 22nd has no particular specialties and includes a wide variety of 'Mechs.

33rd Avalon Hussars RCT

The 33rd is a heavyweight RCT unit that includes an extra battalion of Assault 'Mechs.

39th Avalon Hussars RCT

The 39th is a lightweight RCT unit currently missing the normal number of AeroSpace Fighters.

41st Avalon Hussars RCT

The 41st RCT has a number of LAMs and more than twice the normal number of AeroSpace Fighters.

42nd Avalon Hussars RCT

The 42nd was recently reorganized out of the remainder of the 34th and 36th Avalon Hussars. This has given the 42nd five 'Mech battalions, an extra tank regiment, and two extra infantry regiments. Though well-equipped, this RCT lacks coordination and a command staff experienced enough to handle an extra-large complement.

SYRTIS FUSILIERS

The Syrtis Fusiliers was once the private army of the Hasek family. When the AFFS was reorganized after the era of the Five Principalities and the Civil War, the Syrtis Fusiliers was purged of any member disloyal to the Davions and then was incorporated into the military. Since that time, the Syrtis Fusiliers have drawn a majority of their troops from New Syrtis and other worlds in the Capellan March, but most of its command staff must be from the Crucis March. This structure was supposed to keep the Syrtis Fusiliers loyal to the Federated Suns, but that has not been the case. The three RCTs are all extremely loyal to Michael Hasek-Davion, Duke of New Syrtis, and will follow AFFS orders only if the Duke approves them.

The Fusiliers have a long history of brave and gallant actions against the forces of the nearby Capellan Confederation. This has given them enough reputation that some Liao troops are gunning for them, but it has not kept supplies flowing to them smoothly enough. The unit sometimes suffers long waits for replacement parts, making tensions flare between the Fusiliers and the AFFS bureaucracy. Some recent evidence indicates that Duke Michael Hasek-Davion has purposely created paperwork problems to delay the flow of supplies so that these regiments will become even more loyal to him.

The Syrtis Fusiliers favor dark-green paint for their vehicles. The unit is one of the few that does not wear the Federated Sunburst, preferring to wear its RCT symbols instead.

Fifth Syrtis Fusiliers RCT: The Evil Eye

This RCT is known for the ferocity with which they put their heavy 'Mechs and extra fighter protection to good use. The unit is so despised in the Capellan Confederation that some of the Warrior Houses in the Liao military have already sworn a Blood Vendetta against the Fifth. The RCT patch is similar to the Hasek's All-Seeing Eye, except that the pupil is blood-red instead of white.

Eighth Syrtis Fusiliers RCT: The Snow Cobras

The Eigth Syrtis Fusiliers RCT is a light and highly mobile unit. It uses light 'Mechs and numerous hovercraft in conjunction with infantry regiments supported by a unit of VTOL craft. The unit symbol is a white cobra.

Sixth Syrtis Fusiliers RCT: The Duke's Own

The Sixth Syrtis RCT is a combat unit oriented to defense. It has extra combat engineers, heavy assault 'Mechs and tanks, plus plenty of heavily armed soldiers. The RCT is stationed on New Syrtis and acts as Duke Michael's private guard, manning the planet's many defensive installations. The unit's symbol is the All-Seeing Eye, the Hasek family crest.

DENEB LIGHT CAVALRY

At the height of the Star League, its Regular Army boasted a group of 'Mech regiments whose ability to respond quickly anywhere they were needed was legendary. These quick-response troops were officially known as the Rapid Deployment Mixed Arm Forces, or RDMAF. The general public, however, knew these elite troops—because of several highly romanticized holoplays—as the Deneb Light Cavalry, after the headquarters of the RDMAF on Deneb Kaitos and the pony on their banners.

When the Star League collapsed, only one RDMAF unit refused General Kerensky's orders and stayed on Deneb Kaitos. The Fourth Deneb Light Cavalry was commanded by General Eveline McDennis, who firmly believed it would be an act of cowardice to abandon the Inner Sphere. Her original intention was to remain on Deneb Kaitos to protect that world somehow from the warfare that was sure to come. As trade to the world collapsed and the population began to starve, General McDennis was forced to consider offers of employment from all five Successor Lords. After hearing them out, McDennis replied that she and her men would work for whichever House could guarantee that the people of Deneb Kaitos would have food until they could grow enough on their own. It was a condition that only the Federated Suns would accept. As the desperately needed shipments of food and agricultural equipment made their way to Deneb Kaitos, the Fourth Deneb Light Cavalry packed up and prepared to become part of the AFFS.

During the First Succession War, the Deneb Light Cavalry played only a minor role. This was because the Davions preferred to use the unit's experienced members to help organize other regiments and RCTs into their rapid deployment abilities. The Fourth and the new Deneb Light Cavalry units made their abilities felt in the Second and Third Wars, however. Both are known today as important fighting forces equally able to respond to enemy offensives or to play a role as dangerously swift offensive forces.

Deneb Light Cavalry regimental teams are organized slightly different than other RCTs in the Federated Suns. Around the one 'Mech regiment, which usually has light to medium 'Mechs, there are two hovercraft regiments and two jump infantry regiments. Also, there are three times the normal number of AeroSpace Fighters present, plus enough DropShips and JumpShips to transport them.

The Deneb Light Cavalry has always preferred to use a light-tan paint scheme to show their pride in their Regular Army roots, even when other units would adopt camouflage paint schemes. The Deneb Light Cavalry insignia is a rectangle of black surrounding the bust of a white horse upon a square of gold. The unit number is written in roman numerals within the golden square. Most of the units are named after swift predators.

Fourth Deneb Light Cavalry RCT: Eveline's Originals

As the unit is descended from the one commanded by General Eveline McDennis, the Fourth Deneb Light Cavalry is the most tradition-bound and reverent of its Star League roots. The unit goes so far as to seek to recruit only descendants of the unit's original personnel to keep the memories alive. The unit has consistently been the best among the Deneb Light Cavalry. The 'Mechs and vehicles of the Fourth show a silver general's star, representing Eveline McDennis.

Fifth Deneb Light Cavalry RCT: The Sprinter-Killers

The Fifth Deneb Light Cavalry was the first Deneb Light Cavalry unit created by the AFFS. Its history is almost as impressive as the Fourth's because of major victories in the Second and Third Wars. The Fifth's nickname comes from a predator native to the plains of New Avalon.

8th Deneb Light Cavalry RCT: The Green Leopards

The Eight was the second Deneb Light Cavalry unit created by the AFFS. It is named after the Leopards of Kigamboni, whose fur is green because of the animal's symbiotic relationship with a native animal.

10th Deneb Light Cavalry RCT: The Cheetahs

When the Ninth Deneb Light Cavalry was lost during the early stages of the Third War, the AFFS decided to disband the unit permanently and use the survivors as a core to build the Tenth. Recent losses and the influx of a large number of recruits have made this the most inexperienced of the four Deneb RCTs.

12th Deneb Light Cavalry: The Blur-Snakes

The Twelfth is a 'Mech regiment only, and so has no support from other units. Midway through the Third War, the AFFS High Command saw the need for a reserve of unattached 'Mech regiments that could easily be sent about the realm without disrupting a RCT. The Twelfth is one of those so-called "Free 'Mech regiments."

15th Deneb Light Cavalry: The Whip Bears

The 15th is another free 'Mech regiment. The unit is unusual because it carries a fourth battalion of heavy 'Mechs. On several occasions, the 15th has caught its enemies flat-footed because they assumed that all Deneb 'Mech regiments were composed of only light 'Mechs.

CRUCIS LANCERS

The origin of the first Crucis Lancers unit resembles that of the Deneb Light Cavalry. When General Kerensky ordered the exodus of the Regular Army from the Inner Sphere, many soldiers could not obey, particularly those who had grown fond of a particular world and its people. The Federated Suns invited those who had stayed behind in the Crucis March to join the AFFS. Most accepted.

These were gathered into two Regimental Combat Teams newly named the First and Second Crucis Lancers. The Crucis Lancers' superior training and the weaponry they had obtained from stores left behind by General Kerensky made them a much-feared fighting force during the later stages of the First Succession War. This reputation has continued through the Second and Third Wars, despite the regiment's difficulties in obtaining supplies and the general decline in technology. Today, the Lancers recruit soldiers directly from the administrative sectors for which the individual units are named.

In 2930, the Crucis Lancers underwent a major reformation and restructuring. Though most of the changes were merely paperwork alterations such as renumbering the regiments, the soldiers did not like the break with their proud traditions. The changes became so controversial that the issue almost turned into a violent protest by the soldiers. It was only after a battalion of the 33rd Crucis Lancers, now the 7th, was lost because of the outdated paperwork procedures that the near mutiny evaporated like smoke.

The symbol of the Crucis Lancers is a stylization of the baton sometimes carried by Generals in the Regular Army. In memory of this tradition, the commanders of the various Lancer regiments carry gold batons with flails made of Avalon Stick-Vine. Units of the Crucis Lancers are named after regions in the Crucis March and carry the initials of that region just below the unit insignia.

First Crucis Lancers RCT: New Avalon Lancers

The regiments that make up the New Avalon Lancers have all been drawn from the New Avalon administrative area. The other, hidden criteria is that all be staunch supporters of the Davion family. It is therefore not surprising that the soldiers of the First Crucis Lancers have always reacted vigorously and speedily against enemies of the Davions. It also explains why the First has so many gaping holes in its regiments.

Second Crucis Lancers RCT: Kestrel Lancers

The Second is another pro-Davion, pro-Federated Suns unit. It has a right to be because so much of the Kestrel administrative area suffered severely during the First Succession War. The unit's ability to fight has been seriously affected by damage from recent actions. Talk around the Fox's Lair is that Prince Davion is considering pulling the Second off active duty for an extended repair, refit, and rest.

Third Crucis Lancers RCT: Achernar Lancers

The Third Crucis Lancers is an accomplished collection of regiments. The effectiveness of the commanders and soldiers drawn from Achernar and environs have given the region the reputation of literally growing fighting men and women. The 'Mech regiment in this RCT has four battalions.

Fourth Crucis Lancers RCT: Point Barrow Lancers

Though it is almost impossible to attribute a particular specialty to an RCT, the Fourth Crucis Lancers definitely have one. The Point Barrow Lancers love a sloppy fight on worlds whose terrain everyone else considers the worst possible place for a fight. This is why the unit is called in either to attack such a fight-poor world or to defend one.

Fifth Crucis Lancers RCT: Remagen Lancers

The Remagen Lancers are one of the few RCTs in the AFFS that completely lack a 'Mech regiment. The RCT lost most of its 'Mech regiment in an abortive attack against the Capellan Confederation in 3021. To make up the loss and to give the High Command time to decide what to do, the RCT was given three additional armored units. To everyone's surprise, especially the RCT's, the arrangement has proven successful. Though they have yet to encounter a concerted attack against them, the extra number of tanks, mostly heavy Manticores, have held their own against Capellan 'Mechs.

Sixth Crucis Lancers RCT: The Tsamma Lancers

Besides being a skilled unit, the Tsamma Lancers are one of the few RCTs in the AFFS to adopt uniforms and vehicle decorations that go specifically counter to the wishes of the High Command. In accordance with a Tsamma native festival sanctioned by the New Avalon Catholic Church, soldiers adorn their bodies and their vehicles with flowers to show their love of peace. What was a yearly festival celebrated by the members of the Tsamma Lancers eventually became permanent. Now Tsamma troopers are never without such flowers, whether stuck behind one ear, pinned on a jacket, or tied to their rifle. The vehicles are similarly decorated, and when real flowers are impractical, the soldiers have taken to painting bright flowers on the sides of their vehicles. The AFFS has been increasingly irritated by the sight of *Atlas* 'Mechs thundering across the battlefield with yellow daisies on their foreheads.

Seventh Crucis Lancers RCT: The Broken Wheel Charioteers

As one of the first Crucis Lancers units formed, the Seventh Crucis Lancers RCT has had a long and illustrious career. They were the unit responsible for the defense of Kentares IV during the First Succession War. Though strong, the unit shattered under the massive Kurita assault with the survivors taking to the hills. For a long time, the splintered, decimated Seventh was a ghost unit, existing on paper and in honor guards, but a nonentity on the battlefield. The unit was slowly rebuilt, however, until its motto of "Anything, Anywhere, Anytime" once again strikes fear in the enemies of the Federated Suns.

Eighth Crucis Lancers RCT: The Islamabad Lancers

The Eighth Crucis Lancers has shaped itself into a hard-hitting collection of regiments. They have gathered together a 'Mech regiment of heavy 'Mechs, armored regiments filled with heavy tanks and hovercraft, and troopers armed with the largest weapons they can carry. Though this RCT may not be the swiftest, its firepower is devastating once it catches up to its prey.

CETI HUSSARS

The Ceti Hussars are three RCTs designed and geared for quick, deep strikes into enemy territory. This means that these RCTs are organized differently than others. Instead of organizing all the regiments around one 'Mech regiment, the Ceti Hussar regiments are divided into thirds, which creates three separate groups for an attack on an enemy world. The arrangement is effective, but these three RCTs have the dubious distinction of having lost more men and equipment than any other unit in the AFFS. Experience has taught the Hussars that their three-group arrangement will not be able to withstand a coordinated enemy attack if they are forced to wait too long for pickup by their DropShips. This is why prompt retrieval of a Ceti Hussar unit is important.

The unit symbol of the Ceti Hussars is a stylized rendition of a JumpShip and two DropShips. The unit's headquarters is on New Earth, a small world in the Emporia star system.

First Ceti Hussars RCT: Three-Headed Serpent

The First Ceti Hussars is perhaps the best known of the Hussars because of their recent efforts against the Kurita worlds of An Ting and Igualada. Part of the unit's success lies in its three fighting units, each slightly different and with a different specialty.

Second Ceti Hussars RCT: Three of Diamonds

The Second Ceti Hussars has suffered many casualties in recent months. Their raids against Scheat over a three-month period proved successful in reducing the store of enemy supplies on the planet but cost the unit many skillful soldiers. It is possible that the second will be pulled off the front lines to rest.

Third Ceti Hussars RCT: Three of Spades

The Third Ceti Hussars RCT has just come back from a prolonged, year-long R & R. Though the units were eager to return to the action, snafus in repairing the Third's vehicles have delayed their return.

ZONE-FIVE
G
LAUBENSTEIN·88

CHISHOLM'S RAIDERS

In 2892, Marshal John Chisholm retired from the AFFS after 40 years as one of the most irritating officers in the Davion military. Though Chisholm is a staunch believer in the natural superiority of the Federated Suns, he was one of those annoying people who criticize and complain too much. No one could really ignore him because the Marshal's complaints were accurate enough often enough to be valuable. When Chisholm complained that the elite Brigade of Guards were "a collection of old women and weak little boys," the AFFS, the Davion family, and the Brigade of Guards finally lost patience and forced Chisholm to accept an early retirement.

Back on his homeworld of Colorado, Marshal Chisholm grew bored with his new life. To relieve his frustration, he assumed control of the Colorado Planetary Guards and began to whip the collection of senior citizens, reservists, and young boys and girls into a fighting force. Soon, the AFFS came to investigate reports of a mad Marshal forcing grey-haired ladies to hike 20 kilometers a day. What they found was a collection of ill-equipped men and women, led by a regiment of rickety 'Mechs. Seeing his chance, the Marshal challenged the AFFS to face his troops in a military exercise. The AFFS agreed, bringing in the Second RCT from the Brigade of Guards. The exercise, codenamed "Cowboys and Indians," proved to be enlightening for both sides. Marshal Chisholm did not get his hoped-for victory, but he did manage to outwit the vastly superior Davion Guards on several occasions. The speed and coordination of the Marhsal's troops took the AFFS completely by surprise.

As a result, Marshal Chisholm's troops were designated the First Chisholm's Raiders and assigned better equipment. Since then, the Chisholm Raiders have always been a trusted, though unorthodox regiment. The unit's symbol is a black triangle representing an ancient Native American shield, behind which six arrows point in six directions.

First Chisholm's Raiders RCT

The First Chisholm RCT is considerably smaller than a normal RCT in the AFFS. There are only two armored regiments and three infantry regiments. The RCT does have almost twice the normal number of artillery pieces, however. A Chisholm regiment stresses mobility, the ability to know and take advantage of surrounding terrain, and the ability to improvise tactics when the enemy's actions force abandonment of the official plan.

Second Chisholm's Raiders: The Apaches

This 'Mech regiment was created in 3001 to act as a mobile reserve for the Raiders RCT. It is composed of fast 'Mechs and has twice the number of AeroSpace Fighters for support.

ROBINSON RANGERS

Robinson Rangers was originally an all-volunteer unit that participated in the defense of the planet Robinson during the First Succession War. When the planet was overrun by the Draconis Combine, the survivors of the force continued the fight as a group of widely scattered small guerrilla units. Eventually, the Kuritans hunted down and killed all the Rangers. During the Second War, the Federated Suns decided to commemorate the courage of those brave volunteers by providing material and technical support for two 'Mech divisions. They would choose pilots for these 'Mechs from among young volunteers on the planet Robinson, who would be trained at the Battle Academy at the Duke of Robinson's expense. The arrangement has worked rather well. Recently, each of the two Rangers regiments have added two platoons of elite anti-'Mech infantry troopers.

The unit symbol of the Robinson Rangers is a single red arrow against a gold shield.

First Robinson Rangers: The Vols

The Vols have recently been refitted with brand new *Enforcer*s and *Valkyrie*s to replace the aging *Shadow Hawk*s and *Stinger*s that they had been using. The regiment is famed for its open-field fighting skills.

Second Robinson Rangers: The Patriots

The Patriots is a regiment that favors and excels at city fighting.

NEW IVAARSEN CHASSEURS

The two 'Mech regiments of the Duke of New Ivaarsen are semi-owned by a nobleman in the Federated Suns. The ducal family of the planet, the Stephensons, have always been rich because of their lands and businesses. This wealthy family are fanatical supporters of the Federated Suns and its liberties because of how much they have profited from them. Because of the fervor of their sentiments, the Stephensons have been allowed to slowly build up their household guards into two BattleMech regiments over the years. The two regiments' 'Mechs and AeroSpace Fighters are all old models but still quite capable of putting up a good fight. Though the MechWarriors and Techs are all native to New Ivaarsen, they are technically members of the AFFS, allowing them to fight with pay in the regiments. The unit insignia of the New Ivaarsen Chaussers is a golden-winged sword against a blue triangle.

First New Ivaarsen Chaussers: The Wings

The First New Ivaarsen Chausseurs have three times the number of AeroSpace Fighters normally assigned to a 'Mech regiment. This regiment favors open-field actions.

Second New Ivaarsen Chaussers: The Swords

This regiment has four battalions, the lightest of which is composed of medium-weight 'Mechs. The regiment favors city actions.

ARGYLE LANCERS

The Argyle Lancers were originally formed from a group of Star League Regulars who chose not to leave with Kerensky. Their small company of 'Mechs played a role in the defense of the Federated Suns during the First Succession War. Their actions were rewarded with titles and the right to recruit graduates from military academies to build up the Argyle Lancers to a full 'Mech regiment.

Today, the regiment is led by a descendant of one of the original members of the Lancers. The symbol of the regiment is the white initial A against a square of black and red.

First Argyle Lancers: The Fast-and-Furious

The First Argyle Lancers trains for mobile battles and has many swift 'Mechs and plenty of aerospace support.

KESTREL GRENADIERS

The First Kestrel Grenadiers is a combined BattleMech and hovercraft regiment. The third company of each battalion is composed of fast, but heavily armed hovercraft that function as the regiment's scouting force. The unit is also known as the Blue Flamers and their insignia is a white torch with a blue flame, the crest of the ducal family of Kestrel.

ARAGON BORDERERS

The Aragon Borderers were New Aragon's planetary guards unit until the unruly population and the almost constant raids by the Capellan Confederation made it obvious that a much larger force was needed to handle these conditions. The Borderers are a mixed-unit regiment, with the third battalion composed of tanks and infantry. The unit insignia is a black rectangle sliced diagonally with a broad white stripe.

First Aragon Borderers: Pride of New Aragon

The Aragon Borderers have a wide variety of 'Mechs and vehicles. The unit is low in aerospace support, however, because of the transfer and retirement of some of their pilots.

CAPELLAN DRAGOONS

The Capellan Dragoons were once the private army of the Duke of Kathil. During the Second Succession War, the unit fought valorously against the Capellan Confederation. Because many of the original members of the Dragoons were defectors from the Capellan military, their knowledge of the enemy's military often gave the Dragoons the upper hand against the Liao forces. Early in this century, the Duke of Kathil became impoverished and without means to pay off his mounting debts. He decided to "sell" his regiment to the AFFS for a sizable chunk of money and the proviso that he or his heirs could buy back the Dragoons if they ever accumulated enough money to do so.

The insignia of the Capellan Dragoons is a rearing golden Lion against a gold and red rectangle.

First Capellan Dragoons: Roar of the Lion

The First Capellan Dragoons is chronically short of parts and is therefore seldom up to full strength. The heavy 'Mechs of the regiment make this unit best suited to slugging matches between two forces.

KITTERY BORDERERS

The Kittery Borderers is similar in design and purpose to the Aragon Borderers and fulfills much the same duties. Because both Kittery and New Aragon sit on the tip of bulges deep into the Capellan Confederation, these planets have a need for planetary guard units strong enough to stand up to the constant Capellan raids. The Kittery Borderers have proven themselves, particularly in a face-off last year with a detachment from one of the Confederation's elite Warrior House units. The unit insignia of the Kittery Borderers is a gold disk against a stylized sunrise.

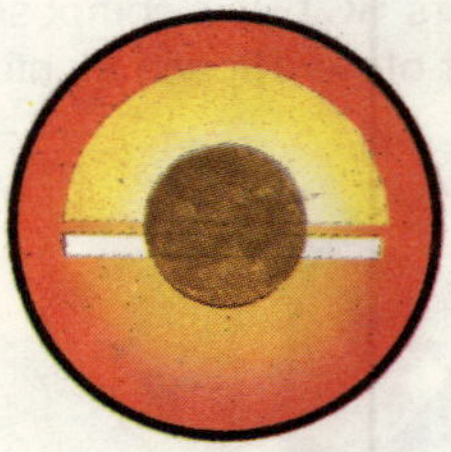

First Kittery Borderers: Pride of Kittery

Through a fluke of military bureaucracy, the Kittery Borderers have recently been outfitted with brand new *Valkyrie*s, making them a much more potent force than previously.

FEDERATED SUNS ARMORED CAVALRY

The First Armored Cavalry is a highly mobile fighting unit consisting of elite MechWarriors and fighter pilots. It was Prince Ian Davion who decided to create a 'Mech regiment so fast and powerful that it could easily overrun most enemy positions before an effective defense could be mounted. To to do this, he assembled a regiment of 'Mechs chosen for their speed as well as their weapons. As a complement to their speed, he organized a wing of heavy Aero-Space Fighters whose pilots were as skilled in air-to-ground tactics as they were against enemy fighters. Prince Ian's idea has proven effective. As a raiding force, the Federated Suns Armored Cavalry has few peers, but their success is often achieved only at great cost in machinery and men, which is why there is only the single regiment.

The unit insignia of the First Armored Cavalry is a variation on the Federated Sunburst. Its nickname is Point of the Sword.

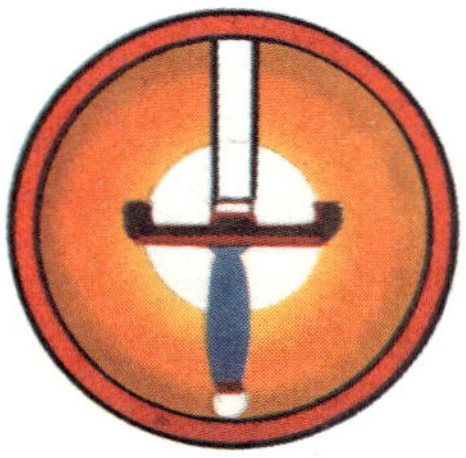

MARCH MILITIAS

As discussed earlier, each March in the Federated Suns has several Regimental Combat Teams to protect it. Each RCT is made up of troops native to the particular region the unit is supposed to protect. The weapons assigned to the March Militias are not the best, but they are serviceable enough to make the Militia unit a formidable foe if attacked. There are ten March Militia units in the Draconis Combine, seven in the Capellan Confederation, and seven in the Crucis March.

NAIS TRAINING CADRE

The College of Military Arts at the NAIS is the most prestigious military academy in the Federated Suns. Its cadets are expected to study long and hard to perfect their skills if they hope to become MechWarriors or fighter pilots. Those that show promise by the end of their third year of studies are transferred to one of the NAIS Training Cadres for training in the field.

Though the NAIS regiments are sent to the front to give the young cadets a taste of battle, they are not intended for the thick of a fight. The cadres are given rear-area assignments and only occasionally conduct combat patrols into areas where they might encounter enemy 'Mechs. If involved in a fight where the enemy numbers come even close to matching the number of cadet warriors, the experienced field instructors leading the cadets have standing orders to pull out as quickly as possible. The Federated Suns is willing to give up a little territory to preserve the lives of their young pilots and MechWarriors. The unit's insignia is a white diamond and black diamond upon a shield of green.

First NAIS Cadet Cadre

The First NAIS Cadre regiment is composed of light 'Mechs.

Second NAIS Cadet Cadre

The Second NAIS Cadre regiment is composed of medium 'Mechs.

Third NAIS Cadet Cadre

The Third NAIS Cadre regiment is composed of heavy and assault 'Mechs.

ALBION TRAINING CADRE

The Albion Military Academy is almost as prestigious and important to the AFFS as the NAIS. Like the NAIS, the Albion fields regiments of its more promising cadets for actual combat training. The unit symbol of the Albion is a white diamond and green diamond upon a black shield.

1st Albion Training Cadre

The First Albion Cadre regiment is composed of light to medium 'Mechs.

Second Albion Training Cadre

The Second Albion Cadre regiment is composed of heavy assault 'Mechs.

REGIONAL TRAINING BATTALIONS

Prince Hanse Davion has recently been looking for ways to train MechWarriors without the expense of shipping young men and women to far-off military academies. His solution was to create a system of Training Battalions on worlds near the front lines. There, experienced MechWarriors use relatively cheap 'Mechs to instruct local young men and women how to operate the machines. The first Training Battalion was created on Conroe in 3024. Experience so far seems to suggest that the MechWarriors trained by these battalions are capable of holding their own, though the lack of technical expertise makes them woefully ignorant of what makes a 'Mech tick. The only real drawback to the Regional Training Battalions idea is the prejudice its graduates encounter from academy-trained MechWarriors.

MERCENARY UNITS CURRENTLY EMPLOYED BY THE AFFS

ERIDANI LIGHT HORSE

CO: Brevet General Nathan Armstrong
Unit Size: 3 Regiments
Skill Level:
- 71st Light Horse Regiment (The White Horse Regiment): Veteran
- 21st Striker Regiment (Winston's Moonrakers): Veteran
- 151st Light Horse Regiment (The Black Horse Regiment): Elite

'Mech Weight: Light/Medium
Fighter Weight: Medium
Armor: Yes **Infantry:** Yes **DropShip:** Yes **JumpShip:** Yes
Financial Situation: Good
Contract Expires: November 3027

LINDON'S COMPANY

CO: Acting Captain Sarah Lindon
Unit Size: 1 Battalion
Skill Level: Veteran
'Mech Weight: Light/Medium
Fighter Weight: Medium
Armor: No **Infantry:** No **DropShip:** Yes **JumpShip:** No
Financial Situation: Fair
Contract Expires: November 3026

MILLER'S MARAUDERS

CO: Major Grissom Miller
Unit Size: 2 Battalions
Skill Level: Veteran
'Mech Weight: Heavy/Assault
Fighter Weight: Heavy
Armor: Yes **Infantry:** No **DropShip:** Yes **JumpShip:** No
Financial Situation: Fair
Contract Expires: July 3025

TEAM BANZAI

CO: Dr. B. Banzai
Unit Size: 1 Regiment
Skill Level: Elite
'Mech Weight: Light/Assault
Fighter Weight: Medium
Armor: No **Infantry:** No **DropShip:** Yes **JumpShip:** Yes
Financial Situation: Good
Contract Expires: January 3030

ILLICIAN LANCERS

CO: General Ellaine Stewart
Unit Size: 4 Regiments
Skill Level:
- 59th Strike Regiment (Bradley's Bad Boys): Veteran
- 21st Illician Rangers (Drop-and-Stomp): Regular
- 4th Illician Rangers (Janey's Lovers): Regular
- 9th Illician Rangers (Roadrunners): Green

'Mech Weight: Light/Assault
Fighter Weight: Medium
Armor: No **Infantry:** No **DropShip:** Yes **JumpShip:** No
Financial Situation: Fair
Contract Expires: January 3026

12TH VEGAN RANGERS

CO: General Carl VanTan
Unit Size: 4 Regiments
Skill Level:
- Alpha Regiment (The Concrete Crushers): Elite
- Beta Regiment (The Victories): Veteran
- Gamma Regiment (Vella's Dancers): Veteran
- Delta Regiment (Noble Nelson's Nasties): Regular

'Mech Weight: Heavy/Assault
Fighter Weight: Medium
Armor: Yes **Infantry:** No **DropShip:** Yes **JumpShip:** No
Financial Situation: Good
Contract Expires: July 3026

THE DIOSCURI

CO: Colonel Timothy Nels, Colonel Michael Nels
Unit Size: 2 Regiments
Skill Level: Regular
'Mech Weight: Medium
Fighter Weight: None
Armor: No **Infantry:** Yes **DropShip:** No **JumpShip:** No
Financial Situation: Good
Contract Expires: November 3026

GRIM DETERMINATION

CO: Colonel Fiona Trent
Unit Size: 4 Battalions
Skill Level: Regular
'Mech Weight: Heavy
Fighter Weight: Heavy
Armor: Yes **Infantry:** No **DropShip:** No **JumpShip:** No
Financial Situation: Very good
Contract Expires: July 3026

REDFIELD RENEGADES

CO: Colonel Tyrell Redfield
Unit Size: 1 Regiment
Skill Level: Regular
'Mech Weight: Medium
Fighter Weight: Light
Armor: Yes **Infantry:** No **DropShip:** Yes **JumpShip:** No
Financial Situation: Good
Contract Expires: November 3026

WACO RANGERS

CO: Colonel Wayne Waco
Unit Size: 1 Regiment
Skill Level: Regular
'Mech Weight: Medium
Fighter Weight: Medium
Armor: Yes **Infantry:** Yes **DropShip:** Yes **JumpShip:** No
Financial Situation: Good
Contract Expires: July 3026

MARKSON'S MARAUDERS

CO: Colonel Richard Markson
Unit Size: 1 Regiment
Skill Level: Regular
'Mech Weight: Light
Fighter Weight: None
Armor: No **Infantry:** No **DropShip:** Yes **JumpShip:** No
Financial Situation: Poor
Contract Expires: January 3026

LEXINGTON COMBAT GROUP

CO: Major General Isabella Rahm
Unit Size: 3 Regiments
Skill Level:
- 32nd Recon Regiment (Tessie's Sprinters): Veteran
- Golda's Golden Hammers: Regular
- Timothy's Gazelles: Green

'Mech Weight: Light/Assault
Fighter Weight: Medium
Armor: Yes **Infantry:** Yes **DropShip:** Yes **JumpShip:** No
Financial Situation: Good
Contract Expires: December 3026

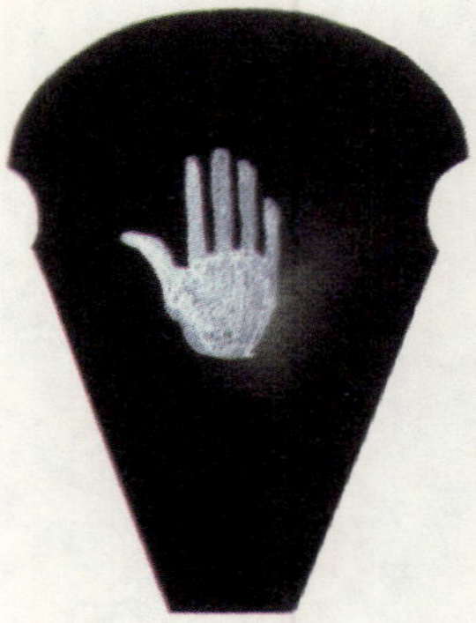

THE FIGHTING URAKHAI

CO: General Harold Greenspan
Unit Size: 3 Regiments
Skill Level:
- 8th Striker Regiment (The Lazy Eights): Veteran
- Candelli's Bar Hounds: Regular
- Jackson's Sluggers: Regular

'Mech Weight: Heavy/Assault
Fighter Weight: Medium
Armor: No **Infantry:** No **DropShip:** No **JumpShip:** No
Financial Situation: Poor
Contract Expires: March 3026

THE KELL HOUNDS

CO: Lieutenant Colonel Patrick M. Kell
Unit Size: 1 Battalion
Skill Level: Veteran
'Mech Weight: Medium
Fighter Weight: Medium/Heavy
Armor: No **Infantry:** Yes **DropShip:** Yes **JumpShip:** Yes
Financial Situation: Good
Contract Expires: January 3026

SCREAMING EAGLES

CO: Colonel Walter Hokala
Unit Size: 2 Regiments
Skill Level: Regular
'Mech Weight: Medium
Fighter Weight: Heavy
Armor: Yes **Infantry:** Yes **DropShip:** No **JumpShip:** Yes
Financial Situation: Poor
Contract Expires: January 3026

WYLIE'S COYOTES

CO: Colonel Enzo Wylie
Unit Size: 1 Regiment
Skill Level: Regular
'Mech Weight: Light
Fighter Weight: Light
Armor: Yes **Infantry:** Yes **DropShip:** Yes **JumpShip:** Yes
Financial Situation: Fair
Contract Expires: December 3026

HARLOCK'S WARRIORS
CO: Colonel Rodney Viele
Unit Size: 1 Regiment
Skill Level: Regular
'Mech Weight: Medium
Fighter Weight: None
Armor: Yes **Infantry:** Yes **DropShip:** No **JumpShip:** No
Financial Situation: Good
Contract Expires: July 3026

DISMAL DISINHERITED
CO: General Ivan Fontral
Unit Size: 3 Regiments
Skill Level:
O'Hare's Plague of Locusts: Regular
Ushida's Crypt Kickers: Regular
Kelly's Hostile LAMs: Green
'Mech Weight: Light/Heavy
Fighter Weight: Light/Heavy
Armor: No **Infantry:** Yes **DropShip:** Yes **JumpShip:** No
Financial Situation: Fair
Contract Expires: November 3027

SWANN'S CAVALIERS
CO: Colonel Sarah Dosimion
Unit Size: 1 Regiment
Skill Level: Veteran
'Mech Weight: Heavy
Fighter Weight: Heavy
Armor: Yes **Infantry:** Yes **DropShip:** Yes **JumpShip:** Yes
Financial Situation: Good
Contract Expires: December 3026

THE BLUE STAR IRREGULARS
CO: General Mona Troy
Unit Size: 3 Regiments
Skill Level:
1894th Light Horse Regiment (The Thoroughbreds): Veteran
Avatars of Painful Death: Regular
21st Rim Worlds 'Mech Regiment (The Drunken Surgeons): Green
'Mech Weight: Light/Heavy
Fighter Weight: Heavy
Armor: Yes **Infantry:** Yes **DropShip:** Yes **JumpShip:** Yes
Financial Situation: Fair
Contract Expires: January 3027

THE CRATER-COBRAS
CO: Colonel Richard Westrick
Unit Size: 2 Regiments
Skill Level:
789th Striker Regiment (Chan's Choppers): Veteran
Westrick's Black Cobras: Regular
'Mech Weight: Medium/Assault
Fighter Weight: Medium
Armor: Yes **Infantry:** Yes **DropShip:** Yes **JumpShip:** No
Financial Situation: Good
Contract Expires: November 3026

WEAPONS INDUSTRIES

The Federated Suns weapons industries are among the healthiest of the Inner Sphere, especially with firms such as Federated-Boeing and Kallon Weapons producing at full capacity and even devoting resources to development of new equipment.

Name: Norse BattleMech Works
Plant Location: Marduk
CEO: Dolan Rastkel, Duke of Marduk
Weaponry Manufactured: GRF-1N *Griffin*, WVR-6R *Wolverine*
Notes: The Norse BattleMech Works is one of the few nearly self-reliant 'Mech factories in the Inner Sphere. It is located deep in a jungle directly over a vast deposit of rare metals that the NBW mines and processes in a smelter system next to the factory. The finished product travels less than a hundred meters before being turned into *Griffin* and *Wolverine* chassis. Other nearby factories make almost all the electronic and mechanical components necessary to build the 'Mechs. Only the tracking system must be imported from offworld. This self-sufficiency has, of course, made the factories extremely attractive to the Draconis Combine, which has attempted to take the planet more than a dozen times.

Name: Cal-Boeing of Dorwinion
Plant Location: Dorwinion, Belladonna
CEO: Count Mitchell Fassel
Weaponry Manufactured: Ferret, Wild Weasel, and Marten VTOL Vehicles
Notes: Cal-Boeing of Dorwinion is a small company made up of engineers and designers who left Federated-Boeing in 2899 over a disagreement about whether the company should expand into new areas of weapons design. The parting was amiable, however, and so Federated-Boeing donated a group of long-empty warehouses on the planet Belladonna to their former employees. The new Cal-Boeing produced its first Ferret in 2901 and won an AFFS contract to build 6,000 of the light scout VTOL. Surprised by their good fortune, Cal-Boeing paid back Federated-Boeing to avoid any nasty takeover attempts on their part, then proceeded to expand their facilities to cope with the sudden demand. For 40 years, the company was so busy with the demand for Ferrets that they could not stop to design any other aircraft. In 2967, Cal-Boeing finally came out with the Wild Weasel, a combat version of the Ferret. The design did not sell, however, except to Periphery kingdoms. Stung by the poor showing of the Wild Weasel, Cal-Boeing did not produce another new design until 3021. This was the Marten, another VTOL scout but with a wiser selection of weapons.

Name: Corean Enterprises
Plant Location: Jeratha, New Avalon
CEO: Count Yethers Corean
Weaponry Manufactured: VLK-QA *Valkyrie*
Notes: Corean Enterprises has been the focus of much attention from the engineers and scientists of the NAIS because its factory at Jeratha is one of the few surviving examples of a fully automated BattleMech factory in the Federated Suns. This has been both a blessing and a curse. On one hand, the factory can churn out 130 'Mechs a year, which is more than a manually operated factory could ever hope to produce. On the other hand, the factory runs so well that its engineers have forgotten how it operates and cannot repair the equipment when it breaks down. Recently, some of the machinery has definitely begun to show signs of wear, which is why the NAIS scientists have decided it is urgent that they rediscover how the factory operates. They hope to apply what they learn to other factories and also to be able to repair the Corean factory when it inevitably breaks down sometime in the near future.

Name: Achernar BattleMechs
Plant Location: Ith, New Avalon
CEO: Countess Rodina Achernar
Weaponry Manufactured: WSP-1A *Wasp*, PXH-1 *Phoenix Hawk*, LCT-1V *Locust*
Notes: The 'Mechs coming out of the Ith factory have been an important part of the Davion military for centuries. Unlike the Corean factory, the Achernar facility is not fully automated, which allows at least patch-work repair when the machinery breaks down.

Recent rumors from New Avalon tell that Achernar BattleMechs might open a small factory on Waimalu. In the past decade, deposits of rare metals have been discovered on this poor world, and so it would benefit the planet should Achernar open a factory there. The two obstacles that might prevent it are the lack of sufficient water and Waimalu Fever, an unconquered virus that kills one out of every two visitors to the planet.

Name: Kallon Weapon Industries
Plant Location: Talon
CEO: Duke Roderick FaCrimeia
Weaponry Manufactured: RFL-3N *Rifleman*, GRF-1N *Griffin*, ENF-4R *Enforcer*, JM6-S *JagerMech*, and the Partisan Heavy Tank
Notes: The planet Talon is a minor world not often noted on star maps. As the smaller sister to the agriculturally important planet of Wernke, it seems to many people just another two-bit rocky world. Talon does have great importance to the AFFS, however. Despite its lack of water and food resources, the Kallons felt that the planet's excellent rare metal deposits justified construction of a factory on the mountainous world.

Since that time, the Kallon BattleMech factory on Talon has been churning out 'Mechs at record speed. Not only is the factory capable of fast production, but the level of technology and quality are phenomenally high. The Kallons somehow manage to build 'Mechs that are almost of Star-League era quality, but no one knows how. Now that they have shown their intention to expand into new areas of weapons design, such as their well-built Partisan Heavy Tank, which appeared in 2961, it is even more important to know their secret. All efforts by our agents to penetrate the Kallon factories have so far been unsuccessful.

Name: Independence Weaponry
Plant Location: Quentin
CEO: Duke Oscar Ivanick
Weaponry Manufactured: AS7-D *Atlas*, JM6-S *JagerMech*, VTR 9B *Victor*, MAD-3R *Marauder*
Notes: Steel Valley, the name given to a heavily industrialized valley on Quentin, was the site of a 'Mech factory for the Regular Army. When the Star League collapsed, the Draconis Combine, the Capellan Confederation, and the Federated Suns fought bitterly for control over the factory. During the later stages of the Second Succession War, House Davion seized the planet and has established what seems to be permanent control over the planet.

Davion has repaired the huge complex of assembly lines that were damaged during the fighting, and these have since become the core of the Davion military's heavy 'Mech production. Due to repairs and almost constant raids by Houses Liao and Kurita, the factories have been operating at only half-capacity. Some of the assembly lines, like those for the *Victory* and *Marauder*, can churn out 20 'Mechs a year. The line for the awesome *Atlas* can turn out only five a year. The numbers of completed *Atlas*es may soon rise because of the recent alliance with the Lyran Commonwealth. A team of engineers from the massive Defiance Industries 'Mech factories on Hesperus II have arrived on Quentin specifically to help increase the number of *Atlas*es built in a year.

Name: Jalastar Aerospace
Plant Locations: Amritsar, Panpour
CEO: Count Edwas Trehala
Weaponry Manufactured at Armistar: SPR-H5 *Sparrowhawk*, *Sabre*, CN9-A *Centurion*, and the Manticore tank.
Notes: Jalastar is one of the Federated Suns's major aerospace manufacturers. Early in the Star League days, the company acquired the rights to manufacture fighters for the Regular Army and so became privy to secret designs. One of these was the *Sparrowhawk*. When the Star League fell, Jalastar did not hesitate to manufacture the swift yet potent light AeroSpace Fighter for the Federated Suns.

Jalastar is currently in a state of confusion. An unknown financial giant (suspected to be Federated Boeing) is attempting to buy up a controlling share of the company. The effort to stave off the takeover has drained money from the company's many research projects at just the moment when a new lighter engine for the *Sparrowhawk* was near completion.

Name: Lycomb-Davion IntroTech
Plant Location: Ultron and Novais, New Avalon
CEO: Countess Tyrell Jaffey
Weaponry Manufactured at Ultron: STU-K5 *Stuka*, *Lightning*, and *Hellcat* AeroSpace Fighters
Weaponry Manufactured at Novais: Armored Personnel Carriers, Infantry Weapons, Artillery Pieces
Notes: When the Star League collapsed, the Capellan Confederation was quick to destroy the complex of factories on Demeter that produced the large and effective *Stuka* AeroSpace Fighter. What House Liao did not realize was that Lycomb IntroTech had begun construction of a second aerospace manufacturing facility on New Avalon and that many of the company's vital engineers had been on New Avalon when the Confederation attacked Demeter. This new factory soon began to churn out the *Stuka* fighter as well as other more common aerospace fighter types.

In the last few decades, the company, now partly owned by the royal family, has been attempting to diversify. Their second plant at Novais produces rifles, hand-held laser weapons, and small artillery pieces.

Name: Johnston Industries
Plant Location: Addicks
CEO: Count Davion Johnston
Weaponry Manufactured: Goblin Medium Tank, Portable Lasers, and Particle Guns
Notes: When the Star League fell, the wealthy Johnstons of the Federated Suns purchased the Tauken Weapons System manufacturing facility on Addicks, despite the fact that the planet would most surely become a battlefield once the fighting started. When the war began, the Johnstons had to sit on their hands and watch as forces from the Federated Suns, the Draconis Combine, and the Capellan Confederation made their bids for the planet. By the time the Federated Suns finally secured the world, over half the factories had been destroyed. The only factories that had managed to survive were those manufacturing trooper-portable laser and particle weapons and the one producing the Goblin medium tank. Fortunately for the Johnston family, the Goblin's ability to carry a squad of infantry make it a favorite of AFFS mechanized regiments. There are rumors that Johnston Industries is considering opening a new facility on New Syrtis.

Name: Wangker Aerospace
Plant Location: Axton
CEO: Baron Tevile Wester
Weaponry Manufactured: CSR-V12 *Corsair*, *Eagle*, and *Thunderbird* Fighters
Notes: At one time, Wangker Aerospace had five separate factory complexes on five worlds, producing many kinds of AeroSpace Fighters. The pride of Wangker was the *Corsair*, a design they had created. The First Succession War, with its widespread destruction, demolished all five factory complexes. During the Second War, Wangker had salvaged enough from its earlier plants to construct a fighter manufacturing center on the planet Axton. Since then, the factory has been producing fighters for the AFFS, including the company's famed *Corsair*. The firm has been doing so well that they are thinking of expanding, either into the manufacture of another fighter design or combat vehicles.

Name: Aldis Industries
Plant Location: Terra
CEO: Paula Carlton
Weapons Manufactured: Demolisher Heavy Tank, Schrek PPC Carrier, Behemoth Heavy Tank, Rolling Thunder LRM and SRM Carriers
Notes: Aldis is the largest independent manufacturer of weapons in the Inner Sphere. When the Star League fell, Aldis Industries was just a small company. The exodus of the Regular Army and the declaration of Terra's neutrality by ComStar caused a collapse of all weapons manufacturers on Terra. All except Aldis Industries, that is. The company drew on some unknown source for huge sums of money to buy out the other weapons company on Terra. They correctly guessed that ComStar would have no real objection to weapons being manufactured there. Since then, Aldis has grown tremendously. They concentrate their efforts on huge battle tanks, which they believe are the most profitable weapon type, though they also own the 'Mech factories at Liverpool and the aerospace facilities at Tokyo.

MILITARY EDUCATION

In the Federated Suns, it is the responsibility of a planet's nobility to encourage promising young men and women to enter the military. Those young people who show exceptional potential, either because they have scored high on tests given regularly by the AFFS or because of their athletic abilities, are offered scholarships by the planet's nobility and government to attend the nearest accredited military academy. If the young person would like to join the AFFS but cannot voyage to an academy because of a lack of money, he may join an ROTC corps and become a reserve officer with the local planetary guard units. If his career in the PUG is notable, the officer might be asked to join a front-line unit in the AFFS.

A cadet's first two years at an academy are spent learning military fundamentals as well as getting a heavy background in sciences. The student is also encouraged to think about what part of the AFFS he or she would like to join. Once the cadet has decided, most of the final two years of academy life are devoted to his chosen specialty. To ensure that every soldier understands at least generally how his role fits in with the rest of the effort, even aspiring MechWarriors and AeroSpace Pilots must learn what it means to be a common infantryman, a clerk, and a Tech.

There are ten major military academies in the Federated Suns today, as well as five lesser academies. Though some provide training in all disciplines, most specialize in a particular group of disciplines. The AFFS administers the academies to ensure that the curriculum will prepare the cadets for what the Davion military requires of a new officer.

Recently, the AFFS has introduced a new and radical program designed for training MechWarriors outside the military academy system. The basis of the program are Training Battalions, a small staff of teachers and instructors who travel to worlds, particularly poorer worlds on a hostile border. Once there, they provide 'Mech training for young recruits who would otherwise be too poor to attend a military academy. Many AFFS MechWarriors believe the Training Battalions offer training that is skimpy at best, and dislike them intensely. There are currently three Training Battalions in the Federated Suns.

ALBION MILITARY ACADEMY, NEW AVALON

The Albion Military Academy is is the oldest and one of the most prestigious of all military academies in the Federated Suns. It began its life in the 2400s as an academy for officers entering the New Avalon Militia. The school, which is some 30 kilometers south of Avalon City, was designated an official Star League Military Academy in 2604 after extensive renovation and expansion of the facilities. Those attending were guaranteed the finest possible military education. After spending five years in the Regular Army, any graduate of the Albion could expect to come back to the AFFS and rise quickly in the ranks.

Many of the Academy's instructors chose to stay with the school rather than heed General Kerensky's order to abandon the Inner Sphere. This was fortunate for the Federated Suns because the skills of several key soldiers, all trained at Albion, were what permitted the Federated Suns to survive the ravages of the Succession Wars. Today, the school trains MechWarriors and AeroSpace Pilots, as well as infantry, armor, and artillery unit officers. Strategy and tactics are heavily emphasized at Albion.

The NAIS College of Military Sciences is the Albion's rival. Yearly, the BattleMech Training Cadres of the two schools meet on Jasper Plain a few hundred kilometers away to challenge one another in a series of wargames. The winners are declared "the young lions of New Avalon."

The Commanding Officer of the Albion Military Academy is Colonel Fethu Simons.

NAIS COLLEGE OF MILITARY SCIENCES (NEW AVALON MILITARY ACADEMY)

While the Albion Academy may be the oldest training institution for officers in the Federated Suns, the nearly forgotten origins of the NAIS College of Military Sciences make it a close second. Few people seem to remember that the shiny new buildings of the largest university in the Inner Sphere stand on the location of what was once the New Avalon Military Academy. The NAMA was built a decade after the Star League took over the Albion Military Academy so that New Avalon might continue to contribute its fair share of soldiers to the armed forces of the Federated Suns.

Besides getting their training, the cadets of NAMA received a heavy dose of political training and were closely scrutinized by the MIIO. This was to ensure that the school's graduates would form a loyal core of soldiers in the AFFS who could be counted on in times of internal strife. This tradition of unquestioned loyalty was to continue for many years.

After the fall of the Star League, the NAMA fell into disfavor, even though its graduates continued to be loyal soldiers. For awhile, the NAMA and its various 'Mech regiments that did tours of duties on the front lines become the object of public scorn because the people thought they could not measure up to their flashier counterparts from the Albion Academy. When the number of cadets requesting to attend the academy plummeted, some government and military leaders suggested tearing down the academy or allowing Albion to annex it.

It was Prince Ian who saved NAMA and set it back on the road to prominence. As a graduate of NAMA, he showed his gratitude to his old school by ordering it rebuilt and restaffed when he came to power. Later, Prince Hanse Davion, looking for a way to justify his desire to build a huge and expensive university, decided to make the New Avalon Military Academy the physical center of his

Pictured here is a young MechWarrior cadet. Though the official age of entry into the AFFS is 16, some special circumstances permit a boy or girl to join as young as 14. The Cadet is wearing a cooling undershirt, a garment made of tiny tubes through which water can be circulated to cool the wearer. The white epaulettes on his shoulder show his status as a student in one of the many MechWarrior academies in the Federated Suns.

new university, which would give the institution both military and scientific importance. Though the nobility had objected to Davion's proposal to build a new university, they welcomed the idea of a new academic institution built around the military academy. That Hanse Davion foresaw their reaction is just another example of his skill as a politician.

The College of Military Sciences, as it is known today, may be the most expensively furnished military academy in the Inner Sphere. It lacks nothing. There are plenty of training 'Mechs and fighters as well as simulators of any conceivable 'Mech design known in the Inner Sphere. There are three Gauntlets (booby-trapped mazes designed to test a cadet's responses) as well as plenty of open land for practicing the art of war.

The College of Military Sciences retains many of the traditions of the NAMA. One of the most highly prized is the tradition of loyalty to the Federated Suns and the Davion family. The Commanding Officer of the College of Military Sciences is Major-General Jeremiah Davion.

WARRIOR'S HALL, NEW SYRTIS

Warrior's Hall was originally an officer's club and VOQ (Visiting Officer's Quarters) for a nearby Regular Army base near the capital city of New Syrtis. During the early years of the Star League, the Hall served both Regular Army officers and those from the Davion military as a place to mix and mingle. Whenever these two groups met, they inevitably disputed the question of which military was the best, the Regular Army or the AFFS. The heated discussions, arguments, and drunken brawls soon convinced the commanding officers of both militaries to give their men a chance to settle the question. The two groups raised enough money to build a maze of obstacles and booby-traps, similar to the Gauntlets used today, and gave the soldiers old 'Mech trainers to race and fight in. Now, whenever the arguments inside Warrior's Hall got too heated, the officers could jump into a trainer and either fight it out or simply compare times through the obstacle course.

It was a short step from this to using the mazes and trainers to teach 'Mech skills to the promising young men and women in the nearby village who were too poor to attend a military academy. Soon, both the AFFS and the Regular Army realized the benefit of the Warrior's Hall and jointly began to expand the Hall's capacity to train would-be MechWarriors. By the end of the Star League era, Warrior's Hall was a fully operational military academy capable of teaching every military discipline except for JumpShip operations.

In recent years, the Warrior's Hall has become the center of pro-Michael and anti-Hanse sentiment. No doubt the Duke of New Syrtis's agents promote this attitude, for Hasek-Davion likes to fill vacancies in his elite Syrtis Fusiliers with the top graduates of the Hall.

Today, Warrior's Hall is commanded by Colonel Kathleen Sullivan.

BATTLE ACADEMY ON ROBINSON

The Battle Academy has just recently come back from years of disuse and war damage. At the start of the First Succession War, the Battle Academy had just graduated its first class of officers. When the Draconis Combine took Robinson, the conquering soldiers gave special attention to the Battle Academy. In a series of actions intended to demoralize the people of Robinson, the Occupation Administrators of the Draconis Combine forced the few cadets who had not perished in the invasion to tear down the buildings of the Battle Academy with their bare hands. That it was impossible did not matter to the Kuritans.

It has taken these many years for the Battle Academy to be fully rebuilt and restaffed. In 3020, with the Robinson economy entering a prosperous period, the Sandoval family, Dukes of the planet and nominal controllers of the entire Draconis March, decided to reconstruct the school. They spared no effort to make it equal to anything at Albion or the College of Military Sciences on New Avalon.

The Battle Academy is currently commanded by Colonel David Sandoval.

SAKHARA ACADEMY

The Sakhara Academy is one of the few private military academies in the Federated Suns. Created by a group of retired Star League officers in 2613, the Academy has willingly accepted anyone who can afford the tuition. During the First Succession War, the Sakhara Academy distinguished itself when its staff and students, about six regiments of mixed weapons, withstood and even repulsed three separate attacks by the vaunted 3rd Sword of Light 'Mech regiment and three other veteran Kurita regiments. This action, known in the academy as the Days of Jousting at Dragons, made the academy's reputation.

Today, though Sakhara can only turn out 20 MechWarriors and an equal numbers of pilots, armor, and infantry officers every year, the traditions and history of the school still carry much respect among the AFFS. Because the school is so small and so tradition-bound, graduates grow to love it and consider the four years there as the best years of their lives. Yet, the training is extremely tough, both mentally and physically.

The current commander of the Sakhara Academy is Commandant Lindsey Trevellis, a retired AFFS Colonel.

ARMSTRONG FLIGHT ACADEMY

Located on Galax, the Armstrong Flight Academy was built during the First Succession War to train DropShip and JumpShip crews. The reason the AFFS chose Galax was because Federated-Boeing, the Inner Sphere's largest DropShip manufacturer, was also located on and around the planet.

Today, the teaching staff at the Armstrong Flight Academy are also employees of Federated-Boeing. Their expertise makes graduates of this academy some of the best DropShip and JumpShip crewmen in the AFFS.

Colonel Jaswa Punalla currently commands the Armstrong Flight Academy.

POINT BARROW MILITARY ACADEMY

The Point Barrow Military Academy is currently under investigation for supposed anti-Davion activities. Several anti-government bombings in the nearby capital city have been traced back to some cadets in the academy. ComStar finds the whole matter quite puzzling and it is still under investigation by ROM agents.

The academy is otherwise nondescript. It cannot teach fighter pilots, but is known for the quality of its Techs and Engineers. Command of the academy is currently shared by Colonel Calvin Froshith and Major Claudia Saunders of Military Intelligence.

WAR COLLEGE OF GOSHEN

The War College of Goshen has been quietly training its students in the fine art of piloting a 'Mech or a fighter for the past 70 years. Its training cadre, now disbanded, was once a key part of the local defense structure in the Coreward Theater of the Capellan March.

Colonel Francine Maryvelt commands the academy.

KILBOURNE ACADEMY

The Kilbourne Academy is one of two military academies located near the Periphery. As the representatives of education in a very ignorant section of the realm, admittance is not limited to those who plan to join the AFFS. Citizens are allowed to attend the academy's non-military classes.

The Kilbourne Academy, though it trains MechWarriors and AeroSpace Pilots, actually specializes in training Techs and Engineers. Colonel T. R. Graham commands the academy.

FILTVELT MILITARY ACADEMY

The recently built Filtvelt Military Academy also permits civilians to attend its non-military classes. Most of the graduates of Filtvelt and of the Kilbourne Academy become incorporated into Periphery units of the AFFS. Commanding the Filtvelt Military Academy is Colonel Katie Graham, sister of Colonel T. R. Graham.

MEDALS AND DECORATIONS

The Armed Forces of the Federated Suns is deliberately stingy about handing out medals to its soldiers and officers. Because tradition is so important in the AFFS, every soldier knows that any heroic deed helps to glorify his regiment's honor. To give a soldier a medal every time he does something he probably considers all in the line of duty might be an insult to his honor. This is why only a trooper's regimental commander can nominate him for a medal. If the regimental commander, an almost priest-like holder of the regiment's traditions, thinks the actions of one of his soldiers is heroic, then the Field Marshals and Prince Davion have no doubt about the trooper's worthiness.

Another reason that the AFFS hands out medals only rarely is because the Dukes of the Federated Suns are allowed to use their own medals and ribbons to honor AFFS troops that have defended their worlds. One saying among the soldiers of the AFFS is that if a soldier's dress uniform isn't heavy with medals by the end of his second year in the service, then he's either lucky, cowardly, or dead. The federation government does not wish to dilute the honor and prestige of its own medals by becoming too generous with them.

THE MEDAL EXCALIBUR

The Medal Excalibur is the highest honor the Federated Suns gives to individual soldiers. It is reserved for soldiers whose actions have saved his regiment, a Davion world, or whose actions directly and obviously resulted in some military or social advancement of the Federated Suns. In the centuries since the collapse of the Star League, the Medal Excalibur has been awarded only 30 times.

The first recipient was Lieutenant Colonel Jessica Dewitt of the Third Avalon Hussars. She and a small group of volunteers are credited with holding back the advancing Kurita forces long enough to allow the orderly evacuation of the people of Robinson. A recent recipient of the Medal Excalibur was Colonel Ardan Sortek, friend and confidant of Prince Hanse.

The medal itself is a simple gold disk with the legendary sword of King Arthur imbedded in its anvil, with the Federated Sunburst glowing behind it. This is also one of the few medals that tradition demands be worn on the dress uniform instead of pinned to the Sunburst Vest with other medals. The medal also bestows the title "Knight of the Realm" and the grant of a parcel of land of the recipient's own choosing on his homeworld. Members of the Davion family are not eligible for this medal.

DIAMOND, GOLDEN, AND SILVER SUNBURSTS

The Sunburst Medal is an honor given to soldiers and civilians alike for exceptional duty, devotion, and sacrifice in the name of the Federated Suns. It is usually given to someone whose actions have been crucial in stemming an enemy attack, but the heroic act need not be military. There have been recipients who have won the medal for a particularly brave political or even economic action that proved to have realm-wide importance.

The medal is a large black disk with—in increasing order of importance—a silver, gold, or diamond sunburst on it. Each medal has a corresponding ribbon for use with the breastplate of the dress uniform, but because of the great honor associated with the Diamond Sunburst Medal, recipients usually wear it on the right side of their uniforms.

SYRTIS MEDAL OF HONOR

The Syrtis Medal of Honor is the only medal that Duke Michael Hasek-Davion, as Duke of New Syrtis and Commanding Officer of the Capellan March, can award to troops under his command. The medal is made from gold, silver, and polished black BattleMech armor and is shaped into the All-Seeing Eye, the family crest of the Haseks. Recipients of this medal also take the title, "Knight of the Capellan People".

Though any unit currently defending the Capellan March is technically eligible, the Syrtis Medal of Honor has become an extremely political medal. Ever since the rise of Michael Hasek-Davion, nearly all recipients of the Syrtis Medal of Honor have also been fanatic supporters of the Duke. Few, if any, soldiers or officers loyal to Prince Hanse have received the medal.

ROBINSON MEDAL OF VALOR

The Robinson Medal of Valor is similar to the Syrtis Medal of Honor. The Duke of Robinson, as Commanding Officer of the Draconis March, bestows this medal upon those who have done more than duty or honor requires to help defend the Draconis March. Recipients of the medal also receive the title, "Knight Defender of the Draconis March".

The medal is a burnished red disk with a gold Star of David beneath the Federated Suns Sword.

KNIGHTS OF THE FEDERATED SUNS

The Knights of the Federated Suns was created just before the start of the Succession Wars to honor 100 of the realm's best, brightest, and most heroic citizens. The Knights was an obvious Davion imitation of the prestigious Order of the Star, a group of illustrious soldiers and citizens brought together by the Cameron family during the best years of the Star League. The Order of the Star vanished with the League, but the Knights FS have only grown in importance and respect since then.

The Order never has more than 100 Knights FS. Because membership is for life, a Prince only rarely has occasion to choose a candidate for the order. A candidate must be approved by the Royal Court. If accepted, a candidate is endowed with land and a yearly grant of money. Membership is open to any citizen of the Federated Suns, no matter what his status or occupation. All that matters is that the candidate has shown unquestioned expertise in his profession and an unswerving devotion to the Federated Suns. This is why soldiers, though the most numerous with 71 members, are not alone in the Order. There are poets, artists, economists, politicians, and even a former prostitute currently among the members of the Order.

When formally meeting in the Cathedral of Notre Dame at Avalon City, the Knights FS wear long capes of ermine and fox fur. They also carry a sword, representing the Federated Suns, in one hand while carrying the holy words of each member's particular religion in the other. [EDITOR'S NOTE: The Knights of the Federated Suns have taken to carrying the Unfinished Book during these formal occasions.] The official badge of the order is the Federated Sunburst, but with the silhouette of a man (or woman, depending on the recipient) holding up the sword.

ORDER OF DAVION

The Order of Davion is similar to the Knights FS, but it is a more private group. Members of the OD are people who have shown great loyalty to the Davion family. There are currently just 23 members, all of whom are from some branch of service in the AFFS. The badge of the order shows the Davion family crest: the Davion Fox crouched upon a blood-red shield.

The Order of Davion was created in 2965, ten years after the MechWarrior Cabal attempted to mutiny against Prince Peter Davion because he was showing preference to the AeroSpace Fighter Pilots. The Order replaced the older Crucis Legion and the Order of Valor awards, which many felt tended to create the kind of interservice rivalries that led to the MechWarrior mutiny.

DRAGON SLAYER'S RIBBON

The Dragon Slayer's Ribbon is actually a unit commendation that the Lyran Commonwealth awards to its military units that have shown exceptional performance in combat against the Draconis Combine. In 3024, Prince Hanse and the Archon Katrina decided that even though the Federated Suns has a similar unit commendation, the Sword of St. George, it would be good for the alliance to create a single commendation for the armed services of both Houses. This is why the Dragon Slayer's Ribbon, a green and white ribbon with the Draconis Dragon black-striped, is being given to AFFS units. It helps boost the morale of AFFS troops to see the ribbon and be reminded that they have allies, and it might also demoralize the soldiers of the Draconis Combine to see units on both their borders flying the green ribbon.

LIMP SWORD FLAG

Long, long ago some artistic soldier in a Davion regiment scrawled the first Limp Sword on the side of a Capellan 'Mech that he had destroyed. The drawing, with its drooping sword held by a broken wreck of a 'Mech arm, is a sarcastic parody of the official House Liao symbol. [EDITOR'S NOTE: Not only is it an insult to the Capellan fighting abilities, but the drooping sword is believed to have been a slur against the potency of the Liao Chancellor, who had apparently been unable to sire a child at the time.] Soldiers of the AFFS quickly took to the cartoon, and began leaving their own versions of the drawing on the sides of every destroyed Capellan 'Mech and tank as a sign of scorn for their opponents.

During the First Succession War, the Federated Suns officially adopted the Limp Sword Flag as a unit commendation for an AFFS or mercenary unit that has performed particularly well against Capellan forces.

STRENGTHS AND WEAKNESSES

The Armed Forces of the Federated Suns represent the strongest military in the Inner Sphere. Its soldiers, from the lowliest trooper to the best MechWarrior, are a highly trained and highly motivated force that prizes the fact that they enjoy more freedoms than either of their neighbors, the Draconis Combine and the Capellan Confederation. The equipment used by the AFFS, though never sufficient to totally satisfy the needs of the military, is of high quality and dependability. With 119 BattleMech regiments and more then ten times that number in aerospace fighters, tanks, infantry, and artillery units, many people ask why the Federated Suns is not blazing its way across the Inner Sphere, absorbing any House that dares defend itself.

The most obvious reply to that question is that space is immense and the borders the AFFS must protect are longer, more convoluted, and more filled with dangerous bulges and salients than in any other realm. The simple task of stationing troops along these borders to protect the vital inner worlds around New Avalon takes up more than two-thirds of the AFFS regiments, leaving precious few regiments for use in an offensive.

The immensity of the Federated Suns creates another problem. The task of coordinating military actions between forces separated by light years is difficult and has caused the breakdown of many a fine AFFS offensive. Unless an action is meticulously planned and everyone involved has been fully briefed, the only offensive actions the AFFS is capable of at any time are regiment-size raids.

There are two other problems that might hamper future military efforts made by the AFFS. One is the realm's tenuous economic situation, with the Federated Suns attempting to balance the needs of the AFFS with the needs of its citizenry, who, unlike the common folk in the neighboring realms, demand more than just the bare essentials to live. This balancing act, though successful so far, satisfies neither the AFFS nor the average citizen. Any sudden shift in the balance could mean that the slighted side would become angry and potentially dangerous. This is why the alliance with the product-rich Lyran Commonwealth has been so eagerly received.

Another problem facing the AFFS is the inter-service, inter-regimental, and inter-regional rivalry that threatens the stability of the realm itself.

Regional rivalries may be inevitable, for none of the other Houses seems immune to territorial disputes among its own people. When a military allows regiments to be formed almost solely of men and women native to the a region, it is asking for trouble because those troops become instantly involved in the politics of the region. The Syrtis Fusiliers are the most obvious example. Because of their connections to New Syrtis and the Capellan March, the unit has become a staunch backer of Duke Michael Hasek-Davion, chief rival of Prince Hanse Davion. Encouraging such regional affiliations makes the possibility of another Federated Suns civil war very real.

Inter-service rivalry is one of the more hidden problems of the AFFS. Most of the Davion family have been MechWarriors and thus have shown such a preference for this branch of service that AeroSpace Pilots, the other elite service in the AFFS, have felt spurned and ignored. Moreover, the rest of the service branches believe that they are the ones left out, with MechWarriors and pilots getting all the adulation. Though this kind of inter-service rivalry often resolves itself in the traditional arena of local serviceman bars, other occasions of the animosity between services have resulted in more damaging effects.

To combat these rivialries, the AFFS Marshals have actually gone out of their way to promote inter-regimental rivalry. The rationale is that if soldiers think of themselves as belonging to a proud RCT or regiment, then inter-service problems might be lessened the instant another regiment comes along and proclaims itself the best in the AFFS. Perhaps this idea is not so farfetched, for inter-service problems have diminished since the 2850s, when they were at their worst.

Despite these difficulties, the future of the AFFS is bright, and morale is high. Most of the optimism permeating the regiments of the Federated Suns is due to the alliance with the Lyran Commonwealth. The prospect of such a strong and rich ally willing to coordinate efforts against the hated Combine gives even the most pessimistic Davion trooper something to look forward to.

DAVION
LAUBENSTEIN • 88

EDUCATION, CULTURE, AND ARTS

EDUCATION

It has always amazed me how the government will instantly respond to a world's call for arms, yet become suddenly deaf and dumb when that same world makes a plea for teachers. If nothing else, it gives us a good idea of where Prince Davion's priorities lie.

—From "A Criticism of the Davion Government's Education Policy", by Baroness Robin DeCaster, Report to the High Council, 3020

Hanse Davion is an excellent politician who knows how to handle his most vocal critics. Take the case of the Baroness DeCaster, who protested so loudly about how poor was the education system. What did Prince Davion do? He made her the Minister of Education! Now she is the one responsible for the government's education policy and she's the one forced to defend herself against the critics. It's a beautiful example of how to handle your opponents.

—From *The Political Styles of the Davions,* by Thomas Hufill, Galax Political Press, Galax, 3023

VAGABOND SCHOOLS

Providing every citizen of the Federated Suns with even the most basic education is currently impossible. The hundreds of planets and colonies and the vast distances between them would require far more resources than the Davion government is able, or willing, to field. If the Federated Suns were a dictatorship, an uneducated citizenry would be a definite advantage. The Davion government, however, has always operated on the principle that citizens have the right to improve their minds as much as they wish, which makes the current lack of an effective education system a dangerous political embarrassment.

The person most responsible for education policies in the Federated Suns is the Minister of Education. It is she who must find some way to fill the countless demands for teachers and schools from the meager resources at her disposal. When the budget is insufficient to these demands—which is almost always—the Minister must devise ways to spread the limited number of teachers throughout the Federated Suns.

The creation of the Vagabond Schools has been one of the Ministry of Education's most effective policies. The Ministry purchased freighter JumpShips that could no longer perform their hauling duties, and had them repaired and refitted for a new life as floating schools. The Vagabond Schools travel to areas where educational facilities are poor, which means they spend most of their time in the Outback districts. The region's 12- and 13-year-old youths are transported to the JumpShip, where they receive nine months of intensive education before returning to their homeworlds.

There are currently ten Vagabond Schools in the Federated Suns. Six of these—the *Cambridge*, *Washington, Oxford, Sorbonne, Notre Dame,* and *Trondheim*—are located in the most distant reaches of the Crucis March. Two, the *Dakar* and *Krakos,* are in the Capellan March, while the *Strasbourg* is in the Draconis March. The *Salzburg* is currently undergoing repairs and is expected to return to duty within the year. All ten ships are well-staffed with dedicated teachers and experienced crews. The Vagabond Schools are a proven success, for research indicates that most graduates continue their education or return to their homeworlds to become useful members of the community.

The main problem with the Vagabond Schools is that the ships are so old that breakdowns are frequent, making them somewhat dangerous. The last incident occurred on the *Sorbonne* last year when an inter-compartment seal broke, compromising the ship's atmosphere in one section. Ten people died, including two students and a teacher. The worst accident involving the Vagabond Schools occurred twelve years ago, when the *Kennedy* vanished with over 500 students, teachers, and crew.

Though the Vagabond Schools have produced impressive results, their efforts have not been enough to raise the literacy levels on the worlds nearest the Periphery. There still exist too many Davion planets where illiteracy is the norm and the intelligent are starved into stupidity because of a lack of teachers. In several recent decisions, Hanse Davion has pledged that the Ministry of Education will have a larger share if the economy takes the upward turn it promises. Until a stronger Ministry of Education becomes one of the government's priorities, however, Prince Davion will have a long wait before his dream of an educated citizenry is fulfilled.

UNIVERSITIES

There are currently about 50 institutions of higher learning scattered throughout the Federated Suns. Most are clustered around New Avalon and the Golden Five worlds, with only a handful near the Periphery border.

Several times in recent history, the Davions have attempted to build more educational facilities in the outworld border area, yet some unexpected obstacle always seemed to halt construction before it was even half-done. Even if the planetary governments had been able to complete these construction projects, they would have lacked the teachers to staff them. As a result, there are several outlying worlds whose half-built schools and universities sit like mute indicators of the political as well as interstellar distance separating them from New Avalon.

Prince Hanse Davion is also trying to involve major industries in his plan to provide educational opportunities for citizens of the farflung outworlds. In many cases, he has made personal requests to major business leaders, especially the heads of trading companies, to donate either money or products to the effort. In return, they can hope for the long-term economic stimulus of better-educated populations in their spheres of influence.

Some of the better-known universities in the Federated Suns today are described below.

University of Saso, New Syrtis

The University of Saso is a complex of four colleges just outside the second-largest city on New Syrtis. Three of these are liberal arts schools. The fourth, Hasek College, specializes in the hard sciences, especially Kearny-Fuchida interdimensional mathematics. Duke Michael Hasek-Davion has been pouring money into Hasek College in an effort to rival his brother-in-law's extremely successful New Avalon Institute of Science (NAIS). The Duke's efforts seem to have yielded some results. Two of the college's most prominent professors, Mitchell Colophon and Jasmine Rhinesla, have recently published a paper that explains, in theory, the workings of a key aspect of the hyperspace drive. If their work can be translated into practical engineering, it may be possible to bring back to life many JumpShips long thought to be beyond repair.

Dean Tyler Escribe, Count of Saso, is the current president of the University of Saso.

Tiskilwa University

Tiskilwa University is the cornerstone of Federated Suns' legal and financial education. Indeed, one popular saying claims that the only chance for a good trial is one where the judge, the lawyers, and the accused are all alumni of Tiskilwa. The university was named for Aaron Tiskilwa, first Chief Magistrate of the Bailey Street Court.

Tiskilwa University was erected just prior to the First Succession War. Because of New Syrtis's location so deep into the realm's interior, the university has escaped physical damage from the war. Indeed, its soaring marble architecture is widely admired as a rare example of Star League architecture.

Dean Deborah Henders is the current president of Tiskilwa University.

The Colleges of Savonburg

The Colleges of Savonburg are among the few fully staffed institutes of higher learning in any area of Davion space bordering the Periphery. The ten separate colleges are scattered throughout Savonburg's one habitable continent and offer a wide range of specialties. The College of Medicine at New Cristos is especially important, for most of its graduates go on to serve in the military or in outworld hospitals.

In a recent effort to upgrade the colleges, some funding became available to build a large science facility at Carlsruth, the planet's largest city. The project fell through, however, for lack of skilled construction workers to handle the special needs of the facility's twelve-kilometer subatomic acceleration ring. This proved disastrous for Dean Tilth Refol, who lost his post as planetary Minister of Education when his inability to raise the additional support turned the project into a fiasco.

Gogh-Bukowski University of New Avalon

What the NAIS is to the sciences, the Gogh-Bukowski University is to the arts. It was named after two of the the 20th century's greatest artists, Vincent Van Gogh and Charles Bukowski, whose works are admired to this day. The university is composed of three separate colleges: the School of the Written Arts, the School of the Visual Arts, and the School of the Musical Arts.

The New Avalon Theater Group's first performance of Shakespeare's *Hamlet* on the Outback world of Benedict came to a screeching halt early in the play when the audience fled the auditorium in terror. As the ghost of Hamlet's father appeared on stage in the first act, many in the audience of mostly poorly educated farmers gasped in astonishment and seemed unsure whether to hide under the seats or to run. They quickly came to a decision when the ghost drifted through a wall and spoke with a voice from the grave. In one huge wave of fear, the audience threw itself toward the exits, leaving behind a stageful of astonished actors.

I found out later that the audience had never even heard of holographic projections, let alone seen one.

—From *My Life as an Interstellar Roadie*, by Elton "Effect" Mosh, Layover Entertainment Books, 3001

Gogh-Bukowski University is a seething cauldron of political passions, where the students often mount protests for or against the latest good cause. Some of the demonstrations have been aimed at the NAIS, the university's rival. Though most past demonstrations have been peaceful and often brought constructive reforms, there has lately been an upsurge of violence among the protestors. This is because two underground movements, one for the alliance with the Lyran Commonwealth and one against it, have taken to using the university as their battlefield. If the violence rises above the occasional mugging and bombing, it is likely that Dean Uvonda Trimal, Countess of Melibourne, will have to ask assistance from the Melibourne city militia.

New Avalon Institute of Science

The New Avalon Institute of Science (NAIS) is the biggest and best university of the sciences in the entire Inner Sphere. Formed in 3015, the Institute has already made its presence felt through several key scientific rediscoveries. A number of these successes have repaid the sizable investment that creation of the NAIS demanded from the people of the Federated Suns. In addition to the continuing efforts to discover and develop new technology, our latest intelligence suggests that professors from the NAIS helped to unmask the imposter that House Liao had planted to replace Hanse Davion in a recent elaborate plot.

The NAIS is located about 30 kilometers outside the capital of Avalon City. The site of the university used to be part of the Davions' personal hunting preserve. Where formerly there stood thick forests and open meadows, there now stand 200 or so university buildings, with BattleMechs occasionally stomping through the open fields and DropShips coming and going from the Institute's private spaceport.

The NAIS is a large and attractively designed university, with most of the colleges arranged around the huge Lucien Memorial. To the east of the memorial are the spaceport and combat simulators for students in the military sciences. To the north is the large Administration Building. To the west is the large library and computer sciences complex, and to the south are the many dormitories, sports arenas, eating facilities, and tree-lined parks for the students.

The NAIS is composed of seven separate colleges. The College of Biology/Medicine is headed by Doctor Jennifer Shali, a Colonel in the Medical Corps and a noted researcher of microsurgery techniques. Graduates of this college serve the Federated Suns as doctors in the Medical Corps or as scientists in the Bio-Weapons Division. The College's reputation for teaching advanced medical technology puts its graduates in great demand. Most members of the upper classes can boast NAIS alumni as their personal physicians.

The College of Engineering is headed by Professor Sabina Carlton, a Colonel in the AFFS Engineering and Technical Corps. Professor Sabina, a famed technician during her younger days, leads her staff in the effort to rediscover lost bits of engineering science. Most of her research focuses on military engineering, but several other projects of equal importance are also in place. Indeed, one of the main goals of NAIS engineering research is to recover information that will allow House Davion to begin manufacturing water purification facilities at a faster rate. The teaching staff of the College of Engineering is often bolstered by members of the enigmatic Team Banzai, a band of fighting cryptic-geniuses. When not in combat, members of the group teach several more advanced courses and serve on design teams working on all manner of devices from BattleMechs to children's toys.

The College of Mining/Metallurgy is headed by Professor Rajiv Malajah. The purpose of this college is to rediscover some of the lost technology needed to make the high-strength, low-weight metals essential for BattleMech construction. Though mankind still has the knowledge to build BattleMech frames, certain metal alloys used in the actuators of 'Mechs have been lost, forcing Techs to use copies made from inferior metals. If the formula to those lost alloys were ever rediscovered, it could have a major effect on both the military and the economy of the Suns.

The College of Chemistry is in turmoil ever since Dean Deflin Saunders had to step down because of the scandal over the sale of his college's scientific secrets to private companies. There has been no official replacement for Professor Saunders, though it is assumed that the interim leader of the college, Professor Sandina Vasly, will assume control. One of the school's key projects involves the development of a new water-purification process. This process is said to use something called filtrum, a sand-like substance that strains impurities from the water. The chemical composition of filtrum is what the NAIS chemists are trying to refine. If the formula could be perfected, it appears that the filtrum system would be easy and economical to use. It is assumed that the College of Engineering is also heavily involved with this project.

The College of Military Science comes a close second to the elite Albion Military Academy. It lacks nothing in the way of equipment or facilities, with plenty of training 'Mechs, fighters, and simulators for any conceivable 'Mech design known today. The College also boasts three Gauntlets, special 'Mech training grounds. Commanding Officer of the College of Military Sciences is Major-General Jeremiah Davion.

The College of Finance is the section of the NAIS where economists and business mathematicians are trained. The conditions of admittance require most graduates to serve ten years in the government's Ministry of Ways and Means. Those that are gifted and work hard often rise in the Ministry and then step easily into politics at the end of their decade of service. Professor Samuel Muyskens is dean of this college.

The College of Political Administration trains two types of students. The sons and daughters of noblemen come here for training in effective government. The student body also includes those who will eventually join the vast bureaucracy of the Federated Suns. Most Davions attend and eventually graduate from this school as well as from the College of Military Science.

THE ARTS

The way a society treats art and artists is an important measure of how it regards its citizens and their personal freedoms. True to its professed belief in individuality, the Federated Suns treats its artists quite well. On the more advanced worlds, the arts are subsidized by the government, private companies, or wealthy individuals. Citizens on these more advanced worlds tend to take an active interest in the arts, which makes them knowledgeable patrons. Colleges like the Breckly School of Fine Arts on Steeles and the Vandros University on Ipava are renowned for the skill of their graduates.

Artists suffer on the poorer worlds, but the arts somehow manage to survive. Musicians, particularly those who play native instruments, often gain enough fame to leave behind the poverty of their native worlds. Poor worlds like Fincastle and Cambiano are famed for other arts, such as pottery and textiles.

Most artists of any fame will eventually make their way to New Avalon. In Avalon City, the bustling metropolis clustered about the Royal Palace and Court, there are four major centers of artistic expression. In the south part of the city is Mission Way, a long avenue lined with many famous theaters and cinema houses. Sculpture Park is a large, tree-lined grove dotted with numerous sculptures. Facing the park are a number of large museums and galleries where the work of the best of the Federated Suns visual artists is exhibited. The Ink District is a small section in the north end of the city where many of the realm's writers and poets live. The fourth center of art is the Red Block, near the Ink District. It is here that the realm's more scandalous avant-garde performers tend to live and show their work.

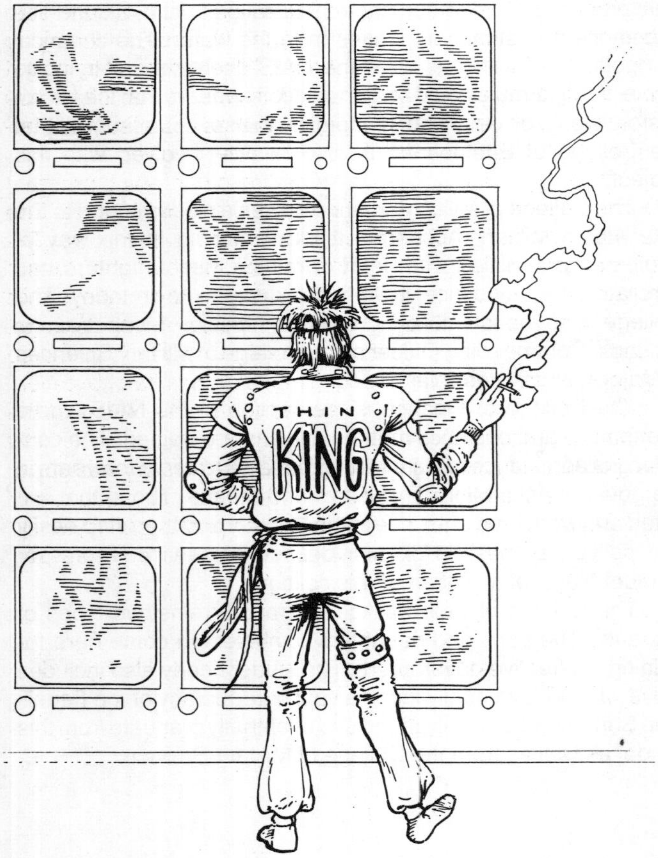

SON OF THE HOUSE OF ATREUS

The argument over who was the greatest composer will never be settled, nor should it be. While names like Beethoven, Bach, Renstedt, Ramananda, Mozart, and Maclean cannot be ignored, my personal choice for the title of greatest composer is the lesser-known Domenicos Atreides (2619-2701), whose very name is musical.

Atreides' *magnum opus* was an extraordinary score for the *Oresteia* of Aeschylus, a cycle of ancient Greek tragedies whose original music was lost centuries before Atreides' time. Atreides was saddened that only the script of this great theatrical piece was all that survived for posterity. Although written in great poetry, the piece was intended, like all Greek drama, to be a total theatrical event that combined music, dance, singing, chanting, masks, and other theatrical effects. Atreides felt that simply reading the poetry in a book was a pale substitute for the total experience Aeschylus had intended. He therefore set about bringing what musical splendor he could to the tragedy, succeeding beyond anyone's dreams by composing music that ranged from primitive chants and swelling, graceful entrance and exit motifs to elegiac delicacy. I submit that no other known musical work ever conceived approaches the emotional range of the *Oresteia*.

Having created this magnificent work, he went on to score portions of other plays by other Greek tragedians, including scenes from *Oedipus at Colonus* and *Philoctetes* by Sophocles, and several works of Euripedes, including *Rhesus, Iphigenia at Aulis,* and *The Cyclops.* From this, we infer that Atreides had intended to create a sweeping, comprehensive lifework that was far from complete at the time of his death.

—From "The Nocturnal Composers and the Gypsies' Revenge," in *Essays Short and Long,* by Carmen Okut, Fine Arts Press, New Avalon, 2982

DAILY LIFE AND CULTURE

The sheer immensity of the Federated Suns and the uneven distribution of technology are the main factors affecting the quality of life for the ordinary citizen of House Davion.

The average citizen of one of the Golden Five or any other advanced-technology world will surely enjoy a life of relative comfort and leisure. He is generally well-educated and politically conservative, though personally tolerant of differing opinions. The typical family has one child, who is quite likely to scandalize his parents during adolescence with displays of rebellious behavior and unconventional friends. Most parents are educated enough to understand that this is no more than a phase in the child's progress toward maturity. On such worlds, dependence on technology is total, to the point where people rely on machines to keep them physically fit. Because the government attempts to keep technological incongruities and the fragility of current technology hidden, this gives the average man or woman the illusion that he or she lives on a superior world in an age of technological ease.

For people living on one of the Skid Row worlds in the Outback of the realm, the situation is drastically different. On such planets, life is generally tough, mean, and often damnably short. Education usually means knowing how to maintain a 500-year-old tractor with an internal combustion engine and keeping it fueled with the methane the owner has obtained from animal dung. Education on these worlds may also mean knowing how to field-strip a slug-throwing rifle, and where to find the nearest bunker in case of an enemy raid. Such an individual's idea of fine art might be the sight of a Kurita 'Mech exploding as it falls to the ground on his neighbor's land instead of his. He might also think of art as Tholimas Jones and his banjo, a hot dance in a dark barn, or a rough kiss and grope with a woman from the next town. Muscles are not a status symbol or expression of physical beauty, as on wealthier worlds, but a necessity for sheer survival. On such less-developed worlds, a man's political views tend to be limited to the idea of protecting oneself and trusting no one. Families average ten children, who will work in the fields or the mines alongside their parents.

PUBLIC MEDIA

The government of the Federated Suns does not believe in complete censorship of the public media. Though it keeps sensitive information secret, the Davion leaders tend to maintain a good relationship with the many public media services in the realm. On the other hand, the fact that the central government is on good terms with the media does not mean that all the planetary governments are. Indeed, the leaders of many worlds often have their own reasons for restricting what is said and not said about their planet. This can create tension between the news services and planetary governments, and sometimes with the Davion central government, which does not approve of overt repression.

The one thing the Davion government will not tolerate is the publication of secret military information. On those occasions when such information has leaked, the government has prevented the broadcast of sensitive military secrets by ransacking a reporter's office or home to steal back the information. Though the news media hate this censorship, reporters and broadcasters have decided that it is better to survive in their profession than to flaunt the government's need to protect certain information.

Federated Independent Broadcasters

The FIB is a loose affiliation of some 50 planetary broadcast and entertainment companies spread out over 20 worlds in the Crucis March. Members of the FIB pool their news-gathering resources and put on their own news programs in direct competition with the more widespread news shows of the Federated News Services. The FIB has also recently begun to produce entertainment programs that have become quite popular throughout the Federated Suns. The most successful of these is a romanticized portrayal of a MechWarrior family, which is entitled *The Fighting Falcons*. The series has just been sold to the Donegal Broadcasting Company for broadcast in the Lyran Commonwealth.

Three Houses Entertainment Agency

Three Houses is a large entertainment agency specializing in music groups and live acts. Headquartered on the planet Addicks, the firm is one of the few Davion agencies that regularly sends acts into the Draconis Combine and the Capellan Confederation.

Because live entertainment is not on the list of items forbidden by the Economic War Powers Act, the Three Houses entertainment troupes move back and forth into enemy realms. Though all our evidence indicates that Three Houses is actually gathering information for both the Department of Military Intelligence and the M17-MIIO, both agencies keep the firm under heavy surveillance to prevent becoming the victims of a double-cross.

At the moment, Three Houses has two equally popular groups under contract. One is the Kyotan Native Instrument Band and the other is the Jameson Troubadours, both groups whose music appeals to the people of all three realms.

New Avalon Publishers

New Avalon Publishers is the largest distributor of books, magazines, and public computer databases in the Federated Suns. They have over 50 divisions, each with a specialty ranging from economics and children's books to religious matters and manuals illustrating the art of love. New Avalon Publishers has offices and presses on ten planets, as well as countless distribution points throughout the Suns.

Though the political views of the firm's publishing program are basically pro-Davion, there have been several clashes. The most recent incident occurred in 3002, with the publication of *The Ethics of the Davions*, a book by Countess Ludmilla Fresof. Fresof expressed severe criticism of the Davions' honorable public image, compared to their alleged underhanded private dealings. Prince Ian and everyone else in the Davion family were so outraged by the book that they attempted to halt its publication. The Davions even secretly threatened to raid the offices of the company to destroy the book's manuscript. The publishers held their ground by publicizing the struggle. Once the story was out in the open, the Davions were forced to withdraw their complaint or else generate even more public interest and publicity for the book.

SOCIAL PROBLEMS

The major problem facing the many societies of the Federated Suns, and the cause of all lesser problems, is the enormous span of the realm and the vast distances separating its worlds. The problem is so great that the government is unable, despite its claims to the contrary, to fully provide the services that its member planets require. Military protection is probably the only service that people on distant worlds can count on receiving relatively quickly from the central government.

The list of related problems is a long one. For the Outback worlds near the Periphery, poverty is severe. Especially on those planets not serviced by trading companies, the hardships can be almost unbearable. Prejudice, whether it be of religion, race, or class, is another problem haunting the poorer, distant worlds.

Though members of the Federated Suns, many Outback planets must also cope with a crushing sense of isolation. To those citizens, New Avalon is a place more mythical than real. When one of these isolated worlds is in the midst of rioting or threatened by a tyrant's rule, it is ludicrous for the people to imagine that a mere speck in the nighttime sky can send help across hundreds of light years in time to save the day. Though the Davion government is concerned for the welfare of its more remote worlds, many people suffer under a burden of violence and ignorance because the government cannot react quickly enough in a crisis.

With the limited resources currently available to the Federated Suns, the best anyone hopes for at the moment is a gradual improvement in the general quality of life and prevention of major social problems before they arise. The Davion government's recent emphasis on rebuilding interest in the Outback worlds has definitely attracted more private business money and other civilizing influences to those often violent worlds. Places where the population is experiencing a palpable rise in the standard of living are less likely to fall prey to violent extremes.

COMSTAR AND THE FEDERATED SUNS

It sometimes seems that one is in two different realms when traveling through the Federated Suns. One realm is ultra-sophisticated, with people skeptical and sometimes aggressively hostile to ComStar representatives, whom they view as outright charlatans. The other realm is the primitive Outback, where ComStar has an almost hypnotic sway over citizens who are easily dazzled and easily angered. Before a ComStar representative can prepare for life in the Federated Suns, he must first learn on which Federated Sun he is going to be living.

—Justin Yovas, Precentor of Galax, "Training Guidelines in the Modern Era," ComStar Publications, Terra, 3018

The freedoms enjoyed by member worlds of the Federated Suns have created many distinct societies, each with its own character and style of action. There is no such thing as a typical Federated Suns planet, as there are typical Lyran or Combine worlds. It is for this reason that officials of ComStar preparing for service in the Federated Suns must undergo additional training and study that can range from two weeks to two years, depending on their destination.

Among the Outback worlds, our hold over the local populations is strong. These worlds, with their general poverty and low technology level, are easily impressed by our abilities. In such places, the greatest danger we face is the stupidity of committing a disastrous cultural blunder. Knowing who holds the power, what are the common courtesies, and what are the local taboos is vitally important for the successful completion of ComStar duties on these planets.

On the more advanced worlds, however, we face stiffer challenges. The citizens of these worlds are more skeptical of any technological special effects we might be able to produce. Indeed, citizens on worlds like Galax and New Avalon are almost as difficult to win over as are the people in the Lyran Commonwealth. Their higher level of education, coupled with the unusual religious phenomenon called the Unfinished Book, have recently combined to produce some frightening anti-ComStar attitudes. Fortunately for us, prejudice against our Blessed Order has been confined mostly to bitter articles and speeches. What is truly worrisome is that the Davions seem to have no interest in repressing such scandalous assaults on our Order. A mere 70 years ago, the governments of all five Successor States would have jumped to punish these same attacks on our holy cause.

The New Avalon Institute of Science, which is staffed by some of the best minds in the Inner Sphere, is the major threat to our position as a communications conglomerate and unofficial political power. According to information gathered by our agents, there are three heavily funded research projects currently under way in the College of Engineering that represent a direct threat to our communication technology.

One project, a study into Kearny-Fuchida hyperpulse generation, is so close to a major breakthrough that our ROM agents have suggested a termination order for the project's director, Professor Greth Tefler. Even if Professor Tefler were eliminated, however, the NAIS would still have many ways to challenge our powers. Indeed, this threat can only grow larger as the NAIS begins to openly exchange its knowledge with advanced universities in the Lyran Commonwealth.

Politically, our problems are fewer, though troubling nevertheless. The Davion family has always been an inquisitive group, eager to see how we operate and how we react under pressure. Prince Hanse Davion is no exception. It is probably the reason that he has allowed some of the more radical members of the Inter-Faith Council, publishers of the Unfinished Book, to attack us in the press. We must react to the lies and insults in only the most controlled manner, however, or else risk losing face with Hanse Davion.

Among the common people, our influence is still fairly strong, depending on what part of the Federated Suns is under discussion. In some areas, the local population remains spellbound by our mystique and our technology, while in others, we are seen as little more than glorified mail-carriers. This makes it difficult for us to create a realm-wide campaign to sway public opinion for uniform results. In many ways, the common people in the Federated Suns show the worst characteristics of both advanced- and low-technology societies, which presents our Order with one of its most difficult current challenges.

SOCIOECONOMICS

INTRODUCTION

Though the Federated Suns is not as well-developed as some of the other Successor States, its sheer size makes it a potential economic powerhouse. Recent events have shown that Prince Hanse Davion is eager to exert the economic muscle of the Federated Suns, despite political and emotional hindrances. If his wishes are fulfilled, it is possible that the growth of the Federated Suns economy will usher in a new socioeconomic order in the Inner Sphere. Whether that bodes good or ill for ComStar remains to be seen.

RESOURCES AND DISTANCES

The Federated Suns is the largest and most planet-rich of the five Successor States, with 500 inhabited star systems, many with two or three worlds, and dozens of lesser star systems. All that prevents House Davion from exploiting its vast natural resources is the immensity of the realm, which makes it difficult, sometimes impossible, to quickly transport resources to distant worlds in need of them. While other Successor States also suffer from this problem, it is more acute in the Federated Suns.

Though the Davion realm has the largest commercial transport fleet in the Inner Sphere, even its many trading companies and independent haulers are barely enough to keep the realm fed. The pressure to deliver essentials to hungry and thirsty worlds makes it desirable for the shipping companies to establish trade routes outside the realm, yet too much contact with other Successor States is discouraged.

The ruling Davions have long believed that an intelligent citizenry is a valuable resource and that an educated citizen is the one most likely to make a contribution to the realm. Here again, though, the immensity of the realm hinders growth. Though there are many universities, they are not enough and are too widely scattered to provide everyone a quality education. Many intelligent young people are forced to give up their dreams of a better life because they cannot afford to travel to a college. Even those lucky enough to graduate from a university are often condemned to poverty when the job they trained for turns out to be light years away. Prince Davion also knows that there are few things more dangerous to a government than the educated unemployed.

GOVERNMENTAL ATTITUDES AND REGULATIONS

Prince Davion's speech declaring the citizenry more important than profits is a clear and concise statement of the government's monetary policy. Indeed, the Davions have always held staunchly to the belief that a citizen should never be at the mercy of unscrupulous business concerns, which is why the Federated Suns is the only realm with large, vocal labor unions.

Nevertheless, leaders do not always do as they say. Businesses often operate with impunity in the Davion realm, and examples abound of corporations choosing profits over people without fear of punishment. The public will stand only so much, however, before it demands governmental protection. At such times, the Davion government asserts its own control over the economy. The last crackdown occurred in reaction to the Jesup Incident, and resulted in major laws restricting a shipping company's freedom to change their trade routes.

The Chancellery of the Exchequer is charged with enforcing the economic laws. Heading the Chancellery is the Minister of Ways and Means, who is a member of the Prince's Privy Council and a member of the High Council. He is expected to take part in all debates over the yearly budget. The current Minister of Ways and Means is Duchess Joan Davion, a cousin of Prince Hanse. Under her capable administration, the Chancellery of the Exchequer has grown in power, much to the consternation of some of the more ambitious corporations.

The Ministers of Regional Finances administer the three Marches of the Federated Suns. They are, in a sense, the economic rulers of the Marches, holding life-and-death control over businesses in the region. Duchess Michella Vandrag, Duke Kisink Falkers, and Duke Talbot Evans are currently the Ministers of Regional Finance for the Draconis March, Capellan March, and Crucis March, respectively. In reaction to the recent emphasis placed on exploitation of the Outback worlds, a Sub-Minister of Regional Finance has been added to help the Minister of the Crucis March cope with increased economic activity. Count Loren Valetta is the Sub-Minister of Expansion.

The Sheriffs of Economic Policy answer only to the Minister of Ways and Means. Often called the Business Police, these five officials and their staff travel through the realm most of the year, with the sole duty of uncovering violations of the economic laws. The Sheriffs of Economic Policy are empowered to override planetary economic laws and to arrest violators in the name of the Federated Suns. Though this power upsets the leaders of some planetary governments, the Sheriffs are usually chosen for their tact as well as their determination to enforce the economic laws.

POLITICAL AIMS OF THE ECONOMY

Unlike the other Successor States, the Federated Suns attempts to balance the personal freedoms of individuals against the needs of corporations. Though this often slows economic growth, most citizens feel that the preservation of their freedom is worth the sacrifice.

In the Federated Suns, the economy is considered to be a weapon against external enemies. Because of generations of deep hatred against the Draconis Combine and the Capellan Confederation, most citizens of House Davion oppose trade with those realms. This attitude was given legal authority in 2802 by the Economic War Powers Act, a group of restrictive laws relating to inter-realm trade. The Act included a list of items strictly forbidden for trade to House Kurita and House Liao, foremost of which was any item of League-level technology. Everything else that might possibly be traded has been examined for military potential. If an item has any military value at all, it is blacklisted and cannot be exported. Violators of the Economic War Powers Act are tried for treason, and suffer a traitor's punishment if convicted.

The economy can also be an aid to potential allies. This use of the economy has suddenly come to the fore with the alliance between the Lyran Commonwealth and the Federated Suns. Though only four years have passed since the treaty was signed, the level of trade between the two realms has already increased dramatically. The Federated Suns recently proposed the establishment of an Inter-Realm Ministry of Trade to coordinate the anticipated flood of trade between the two realms. The Inter-Realm Ministry of Trade would be a part of both House governments, and may be the first step toward eventual union of the two.

THE JESUP INCIDENT

Jesup is a small, dry world at the outer end of the Crucis March. Despite its location at the center of a large, planetless void, it once had a thriving tin and silver industry, which made it profitable for the Black Ball Trading Company to make the difficult journey there. In 3011, a huge vein of silver was discovered on Jesup's nearest neighbor, the planet Perdido. The size of the new strike and the comparative ease of traveling to the planet caused the mining industry of Jesup to collapse. Acting purely from business considerations, the Black Ball Trading Company cut its losses and stopped service to Jesup. Mining was the planet's only industry and the freighters of the Black Ball Line were its only source of food. Thousands died on Jesup before anyone noticed.

The Jesup Incident, as it came to be known, sparked a storm of protest. Most of the public's wrath was aimed at the Black Ball Trading Company for pulling out of Jesup despite being the planet's lifeline. Prince Ian Davion issued the Jesup Trade Regulations in 3013, severely restricting all trade and freighter services from making major changes in their routes. After the Black Ball Trading Company was forced into bankruptcy, it was bought out by White Swan Trans-Stellar, Inc. Its current operating name is Jesup Traders. Prince Ian's Trade Regulations sparked a wave of tough anti-business legislation that has carried on into Hanse Davion's reign.

—From Re*cent Business Policies of the Federal Government*, by Count R. K. Ismon, New Avalon Economic Press, 3022

CONSUMER GOODS

The level of technology varies widely in the Federated Suns. Among the worlds in and around the Golden Five—the planets New Avalon, Galax, El Dorado, Belladonna, and Torrence—it often seems that the glory days of the Star League never left. This is misleading. The Star League did fall and these worlds, though still impressive with their high-tech items and pristine cities, are only just rediscovering and learning to use technology of the 25th century. Nevertheless, the Golden Five are worlds where even the lowliest citizen can expect sufficient food and a warm, dry place to sleep. Items like holosystems and remote communication devices are considered essential for the average middle-class family.

Planets unfortunate enough to be among the "Skid Row Worlds" are usually located along the borders and are under almost constant threat of attack. Most worlds on the Skid Row have had their industries and agriculture severely disrupted. Even though modern combat avoids damaging industry and agriculture, blockades often have the same effect. The fact that a factory still stands is small consolation to laid-off workers who are going hungry.

Other Skid Row worlds in the Federated Suns are those that are resource-poor and/or located outside the major trade routes. On these worlds, a home is a luxury. To someone from a Skid Row world, "personal transportation" means a horse or mule, and even that is considered something only for the rich.

THE RICHEST MAN ON BLUFORD

William Keller Reese, first Count of Bluford (2508-2594), is unique in the way he made his name and fortune. Reese, whose grandparents and great-grandparents all lived into their nineties, literally gambled his life to make his fortune—he entered tontines with smokers.

This is the way it worked. Several young people would pledge a certain sum of money per year as long as they lived, to be paid into a trust in all of their names. When only one of the individuals was left alive, he or she would receive the entire sum. Named after Lorenzo Tonti, a financier of 17th-century Terra, the tontine was originally an early form of insurance plan, intended to guarantee funds for a person's retirement years. By Reese's time, it was little more than a wager that one person would outlive another.

The soil of Bluford's smallest, easternmost continent was, and is, ideally suited for growing a strain of tobacco developed in the 22nd century that was both deadlier and more addictive than any previous strain. Reese made sure that all of the other entrants in his many tontines were smokers. He was willing to take the calculated risk that he might meet an early death himself, and his gamble paid off. Although he was short-lived by the standards of his family, dying at the tender age of 86, he had, by age 75, amassed a fortune of over 70 million D-Bills. By age 81, a title of nobility was conferred upon Bluford for his monetary and political support of Davion entry into the Star League.

Reese's success touched off a flurry of gambling activities on Bluford, with tontines becoming enormously popular among smokers and nonsmokers alike. For a 22-year period in the early 27th century, the planetary government of Bluford even entered every baby born on any given day into a tontine, which was referred to as a "Life Lottery." A backlash began in the late 27th century, when problems with the system had become evident, and a movement to ban tontines developed. Although never successful, the movement did spawn the opposition Bluford Gambling Association, whose members protested any infringement on their rights to squander their money and gamble with their lives.

Between 2620 and 2840, Bluford had the highest annual murder rate of any planet in the Inner Sphere.

—From *Encyclopedia Mordantica,* edited by Morris Nkawa, Double Decker Press, 2833

CORPORATE PROFILES

FEDERATED-BOEING INTERSTELLAR

New Avalon Stock & Commodities Symbol: FdBoIn
Home Office Location: Lagrange Point Delta, Galax
President/CEO: Aleksandr McCorkendale, Duke of Galax
Founding Date: 9 February 2784

Principal Divisions

Division Name: Port Simon Naval Shipyards
Division Head: Simone McCorkendale, Countess of Galax Malabar
Chief Products: *Overlord, Union, Leopard, Leopard-CV, Gazelle, CargoMaster, CargoKing*

Division Name: Boeing Aerospace Frame Ltd.
Division Head: Javinal Glemma, Baron of Auburn
Chief Product: DropShip Frames

Division Name: Boeing Instrumentation Ltd.
Division Head: Kilroth Simons
Chief Product: Instrumentation for DropShips and AeroSpace Fighters

Division Name: Port Simon Test Flight Facilities
Division Head: Katherine McCorkendale, Baroness of Lagrange Point Alpha
Chief Service: Fitting and Testing of DropShips

Division Name: Grumman-Boeing Engine Corporation
Division Head: Mitchell Defross
Chief Product: DropShip Engines and Maneuver Systems

Division Name: Federated-Boeing Civilian
Division Head: Petruc Misokov
Chief Product: Civilian Cargo DropShips

Division Name: Federated-Boeing DropShip Maintenance
Division Head: Ophelia Crodula
Chief Service: DropShip Maintenance and Repair

History:

The roots of Federated-Boeing Interstellar are ancient. ComStar researchers have been able to trace the company back through Star League, Terran Alliance, and Terran Hegemony days. In fact, our researchers have discovered mentions of a Boeing Aerospace manufacturing fixed-wing fighter craft on Terra as far back as the early 21st century. Whether it is the same company is uncertain, however. What is certain is that the Boeing company, through its many manifestations, has consistently supplied superior quality products.

As of 2781, Boeing Interstellar was a Star League company with two major shipyards orbiting Terra and New Earth. Following the tyranny of Stefan Amaris, the board of directors decided it was necessary to expand their facilities beyond the boundaries of the Terran member-state. The company solicited bids for a major industrial center from the other member state planets. The planet Galax, with its perfect conditions and resources, won easily.

The major economic boost that the vast complex of factories and orbiting shipyards known as the Galax Megaplex gave to the Federated Suns economy was a political coup for Prince Paul Davion, who was facing unrest from within and danger from without his realm.

During the five years it took to build the Galax Megaplex, the Star League Council was dissolved and General Kerensky left the Inner Sphere, along with most of the Regular Army. Though it often looked as though political unrest would halt construction of the Galax factories and shipyards, both the Boeing company and the Davion family, for obvious reasons, pressed to see it completed. The Galax Megaplex was completed in December 2786.

Within a year, the Federated Suns was locked in a desperate struggle with the Draconis Combine. Although the ships that the Boeing facility might produce were sorely needed on the front lines, the factory lines sat idle. The Chief Executive Officer of the company, Frewsol Grevers, was a staunch—some say stupid—company man. He refused the Davion government's pleas, entreaties, and demands to start production, waiting instead for orders from the Boeing home offices on Terra. The fact that the home offices, along with the rest of Boeing Interstellar, were smoldering ruins beneath the feet of competing BattleMechs was immaterial to Frewsol Grevers.

Furious at the delay, Prince Paul Davion sent lances of 'Mechs, companies of Royal Guards, and AeroSpace Fighters to the shipyards and factories of the Galax Megaplex. A young officer delivered a sarcastic message from Prince Davion to the furious Frewsol Grevers, which announced that the Federated Suns was seizing the Galax Megaplex along with Boeing Interstellar's assets. Frewsol Grevers was "invited" to step down from his job and enter the military to fight the Draconis Combine.

Soon thereafter, what remained of Boeing Interstellar was restaffed with businessmen loyal to the Davions and eager to begin operating the largest DropShip manufacturer in the Inner Sphere. The first CEO was the young and enthusiastic Uston McCorkendale. Federated-Boeing, as the new company was called, began official operation on 9 February 2788. The DropShips supplied by Federated-Boeing were vital to keeping the Davion defenders ahead of the enemy in the desperate years that followed, as the Draconis Combine thrust deep into the Federated Suns.

Private Frewsol Grevers proved himself a better soldier than businessman. He died with the rest of his platoon while defending a landing field on Galtor. Perhaps it was guilt for what he had done to Grevers that made Prince Paul hang a gold plaque on the Federated-Boeing headquarters in his honor.

Since the time of the First Succession War, Federated-Boeing has continued to produce DropShips. Except for a failed attempt by Kurita commandos in 2810, the facility has never been directly attacked. The firm has not escaped the overall decline in technology, however. Losses in knowledge and a chronic lack of raw materials have reduced the Galax Megaplex's output by 55 percent. It has also narrowed the number of DropShip types they can produce from 25 to just 10. Even at this reduced output, Federated-Boeing Interstellar is still the largest manufacturer of DropShips in the Inner Sphere.

MESSAGE TO GREVERS

It is with deep humility and thanks that I, Prince Paul Davion, do accept your generous gift. To hand over the entire Galax Megaplex and the remnants of the Boeing Interstellar company to us is a tremendous patriotic gesture by itself. But to go on and enlist in the army as you did! My heart swells with pride knowing the Federated Suns will be protected from the fearsome Draconis Combine by such as you. I hope to see a plaque inscribed with your likeness hung on a wall in the Megaplex so that future generations might know what a true patriot looks like.

—From *One Man Against the Wave of Destruction*, by Countess Rebecca Devon, New Avalon Press 2900

Business Profile:

Federated-Boeing is a classic example of the vertically organized corporation. Every division is concerned with the manufacture of one portion of a DropShip, and nothing else. A saying commonly heard on Galax claims that Federated-Boeing manufactures everything on its DropShips except the guns and the upholstery on the flight couches. This is not quite true, but the company does manufacture 72 percent of all components needed to build their ships. This allows the company to manufacture huge chunks of DropShips and place them in orbit around Galax, if necessary, until parts from other companies arrive.

All manufacturing is done at the huge shipyards of Port Simon. There are twelve separate manufacturing lines where DropShip components are slowly moved from workstation to workstation by huge robotic arms. Though the shipyards are growing old and repairs to the workstations are frequent, the McCorkendales have always been fanatic believers in maintenance. Few manufacturers in the Inner Sphere can claim to have better-kept equipment.

FEDERATED BOEING

Once built, the DropShips are taken to the Port Simon Test Flight Facilities. There, the ships are put through their paces by the company's expert DropShip crews, most of whom also teach at the nearby Armstrong Flight Academy. Test flights usually involve high-G Runs at maximum thrust from the flight range to a jump point, then no-atmosphere landings on Galax's moon, Malabar. The final test is a combat-style reentry down to the Federated-Boeing spaceport at Auburn City on Galax.

Federated-Boeing also trains maintenance personnel to service their DropShips throughout the Federated Suns. A military maintenance crew from the company is almost always attached to naval stations. The military engineers often look to the maintenance crew as sources of information, to the point of allowing members of Federated-Boeing Maintenance to make command decisions about DropShip repairs. Only the best engineers are allowed to service the company's pride and joy, the *Running Fox*, Prince Hanse Davion's personal *Overlord* Class DropShip.

Federated-Boeing has an impressive record of workmanship. The company has lost a total of only 13 DropShips from faulty construction since creation of the company during the First War. Seven were lost during testing around Galax, and the remaining six were lost in combat situations. In each instance, the company did not hesitate to accept responsibility or to seek out ways to prevent a recurrence. Despite the general slide in technology, there is little reason to doubt the firm's claim that their ships are the safest in the Inner Sphere.

Profile of CEO:

Duke Aleksandr McCorkendale is a tall, burly man of mixed heritage. His bulk belies the fact that he has spent almost three-quarters of his life in Zero-G. Now that he is approaching his 65th birthday, his body is no longer able to stand the strain of planetside life or space voyages with accelerations greater than a few tenths of a G. This has largely confined him to the Zero-G section of Federated-Boeing's orbiting space station, the *Floating Tower,* which orbits Galax's Lagrange Point Delta.

Duke McCorkendale is, on the surface, a genial man, prone to long stories about his experiences as a crewmember on the DropShips attached to the First Ceti Hussars. His early life was one of grand adventures interspersed with love affairs with some of the most beautiful women in high society. His marriage to Jessica Talos, Duchess of Covington, was a near-fairy tale romance. In public, Duke McCorkendale is still the carefree, spritely gentlemen despite age and declining health.

Underneath, however, is a man haunted by private demons. His wife died in a DropShip accident five years ago. The accident, officially blamed on a collision with some space debris, apparently created deep guilts and fears in Aleksandr. Since then, Duke McCorkendale has become obsessed with safety. His almost daily flood of memoranda and orders to the company about safety have almost brought operations to a grinding halt.

Duke McCorkendale has been advised to see a psychiatrist by his 34-year-old daughter, the Duchess Simone McCorkendale, who was subsequently barred from ever entering the Duke's presence. This suits Baroness Katherine McCorkendale, the Duke's 22-year-old niece, who has been fawning over her uncle for the past year. Baroness McCorkendale obviously wants to inherit the company, but no one is sure whether she is simply taking advantage of her uncle's dementia or actively feeding it.

Whatever Katherine's role, Duke McCorkendale is quickly losing control of his company. Already there are rumors that other firms are tendering offers to the various division heads, hoping to lure them and whole sections of Federated-Boeing away.

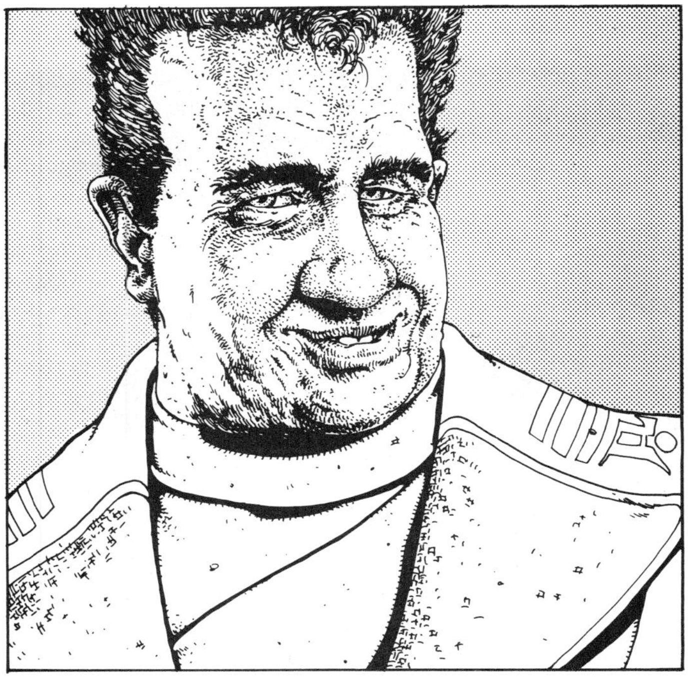

Relations with Federated Government:

The McCorkendales have always been staunch supporters of the Davions, if not of all their policies. Some of the McCorkendales are concerned about the alliance between the Federated Suns and the Lyran Commonwealth, for example. They worry that Federated-Boeing will not be able to participate fully in the anticipated economic boom because of Galax's distance from Commonwealth space. Negotiations are already underway for Federated-Boeing and one of several Commonwealth aerospace companies to build a new DropShip manufacturing plant somewhere in the Commonwealth. If this comes to pass, it will be a major boost both to Federated-Boeing and the Lyran Commonwealth Armed Forces.

Security and Military Forces:

Because so much of the Federated-Boeing manufacturing facilities are in space, the company uses armed DropShips and AeroSpace Fighters to handle most of its security. The security DropShips, usually *Union* Class vessels and *Leopard-CV*s, are manned by combat-experienced crews retired from the Davion navy. The same is true of the AeroSpace Fighter pilots, who usually fly *Corsair*s. It would require a major effort for anyone to fight their way past or to sneak through the Federated-Boeing security force guarding its orbiting factories.

Federated-Boeing has always relied on detachments of Galax Reservists for their first line of security on the planetary facilities. These units are usually armed with slug-throwing weapons and wear standard infantry uniforms modified only with the Federated-Boeing patch on their sleeves. The Galax Reservists have proven reliable over the centuries, with few incidents of disloyalty.

Federated-Boeing relies heavily on electronic means for internal security, including the use of automated security strong points that use lasers and/or lethal gas to discourage unauthorized personnel. Their computer system, also very advanced, is programmed to lock out and trace any unauthorized access. Also, its data storage is kept in a vault buried deep beneath the building in case of physical attack, making it both difficult and hazardous to attempt to interfere with the computer's memory.

Federated-Boeing's Security precautions have proven so effective over the years that there have been only five instances when the company has lost either information or material.

Notes:

Several rumors have been circulating about Federated-Boeing. One has it that the company is attempting to design a completely new DropShip class. Apparently codenamed "Trojan," the ship is said to have an extremely small sensor silhouette that will make it difficult for enemy defenders to detect. Our agents have yet to find any hard evidence to prove this rumor true or false.

Another rumor is that Katherine McCorkendale has been feeding her uncle hallucinogenic drugs to heighten his mental instability. As we do know for certain that Katherine has complete control over the Duke's diet, our agents are investigating further.

INTERCONNECTEDNESS UNLIMITED

New Avalon Stock & Commodities Symbol: IntUnl
Home Office Location: Granite Falls, Rosamond
President/CEO: Elric Von Steffelbus IV, Count of Islington
Founding Date: Unknown
Principal Divisions:
Division Name: Triple-F (Federated Fast Food) Restaurants
Division Head: Josephine Degrassus, Countess of Smithdon
Chief Service: Fast food
Division Name: D. Gently's Detective and Information Agency
Division Head: Dirk Gently IX
Chief Service: Obtaining Information for Divorces
Division Name: Holy Order Press
Division Head: Father James Dreysdon
Chief Product: Philosophy and Religious Books
Division Name: Golden Star Entertainments
Division Head: Mitch Grearson
Chief Product: Holo-films
Division Name: Golden Star Hotels and Casinos
Division Head: Thomasina Grearson
Chief Product: Gambling Casinos
[NOTE: There are thirteen other divisions within Interconnectedness Unlimited.]

History:

Interconnectedness Unlimited is an enigma. For the past 50 years, the company has slowly been acquiring companies in a haphazard fashion. There is no apparent plan behind the company's actions, yet its acquired companies bloom into new productivity as soon as I.U. takes over.

Attempts to establish a history for the company prior to 2971 have proven futile. It seems to have appeared fully grown on the planet Marlette in that year. Its first recorded action was acquisition of the Triple-F restaurant chain, a large and popular group of fast-food restaurants. It seems incredible, if not impossible, that a company with enough assets to swallow up a large and profitable restaurant chain could have blossomed overnight.

Business Profile:

Any attempt to categorize Interconnectedness Unlimited would be futile. A description of some of its various divisions is more to the point.

It is estimated that there are over 5,000 Triple-F Restaurants on over 250 worlds in the Federated Suns. The restaurants are famous for their Triple-F burgers, which are made out of whatever meat animal is common to the area. Triple-F is sometimes awarded food contracts by local military units. The Triple-F logo is instantly recognizable to anyone from the Federated Suns.

The D. Gently's Detective and Information Agency specializes in marital cases. Their many agents follow wayward spouses, deal with kidnappers, and deliver heartfelt sermons on the interconnectedness of all things. This is an homage to the company's founder, who lived about a thousand years ago and believed, quite rightly, that all things in the universe are related. Count Steffelbust is an avid follower of this "philosophy," which is why he acquired the company.

The Holy Order Press is the Federated Suns's third largest publisher of religious books. Though they have no formal affiliation with the Unfinished Book, several noted philosophers from that organization have published dissertations through the Holy Order Press. This has added considerably to the company's reputation, making it I.U.'s fastest-growing member.

Golden Star Entertainments is the largest producer of holofilms in the Federated Suns. Their huge movie studio, located just outside New Hollywood, cranks out at least 50 full-length feature films per year. These movies, of all types, quality, and ratings, are distributed throughout the realm.

There are an estimated 50 Golden Star hotels and gambling casinos scattered across the realm. Several investigations indicate that the various games are run honestly, as the hotel rooms and food are usually so expensive that there is no need for Golden Star to cheat in the casino. Rumor has it that the infamous Casimir family, a large organized crime syndicate based in the Lyran Commonwealth, has attempted to worm their way into the staff of the Golden Star Casino on Errai.

Profile of CEO:

Elric von Steffelbus was the founder of Interconnectedness Unlimited, having formed the company when he was in his early twenties. He is somewhere near 80 years of age now, though neither his exact age nor his place of birth are known.

Elric von Steffelbus kept a high profile during the early years of I.U. A handsome young man with flamboyant taste in clothes, vehicles, and women, he became the darling of the upper class. It was mainly through the connections made at this time that he acquired his title and many of his companies. Throughout his heyday, von Steffelbus spouted a philosophy about the interconnectedness of all things, which he attributed to an ancient book he had read as a youth. This eccentricity endeared him to people all the more, especially the rich young, who saw him as a kind of religious guru.

About 15 years ago, however, Count Steffelbus went into seclusion and has not been seen publicly since. There are countless explanations why, some of which are pure fantasy. One of the more plausible explanations states that Count Steffelbus is suffering from a rare degenerative disease that requires his isolation. Given his earlier vanity, it is possible that he would not want the public to see him suffering from such a disease. Another theory states that his withdrawal from the spotlight may be just a ploy to keep the company in the public's attention.

Relations with Federated Suns Government:

Interconnectedness Unlimited has always maintained a low political profile. It heeds the laws and pays its taxes, but otherwise keeps out of government or politics. The only exception concerns the alliance between the Federated Suns and the Lyran Commonwealth. As the company name implies, I.U. enthusiastically supports interconnecting the two realms. Not only will the alliance open a whole new market for the company, it will also give Count Steffelbus a chance to find new, and equally unrelated, companies to acquire.

Security and Military Forces:

The companies of Interconnectedness are seldom of a nature to require security any stronger than a few lightly armed guards and the local constabulary.

Notes:

Rumors about a putative wife and son of Count Steffelbus have recently surfaced. According to the rumors, they showed up at the Count's mansion on Rosamond, but were flatly turned away. Our agents should definitely seek to discover their whereabouts.

I.U. appears to be looking to Grehville Plastics, a small manufacturing firm on the planet Edwards, as its next acquisition. The Grehville employees are not happy with the idea of working for an eccentric trillionaire, and are currently preparing to fight off the takeover by buying into the company themselves. Interconnectedness has always had labor problems, and so it will be interesting to see how they handle these employees if they should acquire the company.

OTHER MAJOR COMPANIES IN THE FEDERATED SUNS

Name: Cerulean Waters Inc.
New Avalon Stock & Commodities Symbol: CeW
Home Office Location: Thousand Springs City, Cerulean
President/CEO: Countess Stelfana Trethel

Business Profile:

Cerulean Waters is a firm specializing in water purification systems. It had its origins as a bottled mineral water company that took advantage of the planet's many mineral springs. With the rise of Star League technology, the Trethel family used their riches to open five water purification factories, four of which were destroyed by House Kurita's special forces units at the start of the First Succession War. The remaining factory in Thousand Springs is on the verge of complete collapse because it lacks parts and because of centuries' worth of jury-rigged repairs. Just recently, the NAIS has expressed interest in funding the plant's reconstruction. It is assumed that Countess Trethel will jump at the chance. Whether the NAIS's sudden interest in Cerulean means that they have rediscovered something about water purification systems is unclear and bears close watching.

Name: Apple Computers Interstellar
New Avalon Stock & Commodities Symbol: ApCmp
Home Office Location: New Delos, Macintosh
President/CEO: Count Simon DeGrassy

Business Profile:

Apple Computers Interstellar is an ancient company, with roots extending clear back to the 20th century. Since its birth, the computer company has gained a reputation for always being slightly ahead of its stodgier rivals. During the days of the Star League, the company grew so fat and contented that it actually bought the planet Drekos in the Federated Suns and renamed it Macintosh after one of their legendary computer models. When the Star League fell, most of the company's factories and secrets were lost in the mayhem of the Succession Wars. Only the small personal computer factory on Macintosh survived. Apple continues to manufacture small computers of near-Star League technology, and hopes to open a military computer facility within the next five years.

Name: Kimball's Aquaculture
New Avalon Stock & Commodities Symbol: KblAq
Home Office Location: Jade Sea City, Numenor
President/CEO: Ivan Kimball III

Business Profile:

Kimball's Aquaculture is a recent success story. Ivan Kimball, a professor of ichthyology, left his post at Numenor University ten years ago and retired to his home in the Jade Islands. With the aid of generous investors, he collected a skilled staff and began to cultivate high-paying food fish, as well as other aquatic foodstuffs. His phenomenal success at raising supposedly delicate fish like the famed Mara Trout or the Capellan Nihal Whale-Sardine is responsible for the company's sudden success. Kimball is currently negotiating a trade agreement with the Lyran Free Traders Association, which should give Kimball's Aquaculture a new and avid market to supply.

Name: Basantapur Fine Metals
New Avalon Stock & Commodities Symbol: BSFM
Home Office Location: Jupalpur, Basantapur
President/CEO: Rajah Mohandis Fredericks

Business Profile:
Basantapur Fine Metals specializes in the production of rare alloys and has several large foundries on Basantapur and As Samik. This makes the company extremely important to the Federated Suns' military, as its metals are needed for the internal framework of 'Mechs. Though the company has not been much damaged by the wars, its chronic bad management prevents the firm from living up to its full potential. On several occasions, the Davion government has contemplated taking control of the company, but possible political repercussions in the area have stopped them.

Name: Mendham Electronics
New Avalon Stock & Commodities Symbol: MhmEl
Home Office Location: Liao City, Mendham
President/CEO: Baron Theodore Footh

Business Profile:
Mendham Electronics is a company currently on Economic Probation. The firm was founded when the planet Mendham was still a part of the Capellan Confederation. When Mendham was annexed by the Federated Suns, Mendham Electronics petitioned the Davion government to continue manufacturing its line of medical electronic equipment. They were granted the right, on condition that they agree to a probation period of 15 years. During that time, their every move would be scrutinized for hidden links to the Capellan Confederation. The company is in its last year of probation.

McRae Q.C.

Name: McRae Quik Construct
New Avalon Stock & Commodities Symbol: MrQC
Home Office Location: New Dallas, McRae
President/CEO: Countess Roberta Yeslers

Business Profile:
McRae Quik Construct, with its location near the Periphery, is ideally situated to meet the enormous demand for housing from worlds teeming with new frontiersman. Created just a year after Hanse Davion's famous "Bold New Frontier" speech, the company's line of lightweight yet durable prefabricated housing has been extremely popular. The company is planning to follow up their success by expanding into the city planning and heavy construction fields. That way, when their many temporary buildings begin to wear out, McRae Quik Construct will be there, ready to build more permanent structures.

Name: O'Keefre Importers-Exporters
New Avalon Stock & Commodities Symbol: OkIE
Home Office Location: Drearson City, Fomalhaut
President/CEO: Baroness Nicia O'Keefre

Business Profile:
This massive import-export firm was once one of the Star League's largest firms, almost singlehandedly controlling the flow of public goods in and out of the League. The company's many freighters were a common sight in all parts of the Inner Sphere and Periphery. When the League collapsed, most of the firm's JumpShips and transports were seized by whatever member-state they happened to be traveling through. This left O'Keefre stripped of its ability to make money. The wars, especially the many battles for Fomalhaut, destroyed most of the company's remaining assets, but O'Keefre survived and now has reason to be optimistic. The recent alliance between the Federated Suns and the Lyran Commonwealth puts O'Keefre in an advantageous position to play a key role in the coming wave of trade between the two realms.

Name: Rander Communications Equipment Inc.
New Avalon Stock & Commodities Symbol: RaCom
Home Office Location: Jacksonville, Schedar
President/CEO: Count Ivanalo Tythis

Business Profile:
Rander Communications has been the most reliable supplier of military communications equipment in the Federated Suns for the past 500 years. Rander was originally formed as a subsidiary of a Star League company, and the Rander family and then the Tythis family have seen to the firm's continued survival. Its main manufacturing plant is built completely underground and is defended by a permanent force of AFFS aerospace fighters. They also have a large subsidiary that manufactures electronic items for the general public.

FNS

Name: Federated News Services
New Avalon Stock & Commodities Symbol: FNS
Home Office Location: Federated News Tower, New Avalon
President/CEO: Bron Norburns

Business Profile:
Federated News Services is the largest independent news organization in the Federated Suns. It has offices on all the major planets and contacts with news organizations on the smaller ones. In cooperation with the Federated Broadcasting Corporation, the FNS broadcasts the Interstellar News Hour. The FNS leans heavily toward planets' rights and often slants their stories against the Davion government, which they see as encroaching on the freedoms of the common man. The battles between the FNS and the Davion government are legendary, averaging at least one court case per year. Hanse Davion is secretly pleased with the diligence of the FNS because of their ability to unmask the incompetent and the greedy in his government.

Name: New Syrtis Shipyards
New Avalon Stock & Commodities Symbol: NSySp
Home Office Location: Lagrange Point 4, New Syrtis
President/CEO: Baroness Deborah Venchil

Business Profile:

Since June 3022, the New Syrtis Shipyards, formerly the main repair work site for JumpShips and DropShips, have been sealed off. Moreover, all inquiries about what is happening there are strongly discouraged. Three armed DropShips and a number of AeroSpace Fighters are on constant patrol to see that no unauthorized person approaches the shipyards. There is also a constant flow of coded communications between the shipyards and the New Avalon Institute of Science. Even the wildest speculation as to what is going on at the New Syrtis Shipyards may turn out to be plausible.

PALYMYRA

Name: Palymyra Petrochemicals
New Avalon Stock & Commodities Symbol: PaPet
Home Office Location: Hilder's Hell, Palmyra
President/CEO: Countess Rebecca Graldy

Business Profile:

Palymyra Petrochemicals has played a vital part in the recent push to explore and exploit the Outback worlds. Their many survey crews have been working with Federated Planetary Surveyors to find or rediscover Outback world deposits of petrochemicals. This is important because colonies generally depend on chemical sources of energy for decades before they ever see a fusion reactor. Palymyra is especially adept at finding and exploiting oil deposits, thanks to its aging, but still functional, Star League-era survey satellites and equipment. Once a deposit is found, Palymyra is one of the few petrochemical companies in the Inner Sphere that can provide orbiting fuel refineries to its clients.

Name: New Valencia Agriculture Company
New Avalon Stock & Commodities Symbol: NVAg
Home Office Location: NeuStravros, New Valencia
President/CEO: Baron Mitchell Kochar

Business Profile:

New Valencia Agriculture is in deep financial and legal trouble. Recent discoveries by the planetary government's Business Practices Department indicate apparent gross illegalities in corporate procedure. It is quite possible that the company will be seized by the planet's government, which would almost guarantee years of disrupted services. This is especially discouraging because of rumors that the company's research division was on the verge of developing a wheat-type grain that could grow in soil with high mineral and salt contents. The Davion government is privately considering bailing out New Valencia Agriculture because the firm has always been in the forefront of agricultural sciences.

W.S.T.S.I

Name: White Swan Trans-Stellar Inc.
New Avalon Stock & Commodities Symbol: WhSwTS
Home Office Location: Jameston, New Avalon
President/CEO: Yvonne LaGrassic

Business Profile:

White Swan Trans-Stellar is one of the few remaining passenger ship companies still operating in the Inner Sphere. Its seven JumpShip liners, all luxurious and all built during the heyday of the Star League, still somehow manage to transport the elite of the Federated Suns to and from the richer planets. White Swan recruits the best JumpShip crews from Davion navy retirees and has long been a favorite of the wealthy. Even our Blessed ComStar Precentors have been known to travel via White Swan. The company has recently announced plans to begin service between New Avalon and Tharkad, the capital of the Lyran Commonwealth. White Swan also runs Black Swan Liners, a less expensive passenger and cargo line servicing the pioneer worlds of the Federated Suns Outback, and Jesup Traders, the successor to Black Ball Trading Company.

Name: General Motors of Kathil
New Avalon Stock & Commodities Symbol: GMKI
Home Office Location: District City, Kathil
President/CEO: Cassandra Hek and Donald Traves

Business Profile:

General Motors of Kathil is a major manufacturer of motors and wheeled vehicles. Much of the company's business is with the Davion military, and so the factories on Kathil have been the target of numerous attacks from the Capellan Confederation. The company has managed to survive, however, and with the aid of its new factories on Salem and Talcott, it looks forward to continuing its growth. GM of Kathil also manufactures a line of rugged all-terrain vehicles, luxury cars, and engines for the general public.

Name: Vinson Pharmaceuticals
New Avalon Stock & Commodities Symbol: VnPh
Home Office Location: Kiverson City, Torrence
President/CEO: Baroness Marth Terenson

Business Profile:

Vinson Pharmaceuticals is currently attempting to avoid being nationalized by the Torrence government, which recently went socialist. In its efforts to remain independent, the company has been looking for offworld investors to buy up at least 40 percent of the company's securities, which would make it ineligible for nationalization. Vinson is famed for its ability to isolate the effective agents in old folk remedies such as herbal teas or homeopathic mixtures. Its several powerful, yet cheaply made, painkillers have made the company extremely popular with the Davion military.

Name: Green Star Emergency Services
New Avalon Stock & Commodities Symbol: GCES
Home Office Location: Amsterdam, Arnaudville
President/CEO: Count Jam-el Delanter

Business Profile:
Green Star Emergency represents an experiment by the Davion government to privatize certain emergency services in the frontier region of the Outback. Green Star runs and operates several old freighters and DropShips and controls a large stockpile of resources. When Green Star receives word of an emergency, such as a bandit raid or natural disaster, it quickly negotiates a contract with the stricken world, then speedily transports needed supplies and personnel to the disaster area. If this system proves successful, the Davion government will probably set up similar arrangements in other sections of the Outback.

Name: Federated Arbitrators and Judges Inc.
New Avalon Stock & Commodities Symbol: FAJ
Home Office Location: Inns of Court, New Avalon
President/CEO: Count Edward Defano

Business Profile:
Count Edward Defano, the founder of Federated Arbitrators, was once a member of the Federated Suns highest court, the Court of the Old Bailey in New Avalon. After ten years on the bench, Count Defano decided that he had had enough of the dry and unexciting Old Bailey cases. He wanted to get back to cases that had an immediate and obvious effect on people. The Count retired from the bench and formed Federated Arbitrators and Judges. Since then, he has begun to travel through the realm to hear cases that require the kind of immediate attention that the government court system cannot guarantee. After eight years as chief executive and Chief Justice, Count Defano has seen his company grow to become a de facto alternative judicial system. Federated Arbitrators and Judges handles only civil cases, and confines itself largely to interplanetary disputes.

M MELCHER MEAT ANIIMALS LIMITED

Name: Melcher Meat Animals Limited
New Avalon Stock & Commodities Symbol: MMA
Home Office Location: Thousand Oaks Ranch, Melcher
President/CEO: Baroness Josephine R. Earwig

Business Profile:
Melcher Meat Animals is the largest supplier of meat in the Federated Suns. Cattle, pigs, horses, Galax beefalos, Brockway goats, and many other animals are raised on the company's 110 ranches scattered over the planet Melcher and surrounding worlds. The company is currently in a state of flux because of its recent takeover by Baroness Earwig. The takeover was so hostile that the former owners turned their corporate headquarters into a stronghold and refused to leave. A company of troopers and a 'Mech from the planet's militia were forced to storm the building. The losses on both sides were heavy, and the current atmosphere in Melcher Meat Animals is quite tense.

Name: NAIS Information Network
New Avalon Stock & Commodities Symbol: NIN
Home Office Location: NAIS, New Avalon
President/CEO: Professor Lindaval Serenson

Business Profile:
The NAIS Information Network is the unusual system used to distribute information discovered at the New Avalon Institute of Science. For a substantial subscription fee, almost anyone can access certain areas of the Institute's computer system. Though the rates are high, agents of the other Successor Houses have been benefiting from the work done at the NAIS. Though much of the really important work is locked away behind formidable computer security systems (which ComStar has, as yet, been unable to break), what is available to subscribers is significant. The Information Network was created because of Prince Davion's wish to see the NAIS discoveries benefit a broad section of the citizenry.

Name: Galax Launch Systems and Satellites
New Avalon Stock & Commodities Symbol: GLSS
Home Office Location: Canaveral City, Galax
President/CEO: Baron Justine Von Tire

Business Profile:
Galax Launch Systems and Satellites provides planets in the Federated Suns with a wide variety of satellites and launch systems. The satellites, using a common framework, can easily be customized to handle almost any combination of tasks, such as communications, resource mapping, navigation, or security. The launch systems also vary between chemical or liquid rockets, payload drivers, or orbital mules. All Galax products are built at their factories on Galax, then shipped to the client. Their reputation as reliable manufacturers has sparked interest in House Steiner, which might invite Galax Launch Systems and Satellites to open a sister company in their realm.

GALAX

TRADE WITH OTHER SUCCESSOR STATES

Vast distances and the nature of relations between the Federated Suns and its neighbors have made the realm's leader reluctant to encourage heavy trade with the rest of the Inner Sphere. What the Davions are willing to export is generally of good quality, though technology levels vary considerably. Federated Suns trading companies, well aware of the shortcomings of their products, have seldom been accused of overpricing their products. This has given Federated traders a good reputation among buyers in other realms, who have grown tired of the trickery of unscrupulous dealers, particularly those from the Lyran Commonwealth.

Few traders from the Federated Suns have the savvy to forecast the needs of foreign markets as Commonwealth traders do. This often puts them in the wrong place at the wrong time with the wrong cargo in their freighter's hold. Although the Davion trading companies have made some attempt to coordinate their activity in foreign realms, intercompany rivalry tends to water down the effectiveness of the effort.

It is the realm's inward-looking nature and prejudices, however, that have made the Federated Suns the smallest inter-realm trader. The Suns have been at odds with the Draconis Combine and the Capellan Confederation for so long that deep and long-lasting hatred has built up. Even in times of relative calm, the ugly memories tend to linger on. Businessmen making trips into those realms often face open hostility from their own people when they return. This attitude was codified when the Davion government passed the Economic War Powers Act in 2802, a law so restrictive that it has choked off anything more than token trading with the two neighboring realms for over 200 years.

Trade with the Lyran Commonwealth and the Free Worlds League has been much smoother for Federated Suns firms. The Davion government has tended to see those realms as potential allies, and encourages trade with them. The vast distances involved are often more than many traders can manage, however.

The alliance of House Davion with House Steiner opens up whole new possibilities for Federated Suns companies hoping to export their products abroad. With Lyran traders soon to be loading up their freighters with Federated products and Federated traders distributing high-tech Lyran items, the economic potential is staggering.

FREE WORLDS LEAGUE

Trade between Houses Davion and Marik has been based on the Federated Suns ability to provide raw materials such as metal ores and petroleum-based materials to resource-starved worlds in the Free Worlds League. In exchange, the industrialized worlds in the Free Worlds League would ship the Federated Suns high-technology items.

Pharmaceuticals and chemicals from Marik industries are two sought-after items among Davion traders. Pharmaceuticals usually find their way into Davion military hospitals. The chemicals help to sustain certain key industries within the Federated Suns, particularly the military electronics industry.

The recent Davion-Steiner alliance and the subsequent alliance between the remaining three Inner Sphere powers have affected Davion-Marik trade relations. Now members of opposing alliances, the two realms are no longer so eager for trade with one another. Many trade contracts have been allowed to lapse, with the level of trade slowly but surely declining. The Federated Suns is not concerned about this, because they will have almost free access to the Lyran Commonwealth's riches.

CAPELLAN CONFEDERATION

Few trading companies in the Federated Suns have anything to do with the Capellan Confederation. The general animosity toward House Liao, along with the fearsome Economic War Powers Act, make most Davion businessmen consider trade with the Capellans both unprofitable and unwise.

Small, independent trading companies can find hefty profits, however, if they are willing to take the huge risk of entering the Capellan black market. These traders face considerable danger, sometimes from their own people, when they slip across the border without having their cargos approved by the Davion government.

Legal trade between the Capellan Confederation and the Federated Suns is confined to a very small list of products, most of which are totally useless. If a Davion trader can avoid Federated port authorities and enter the Capellan Confederation with a load of foodstuffs, however, he can make a huge profit. Of course, the same trader has to make it home with the money, which is easier said than done.

Over the years, independent traders have developed the infamous "Canopus Run". If they claim to be heading for the Magistracy of Canopus, they can instead enter the Periphery edge of the Capellan Confederation, where they hurriedly swap their cargo for Capellan items. From there, they can actually head to the Magistracy to sell their Capellan items, then return home richer for the trouble.

DRACONIS COMBINE

Trade with the Draconis Combine is the most difficult of all. The bitter feelings most citizens of the Federated Suns have for this Successor State are stronger than those against House Liao. Border skirmishing is much more frequent near Kurita space, making life for a trader more difficult.

Those that do try to trade with House Kurita—again usually small independents—go through the Outworlds Alliance. There is less tension between the two realms in that region, and so some honest trading can occur, though with a hefty tax levied by the Outworlds Alliance government.

What the Draconis Combine wants most from the Federated Suns are foodstuffs and raw materials. Though Davion businessmen are more than willing to supply the Kuritans with food, they will not sell them anything that might find its way into a Kurita soldier's arsenal.

LYRAN COMMONWEALTH

Simply stated, the Lyran Commonwealth wants every bit of raw material from the Federated Suns that it can lay its hands on, and is willing to pay almost any price to get it. The Federation of Skye has long since stripped most of its worlds of their natural resources, and the lack of raw materials in that industrialized section of the Commonwealth is now almost as big a problem as the damage caused by the wars. The new alliance has given them hope of obtaining raw materials from Federated worlds that are closer to Skye's factories than many resource-rich worlds in the Commonwealth. Many Lyran businessmen are looking forward to having the resources to run their factories at full capacity.

The sudden surge of money into the Federated Suns is expected to stimulate the whole realm's economy, though some economists fear it will cause inflation. In addition to the expected new riches will be the influx of Commonwealth high-tech items. Military industries are already benefiting from the Commonwealth's ability to manufacture complex machinery.

Every major Davion trading company is anxious for the chance to enter the Commonwealth market. The Davion government is taking a rather cautious attitude toward issuing trading visas at this time, however. Until the last few details are ironed out with the Commonwealth government, they will not let down the final remaining legal barriers and turn the Federated traders loose on the Commonwealth.

STRENGTHS AND WEAKNESSES

In the Federated Suns, many planets have the potential of incredible wealth. Yet, the distance between worlds is often so vast and costly to bridge that some of these worlds can never tap their markets and so their potential is meaningless. This is why the alliance with the Lyran Commonwealth is so important. With the major Lyran companies also sending traders into the realm, trade among the Federated planets should become much more reliable and frequent.

PERSONALITIES

Name: Hanse Davion
Title: Duke of New Avalon
Position: Prince of the Federated Suns
Birthplace: Royal Palace, New Avalon
Age: 42
Sex: Male

Distinguishing Physical Characteristics:

Prince Hanse Davion is a tall, broad-shouldered man who exercises daily and keeps up his martial art skills to stay in shape. The combination of height, build, and physical vitality make his presence most imposing.

The Prince closely resembles his mother, Jennifer Campbell, but he also has the Davion red hair, which he wears cropped short. Though he also has the broad Davion bone structure, his expressive face is lit by the same unusually blue eyes as his mother's.

Davion has several noticeable scars. On the back of his neck, near the left shoulder, is a jagged scar left by a piece of shrapnel that hit him during his cadet days in the NAIS Training Cadre. Just above his right eye is a small scar from a childhood fall from a tree.

Brief Personal History:

Hanse Davion was the second son of Prince Andrew Davion and Jennifer Campbell. As the second son, he was not under the same pressure to prepare to become leader of the Federated Suns as was his older brother Ian. Everyone, including Hanse, expected him to grow up to become the commander of either the AFFS or the federal bureaucracy.

Unburdened by Princely expectations, Hanse Davion apparently enjoyed a happy childhood. Because his father was a firm believer that the Davions must understand how others live and work, the young Hanse was exposed to a wide variety of people and places. Many of his earliest and closest friends were unconventional associates such as Kincaid Fessul, a New Avalon fisherman, and Vivian Sawl, an exotic dancer whom Prince Hanse used to call his "Auntie Viv."

The closest companions of his youth were Ardan Sortek and Professor Sharon Bryan. Sortek was the son of a close friend of Prince Andrew. Younger by about ten years, Ardan was an important friend because he was one of the few people with whom the young Davion felt totally at ease. Together they embarked on mischievous adventures, such as trying to swim the Olsen Channel. Professor Sharon Byran was Prince Hanse's private tutor, and she especially encouraged the Prince's natural curiosity. Many credit Bryan with awakening in him the desire to see the Federated Suns take the lead in rediscovering the information and technology lost during the past centuries.

In his teens, Hanse divided his time between school work, a cheerful pursuit of attractive young women, and a growing interest in government. While still a young man, he revealed a flair for diplomacy when he personally negotiated a settlement between the Davion family and the maintenance union of the Royal Palace, preventing what could have been a nasty strike. This brought him to the attention of many government figures, who saw in young Hanse Davion the political savvy that seemed so lacking in his brother Ian.

Prince Hanse attended the Albion Military Academy and graduated with honors. His early military actions were as lance commander with the Third RCT of the Royal Brigade of Guards, during which time he showed little respect for battlefield conventions. He would attack or defend with curious, if not downright bizarre, tactics. Only the fact that young Davion's quirky ideas often led to spectacular successes saved him from the embarrassment of a court-martial. By 3011, he was commander of the Third Brigade 'Mech regiment. Though only 28 years old, he had already lived through enough battles to be called a Marksman (the MechWarrior equivalent of an Ace) twice over. His early tendency toward unorthodox tactics mellowed when experience taught him how often the old ways do actually work.

His political skills had certainly not withered during his military career. In 3011, he was appointed Military Governor of New Aragon. The next year, he was given the task of rebuilding the entire military structure of a portion of the Capellan March. Between the convoluted political atmosphere of New Aragon, with its hostile population, and the logistics of rebuilding a military force, Hanse got quite an education in politics. Two years of administering such a powder keg made him a seasoned administrator by the time news reached him of his brother's death in 3013.

Hanse would need all the skills he had learned as Military Governor before he could take his rightful place on the throne. With little time to reflect on the twist of fate that made him the new Prince, he returned to New Avalon, where he immediately encountered an assassination attempt by a minion of Duke Michael Hasek-Davion.

The early months of his reign were stressful ones. The entire realm seemed to hold its breath and watch while he and Duke Michael Hasek-Davion wrestled for control of key positions in the federation government. When the dust cleared, Prince Hanse had succeeded in assuming almost complete control of the central government and Duke Hasek-Davion was left isolated and much weaker than before.

Since that time, the Prince has lived up to his early reputation as a talented, unorthodox leader. He has been responsible for new policies that have begun to revitalize the long-stagnant Federated Suns economy, particularly in the poorer Outback worlds. Under his leadership, the AFFS is growing stronger almost daily with the aid of new policies and new equipment. In the diplomatic arena, only he was able to respond to the potential behind Katrina Steiner's cease-fire proposal of 3020. Though still a relatively young man, Hanse Davion has done much for the Federated Suns, and seems capable of much more.

Motivations/Desires/Goals:

Prince Hanse Davion is a master at concealing his motives and desires, one of the reasons he has been nicknamed "the Fox." Many believe that his one true desire is to see the Federated Suns safe from attack and his citizens prosperous and contented. Though he has done much to make those goals a reality, it is unlikely that such purely pragmatic politics would satisfy this complex man.

It has also been suggested that he secretly wishes to be proclaimed First Lord, but Davion is too intelligent to entertain so grandiose a desire. He must realize that restoration of First Lordship over all the realms of the Inner Sphere is a dream for the distant future, if it all. The most he can realistically expect from the recent alliance between the Federated Suns and the Lyran Commonwealth is to reduce the danger his realm faces from its enemies.

Whatever truly motivates this man, he will use his considerable wit, intelligence, and intuition to achieve it. Perhaps the only true desire of this enigmatic leader is to make the most of any situation he encounters.

Manner:

Most people are immediately struck by the apparent openness of Prince Hanse when meeting him for the first time. Nothing in his friendly, jovial manner suggests his masterful ability to identify the strengths and weaknesses of others. Davion's psychological insight is based on both his intellect and his uncanny ability to listen deeply to what others are saying. This ability to gauge others is positively eerie at times. Indeed, Ardan Sortek often jokingly asks his friend why he did not become a psychiatrist or a barkeeper instead of Prince of the Federated Suns.

When among close friends, the Prince relaxes and turns off that mental spotlight. When at ease, he is a warm individual who loves nothing more than to tell to long stories about his childhood and the many unusual people he met and admired.

Special Skills/Power:

Thanks to Professor Sharon Bryan, Hanse Davion has a wide range of interests. Some, such as an interest in Medieval Italian politics, are obviously useful to his abilities as a ruler. Others, like his deep love of poetry, are less obvious aids to his ability to govern the Federated Suns. So great was his love of learning and of books that Davion was willing to risk his own life in the raid to steal a cache of books from under the nose of the Draconis Combine in the now-famous Halstead Station Affair.

Name: Michael Hasek-Davion
Title: Duke of New Syrtis
Position: Minister of the Capellan March
Birthplace: Saso, New Syrtis
Age: 49
Sex: Male

Distinguishing Physical Characteristics:

Duke Michael Hasek-Davion is not particularly tall, but his slenderness seems to give him some extra height. He keeps his black hair long, wearing it in a single neat braid when in uniform. Though handsome, his frozen expression and the look in his deep green eyes can be forbidding.

Duke Michael lost his left hand in combat about twelve years ago when a Capellan *BattleMaster* hit the cockpit of his *Marauder* with its fist, exploding the Duke's cockpit. The force of the explosion threw him out of the 'Mech and into a lake, where he almost drowned before being rescued. Though most of his injuries healed, the Duke's left hand had to be amputated and a replacement attached. This incident also left a large scar on the left side of the Duke's neck and a psychological scar that makes him fear large bodies of water and *BattleMaster*s.

Brief Personal History:

Duke Michael grew up in a household whose members viewed the Hasek family as the only reason the Federated Suns is strong today. His father, Duke George Hasek, was an ambitious man whose efforts had won the Hasek family renewed prominence in Federated Suns political circles. Believing in the Hasek destiny, Duke Michael began to struggle for control of the realm even before his father's death.

As an officer in the Syrtis Fusiliers, he showed special talent as a tactician as well as being a superb motivator of people. Few people could resist his dynamic personality or eloquent speech.

In 3003, he married Marie Davion, illegitimate daughter of Prince Andrew. The marriage was an obvious political move on the Duke's part to insinuate himself into the Davion line of succession. It was also his ticket into the federation government.

Using his tie to the Davion family, Duke Michael began to work his way into the Ministry of Intelligence, Investigation, and Operations as leader of the Bureau of Investigations. In that role, he was able to begin recruiting men and women to act as his private spies. Prince Ian was far too busy leading his 'Mech regiments to care, which gave Duke Michael almost a free hand in expanding his spy network and political power base.

Duke Michael realized that of all the Davions, Hanse would be his greatest opponent. When young Davion's success as a Military Governor began to attract attention in high places, Duke Michael was so worried about Hanse's growing prestige that he organized an assassination plot even before Prince Ian Davion's death elevated Hanse to Prince of the Federated Suns. The assassination attempt was a dismal failure, leaving the Duke to face the reality that his most feared opponent was now the Prince.

Hanse Davion was quick to challenge Duke Michael's powers in the federation government. In a kind of administrative war that began in 3014, the new Prince replaced Duke Michael's governmental minions with his own handpicked people. Only the risk of Prince Hanse finding some evidence to link him with the assassination attempt kept Michael from waging all-out political war.

This political struggle left Duke Michael Hasek-Davion stripped of all but ceremonial positions in the central government. Hanse Davion had removed most of the Duke's followers from their government posts and had even imprisoned some for treasonous actions while in office. Since that time, Duke Michael has been biding his time, hoping to find some serious flaw in Hanse Davion's administration of the Federated Suns. Perhaps it has been his failure to find it that has prompted Duke Michael to begin turning his eyes toward the Capellans.

Motivations/Desires/Goals:

Duke Michael Hasek-Davion's single goal is to be the Prince of the Federated Suns. The dream was born early in his life as he listened to the many tales of the illustrious Hasek family. The desire seems to grow stronger with time, particularly since Duke Michael has begun to feel that the Davion government discriminates against him and his family.

Manner:

Much of Michael Hasek-Davion's charisma comes from his deep, resonant voice and the conviction of his belief in the greater glory of the Hasek family. He is capable of hiding behind an amiable, courteous facade that shows him only as the ever-loyal Duke. The charade is surprisingly effective.

In private, Duke Michael is an intense, focused man. He pays careful attention to anything said or done in his presence, particularly when meeting someone for the first time. Because he is such an intelligent yet scheming individual, Michael immediately begins to assess every new acquaintance's potential worth as a friend or his potential strength as an enemy.

Special Skills/Power:

Duke Michael is a serious historian of the Capellan March, a trait he cleverly but subtly puts to good use in speeches intended to create anti-Davion sentiment. He is also a skilled computer operator, a byproduct of his days as head of the Bureau of Investigations.

Name: Aaron Sandoval
Title: Duke of Robinson
Position: Minister of the Draconis March
Birthplace: Castle Sandoval, Robinson
Age: 63
Sex: Male

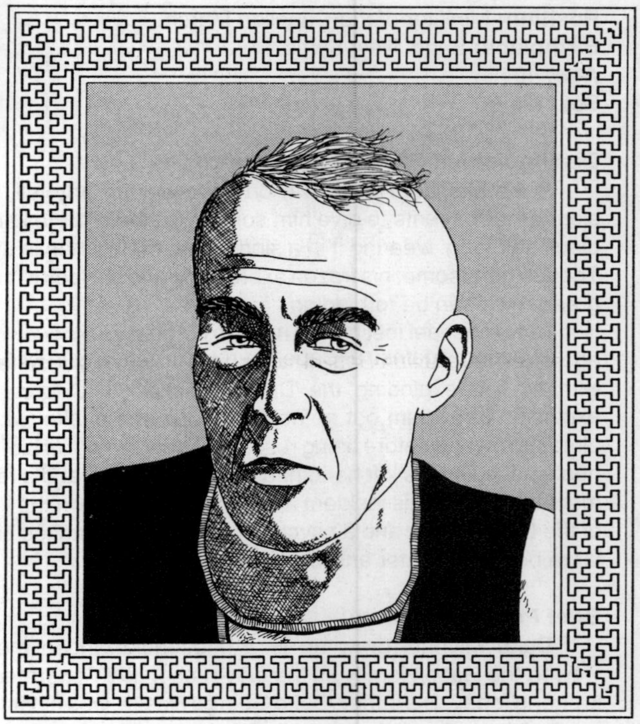

Distinguishing Physical Characteristics:

Duke Aaron Sandoval is one of those fortunate men who age with grace and vigor. He wears his brown hair in a topknot, according to the custom on Robinson. Despite having lost both legs in battle over 40 years ago, Sandoval is in good health and remains quite active. His sharp, handsome features and blue eyes have not been much altered by a few lines and wrinkles. Women still find his appearance and personality so attractive that they make advances to him. Occasionally, he takes one of them up on it.

Brief Personal History:

Aaron was the third son in the Sandoval family. Because his father, Duke Benjamin Sandoval, was such a strict disciplinarian and moralist, Aaron Sandoval grew to resent his severe upbringing at Castle Sandoval. By the time he had reached his teens, father and son were either deep in argument or ignoring one another in stony silence.

At age 16, Aaron ran away from home to enlist in the AFFS under an assumed name. He made it through boot camp and then trained to be a tank crewmember, joining the 159th Galax Heavy Tank Regiment in 2979. After proving himself as an excellent crewmember, he was promoted to the rank of Sergeant Major and command of a Partisan heavy tank.

In 2982, a new driver transferred to Aaron Sandoval's tank. Corporal Jessica Danford was from the distant Outback world of Broken Wheel. As the first member of her poor family to make it into the AFFS, she was sending most of her pay back home. When Aaron and Jessica's superior officers discovered that they had fallen in love, she was assigned to another tank because of a regulation that has since been repealed.

In 2982, the 159th Galax Heavy Tank Regiment, attached to the 15th Avalon Hussars, participated in a military offensive against the Capellan Confederation on Rio. From his hiding place in ambush, Aaron saw a Capellan *Marauder* stalking the Partisan heavy tank that Jessica was driving. As the 'Mech approached the damaged tank and made ready to finish the job, Aaron ordered his own tank to flank speed. In an instant, his Partisan lurched from its hiding spot straight toward the *Marauder*. Stopping directly between Jessica's tank and the Capellan 'Mech, Aaron fired. Most of the shots missed, but the *Marauder*'s reply destroyed Aaron's Partisan. Severely wounded, Aaron slipped into unconsciousness.

He awoke to find himself back on Robinson. His father, now gravely ill, had been searching for him for two years. Just after Aaron ran away, his two elder brothers had died in a DropShip accident, leaving him heir to the Robinson Dukedom. It was only by chance that his father learned of Aaron's heroic stand against the Capellan *Marauder* when the news media covered the action, complete with pictures. To Aaron's surprise, he learned that his father had also brought the wounded, but recovering, Jessica to Robinson.

The long and happy marriage between Aaron and Jessica produced three sons, James, David, and Talbot, and a single daughter, Jessica. Not long after the birth of their first son James, the old Duke died and Aaron became the Duke of Robinson.

Aaron Sandoval reentered the AFFS, this time training at the Battle Academy on Robinson. After graduating with honors, he joined the Robinson Rangers RCT. He saw considerable action and eventually rose to command the regiment, a post he holds to this day.

Jessica Sandoval died in 3001.

Motivations/Desires/Goals:

In some respects, Duke Aaron is as ambitious as Duke Michael Hasek-Davion of the Capellan March. The difference is that Duke Michael aspires to power while Duke Aaron aspires only to protect the Draconis March from the Draconis Combine by whatever means possible. He is steeped in the history of the March and has developed a genuine love for its people during his travels through the March.

Duke Aaron is one of the staunchest supporters of the alliance with the Lyran Commonwealth because his portion of the Federated Suns will benefit most from coordinated attacks against House Kurita. He will do anything to promote the alliance.

Manner:

When meeting someone for the first time, Duke Sandoval is reserved, even slightly cold. His manner definitely thaws after he comes to know and trust a person. Once Duke Sandoval makes a friend, he remains fiercely loyal and expects the same loyalty in return. Because he has been disappointed in this expectation several times, it has made him quite careful about his choice of friends.

Special Skills/Power:

Duke Aaron loves ancient art. He has funded several museums on Robinson and gone out of his way to see precious artworks protected from the ravages of war. He is especially fond of the ancient American painter Norman Rockwell. The Duke is also a connoisseur of fine liqueurs.

Name: Ardan Sortek
Title: Colonel, Knight Excalibur, Member of the Order of Davion
Position: Military Advisor to Prince Hanse Davion
Birthplace: Sortek Hall, New Avalon
Age: 32
Sex: Male

Distinguishing Physical Characteristics:

Ardan Sortek is a slim man of muscular build and rough-hewn, square features. Though strong in physique, he can be quick and graceful, especially in a fight. Sortek wears his brown hair short, which reveals the thin scar above his right eyebrow that he picked up in a 'Mech training accident. His left wrist was injured in an action against a Capellan 'Mech, leaving it with limited mobility.

Brief Personal History:

Ardan Sortek was born into a military family. His father, Major Adriaan Sortek (retired), was a famous battalion commander in the 134th New Avalon Tank Regiment attached to the First Royal Guards RCT. He once saved Prince Andrew's life during an RCT action against the Capellan Confederation. In gratitude, the Davion family gave Adriaan the title of Baron of Crossroads Prairie on New Avalon. Prince Andrew and Adriaan were close friends until the time of the Prince's death in 2999.

Because of their fathers, Hanse Davion and Ardan Sortek also became friends, and at one point, the two were almost inseparable despite the ten-year difference in their ages. Their friendship survived even after Hanse left New Avalon to pursue a military and political career.

As a cadet at the Albion Military Academy, Ardan showed himself to be a natural MechWarrior and tactician. Because of his unusually brilliant actions with the Albion Military Training Cadre, he graduated with the rank of Leftenant instead of Sergeant and immediately took charge of a lance in the 17th Avalon Hussars. He participated in several major raids deep into the Draconis Combine as well as the Capellan Confederation, each time coming away with a major victory.

Ardan Sortek won the Medal Excalibur and command of the Heavy Guards in 3019, when he commanded a raid on the Combine world of Nashira. Once on the planet, he and his three regiments of armor and one battalion of 'Mechs found themselves heavily outnumbered by the planetary defenders, a regiment of Dieron Regulars and six armor and infantry regiments. Cut off from his pick-up points, Sortek engineered a brilliant campaign across the northern continent of the planet, destroying anything of value in the way, including the top-secret Tanadi Computers research facilities.

After six days of hard fighting in the capital city of the world, Sortek unexpectedly turned his forces and lunged at the world's major spaceport, leaving the Kuritans flat-footed. Commodore Nichelson and his DropShips, unable to fight their way through Kurita fighter support, took Sortek's cue and bulldozed through the Kurita fighters to land at the spaceport and pick up Sortek and his forces. From the computer memory seized at the Tanadi research lab, it was later learned that Sortek and his men had destroyed several major military projects that might have given the Draconis Combine a major edge in computer technology.

Since that time, Ardan Sortek has been commander of the Heavy 'Mech regiment in the Royal Brigade. Apparently frustrated by the inaction of the job, he requested and received permission to leave the Brigade to participate in the AFFS action to retake Stein's Folly in 3025.

Motivations/Desires/Goals:

Like his friend Hanse, Ardan Sortek is a complex man. On the one hand, he believes that every person, from the lowliest peasant to the highest noble, is responsible for the well-being of his fellow men. Yet Ardan is a soldier, accomplished at killing and destroying the lives of his fellow human beings. Similarly, he once chided Prince Hanse for manipulating his subjects, yet is himself a strict and forceful commander of soldiers.

His goals are just as complex. Though he apparently wants to become a Field Marshal in command of a major offensive, he often expresses the desire to find a hunk of land and, like Cincinnatus, give up the military life to become a simple farmer. For a man with so much influence over the Prince, his complexity makes him almost as difficult to read as Hanse Davion himself.

Ardan is sometimes prone to brooding and perhaps even depression over the conflict between his ideals and reality. Though this has yet to prevent him from being a loyal and active member of the AFFS and the Federated Suns, it could one day pose a problem.

Manner:

Ardan was not born into high society and so he lacks the inbred mannerisms and duplicity so typical of the nobles of the Royal Court. His preferred company are the soldiers of his regiment. Indeed, he treats most anyone he meets with the same military directness and courtesy he uses with his men. Though elegant social occasions leave him feeling shy and insecure, his natural candor and intelligence usually get Sortek through the event, and even make a good impression.

Special Skills/Power:

Ardan has always been intrigued with computers and has considerable knowledge in the area. He also enjoys tinkering with BattleMech innards with his Techs.

Name: Marie Laura Davion
Title: Duchess of New Syrtis
Position: Wife of Duke Michael Hasek-Davion
Birthplace: The Royal Palace, New Avalon
Age: 54
Sex: Female

Distinguishing Physical Characteristics:

Duchess Marie is atypical of Davion women. Instead of having a tall, strong figure, Marie is petite and slight. She has blonde hair, green eyes, and a kind of perpetually puzzled or befuddled expression. The combined effect of her build and features makes her seem an inconsequential, though beautiful, woman.

Brief Personal History:

Marie was born to Simone St. Claire, mistress of Prince Andrew Davion. Her mother died giving birth to Marie, which resulted in the infant daughter being raised by a series of relatives. For the first ten years of her life, Marie's father refused to acknowledge her paternity. It was not until after the birth of Ian Davion that Prince Andrew claimed Marie as his child and allowed her to use the Davion name. That was the extent of Prince Andrew's generosity, for he had no intention of allowing her to become a legitimate heir to the throne.

The Davion family ostracized Marie because of her illegitimate origins. Though her father provided her with luxuries, Marie's childhood was neither happy nor normal because of her unusual status in New Avalon society. As a result, she grew up to be a shy, unassuming woman.

What she has hidden from everyone until this day is how lively her intelligence is. Though tutored and sent to the best private schools, Marie never willingly excelled in her studies for fear of drawing unwanted attention to herself. Only when alone in a library does she indulge her curious, questing mind. As a result, Marie is a self-taught scholar, a far cry from the addled woman of her public image.

Her career in the AFFS was a dismal failure. Though she seemed to love her AeroSpace Flight training at the Albion Military Academy, Marie purposely failed her courses when her obvious piloting skill began to attract notice. Eventually, she was thrown out of Albion.

From the first, Michael Hasek saw Marie Davion as a way into the line of succession, even though Prince Andrew never recognized her as a legitimate heir. Hasek's original hope was that the Prince would not have any other surviving children and be forced to recognize Marie as heir. Of course, the rule of Ian and then Hanse dashed those hopes.

Why Marie Davion accepted Michael Hasek as a husband is uncertain. What is certain is that she made Hasek wait a full two weeks before agreeing to his proposal. Perhaps she felt marriage to him would take her far from the stifling court life of New Avalon, or perhaps she truly did love the future enemy of the Davion family. Whatever the reason, she fulfils her duty as the Duchess of New Syrtis.

Motivations/Desires/Goals:

There has been considerable discussion about just how soft the "Meek Marie" really is. Our agents have recently collected evidence suggesting that at least some of Duke Michael Hasek-Davion's political maneuvers have been prompted or even planned by Marie Davion. If true, this upsets all theories about the future of the Federated Suns.

What might motivate Marie to betray her brother to further the plans of her husband? The obvious answer is the coldness that the Davion family showed because of her illegitimacy. There were plenty of incidents in Marie's early life on New Avalon that might have nourished a hatred of the Davions. It is entirely possible that the quiet young Marie long harbored a dream of revenge, which marriage to Michael Hasek allowed her to set in motion.

It is also possible that Marie Davion is secretly aiding the Davion family. Michael Hasek was, after all, an accomplished schemer before he married Marie, having easily made his way into the government and seized considerable power for himself. Since his marriage to Marie Davion, however, Duke Michael's plans have either failed or accomplished less than he hoped. Some among our agents believe that Marie has been subtly sabotaging, either by counseling her husband or by direct action, Duke Michael Hasek-Davion's plans. Is it possible that Marie is a true and loyal Davion despite it all?

Manner:

Marie Davion is an outwardly quiet, withdrawn woman who always seems to edge away from conversations and seems frightened of talking with strangers. Her only friends seem to be the librarians at the Hasek palace and an aging handy-man.

Some of our representatives have reported those flashes of a quick intelligence and even cunning in Marie Davion, which give rise to the various theories about her supposed influence over her husband.

Special Skills/Power:

Marie Davion has kept up her interest in aircraft, and she owns and pilots several craft of her own. Perhaps because of her own solitary nature, she has also developed a strong interest in the wildlife of New Syrtis and champions efforts to preserve these native lifeforms.

Name: Yvonne Davion
Title: Field Marshal
Position: Prince's Champion
Birthplace: Thames, Victoria
Age: 62
Sex: Female

Distinguishing Physical Characteristics:

Yvonne Davion resembles nothing so much as a kindly grandmother who is most at home baking cookies for the visiting grandchildren. With her grey hair and matronly figure, she looks somewhat incongruous in her Field Marshal's uniform. Only the glint of uncompromising authority in her brown eyes and a slight limp betray a long career in the military.

Brief Personal History:

Yvonne Davion was born on Victoria in 2963 during the reign of Andrew Davion. She was the eldest daughter of the famous Duchess Matilda Davion, who commanded the Royal Brigade of Guards for 20 years. Because of her heritage, everyone expected great things from Yvonne, too.

Unfortunately, she was born with a hidden heart defect that gradually restricted her ability to participate in physical activities. When she was 16, she agreed to undergo a risky operation that would, if successful, eventually allow her to lead a normal life. The operation was a success, though the recovery was long and difficult for the young Yvonne.

When she had recuperated, the girl began to make up for lost time by building up her body and learning to enjoy the many activities that had been denied her till then. At age 18, she traveled to New Avalon to meet her distant relations, the New Avalon Davions, and to enter the Albion Military Academy. At the Academy, Yvonne relished the physical strains of piloting a BattleMech and she graduated with honors.

After assignment to the Fifth Crucis Lancers, she soon showed that her ability to withstand the rigors of piloting a 'Mech were surpassed only by her innovative tactical skills. Though her actions often got her into trouble with the more conservative commanders, none could deny her genuine ability to devise and lead actions. She rose rapidly through the ranks, becoming Military Commander of the Capellan March in 2994.

During her brief tenure as Commander of the Capellan March, Yvonne Davion attempted to cut away much of the unnecessary bureaucracy that was slowing communications. Unfortunately, these efforts soon ran afoul of the Haseks, who felt Yvonne was infringing on their rights as noble rulers of the Capellan March. In a bitter political struggle, Yvonne was forced out of her post as Military Commander in 3000.

Yvonne almost resigned from the AFFS in disgust. She was particularly upset with Prince Ian for not doing more to help her against the Haseks. It was a very young Hanse Davion, who had always admired Yvonne Davion, who managed to convince her to stay on in the AFFS. For the next 13 years, Yvonne Davion was head of the Analysis and Speculation Department of Military Intelligence, a post she endured but did not particularly enjoy.

When Hanse Davion became Prince of the Federated Suns, one of his first actions was to appoint Yvonne Davion as his Champion. Under this title, Yvonne Davion is the de facto second-in-command of the AFFS. When Prince Hanse is unavailable to make a military decision, it is Yvonne Davion who takes over the authority. Because the Prince learned most of his own innovative tactics by studying Yvonne's career, the two rarely disagree on the proper course of action.

Motivations/Desires/Goals:

Yvonne Davion has never really forgiven the Hasek family for forcing her out as Military Commander of the Capellan March. Though Hanse has asked her to keep her anger and desire for revenge in check, Yvonne has nevertheless used her position as Prince's Champion to strike out at the Haseks. She can show her power in small ways, such as delaying shipments of supplies or sending the wrong spare parts. She does not seem to realize that this plays into the hands of Duke Michael, who uses such minor incidents as proof to the people of the Capellan March that the central government and the AFFS are insensitive to the needs of their region.

Except for her hatred of the Hasek family, Yvonne Davion is a clear thinker and thoroughly loyal member of the Federated Suns. She wants to see the Federated Suns grow to become the dominant force in the Inner Sphere, not as a realm to be feared but as a model to inspire others. For this reason, she fully supports the Prince's alliance with the Lyran Commonwealth. Though she has some private concern about the general competency of the LCAF's officers, she realizes the potential of such an economically strong ally.

Manner:

When on duty, Yvonne Davion is tough and unflinching in doing her duty. She avoids what she calls "the royal fashion show" whenever she can and feels uncomfortable when forced to attend official functions.

Off duty, Yvonne Davion is a kinder, gentler person. Indeed, she becomes as grandmotherly as her outer appearance, often visiting the Royal Orphanage in Avalon City to bring baked goodies to the children.

Special Skills/Power:

Though Yvonne's days of vigorous athletic activity are long gone, she still loves sports and often coaches the orphanage's soccer and baseball teams. Yvonne Davion is also an accomplished cook.

Name: Nelitha Green-Davion
Title: Countess of the Gold Isles on Galax
Position: Minister of Administrative Services
Birthplace: Long Island City, Galax
Age: 38
Sex: Female

Distinguishing Physical Characteristics:

Nelitha is a tall, slender woman with blue eyes and blonde hair. She recently shaved off half her hair in the fashion craze currently popular on New Avalon. Because of an accidental chemical explosion, her left hand and arm were badly burnt during her girlhood. To cover up the nasty scars, she wears clothing with long sleeves whenever possible.

Brief Personal History:

The Greens were originally a powerful family in the Terran Member-State during the era of the Star League. They owned several major companies and wielded considerable political influence in the government of the Star League. When the League fell, survivors of the Green family fled to the Federated Suns. Two years after the outbreak of all-out war, Rebecca Green married Charles Davion. Of all the Greens, only the newly married Rebecca survived the First Succession War. In memory of her family, she and her husband asked permission to use Green-Davion as their surname.

Nelitha grew up on Galax and her childhood was uneventful but pleasant. As she grew older, it became clear that she had inherited her family's flair for leadership but lacked tactical ability. As a soldier in the AFFS, she was transferred from a combat unit to Transportation and Resupply Command, where her administrative skills soon made her the department's second-in-command.

In 3024, Prince Hanse Davion asked Nelitha to leave the AFFS and assume command of the Ministry of Administrative Services, whose incompetence and corruption had made it the laughingstock of the federation government for years. Prince Hanse Davion believed that Nelitha's natural talents and tough attitude would be just the ticket to clean up this Ministry and turn it into a strong and efficient branch of the government.

Motivations/Desires/Goals:

Nelitha is torn between the desire to live a less hectic life away from New Avalon and the desire to fulfill the task Prince Hanse assigned to her. She takes considerable pride in her abilities and realizes that whipping the Ministry of Administrative Services into shape would be the crowning achievement for her career.

Manner:

Nelitha is strict disciplinarian who will not tolerate anyone who does not perform to the best of his abilities. She has been nicknamed "The Whip Minister" for the way she carries her Minister's Baton. Nelitha is also brash and arrogant enough to dismiss people for arriving late to work or for being dressed improperly.

Off-duty Nelitha is another person. She loves to attend social functions at the Royal Court, particularly dances or plays. She looks forward to the end of the day's work as much as her employees so that she can be free to enjoy other activities and escape her nasty reputation, which pains her.

Special Skills/Power:

Nelitha has always been an excellent dancer, and she is skilled in many forms of dance. As a younger woman, she did tour with a prestigious dance troupe for a short time.

Name: Ivan Tevanol
Title: Marshal
Position: Commander of Military Intelligence
Birthplace: Ulsonia Valley, Baranda
Age: 51
Sex: Male

Distinguishing Physical Characteristics:

Having grown up on a very poor world, Marshal Ivan Tevanol has a slim, almost emaciated physique. He is far from weak, however, and daily practices his combat skills with students who quickly come to appreciate his power. His bald head throws into relief the sharpness of his features and his piercing brown eyes. Tevanol's wrists still bear a number of jagged scars, souvenirs from his imprisonment by the Draconis Combine in 3002.

Brief Personal History:

Ivan Tevanol was the fifth son of a minor noble family on Baranda. Though he was a child with grand dreams and desires, his family was as poor as his world. It was unlikely that Baranda would ever give him the chance for meaningful work or a fulfilling life. When Ivan turned 16, he caught a ride on a commercial JumpShip headed for the interior of the Federated Suns.

On Torrence, he enlisted in the AFFS. After his superiors in boot camp discovered that he had a flair for languages and an amazing memory, Ivan was transferred to one of the secret Intelligence training bases, where he was trained as an agent for MI4.

Upon graduation, he was immediately sent out as one of the clerks serving with the diplomatic mission in the Free Worlds League. Once there, Ivan carried out several missions to obtain information on the realm's 'Mech research facilities, for which he received a commendation. Though spying was a bit dangerous at times, Ivan considered it a kind of elaborate chess match. He had not yet encountered the darker side of espionage.

That all changed in 3002. He and a partner, Kelli Reaf, entered the Draconis Combine under the guise of Lyran JumpShip crewmembers. Their objective was to photograph a major aerospace facility on Galedon and to discover whether it was worth the attention of The Rabid Foxes or a BattleMech raid. Before they could accomplish this, Kelli and Ivan were caught and taken for interrogation. In the Draconis Combine, interrogation usually means torture, and Ivan's partner did not survive. Days later, after Ivan told his captors everything he could, they dumped him at the doorstep of a trading company with Federated Suns connections. Ivan barely escaped death and eventually made it back to the Federated Suns to recover.

During that time, he debated whether or not to remain in the MI, for spying no longer seemed amusing. Though he decided to stay, it was because it was the only job he knew. During the next 20 years, he rose slowly through the ranks of the MI, eventually serving in all seven departments.

In 3023, Intelligence Marshal Doger died. After studying the other department heads, Prince Hanse appointed Ivan Tevanol as the best replacement. Ivan, who was head of MI2 at the time, was reluctant at first to accept the post. When he eventually accepted, it was not so much to see the Federated Suns become the dominant force of the Inner Sphere, but to protect agents in enemy territories from having to suffer what he had known at the hands of the Kuritans on Galedon.

Motivations/Desires/Goals:

Ivan Tevanol is not typical of past heads of Military Intelligence, who have been gung-ho types always eager to embark on ambitious covert operations. Before he will even consider a plan, Ivan wants to see it plotted to the last meticulous detail and with plenty of safety measures built in for the agent's safety.

This cautious attitude is not popular with some of the other MI officers, who prefer the spontaneous style of previous regimes that had made the MI so feared and respected. Some colleagues also disapprove of the consideration Marshal Ivan shows captured enemy agents.

Marshal Ivan is well aware of these criticisms. In response, he has begun to draw up a series of intelligence and subversive actions to satisfy even the most ardent blow 'em-up-and-kill 'em proponent in the MI. There has also been a slight but noticeable shift toward more electronic surveillance and the use of agents only when absolutely necessary.

Manner:

Ivan Tevanol still speaks with a bit of his homeworld accent, which embarrasses him. He is reserved in the presence of strangers, watching everyone with great curiosity, as if the scene were a puzzle he had to fit together. Once he has a grasp on another's personality, he becomes more open. In contrast to his initial caution, Ivan is a generous and loyal friend. When drunk, he becomes morose, however, still brooding over the mistakes that led to the death of his partner and his bout with the Kurita interrogators.

Special Skills/Power:

Ivan has a nearly photographic memory and a flair for languages. He is also a good actor who can mask his emotions with ease.

Name: Ran Felsner
Title: Field Marshal
Position: Commander of the Royal Brigade of Guards
Birthplace: Kevin Crossroads, Lexington
Age: 49
Sex: Male

Distinguishing Physical Characteristics:

Ran Felsner is a tall, dark-skinned man whose early life as a lumberjack has left him with a very muscular physique. He is also quite handsome, with short graying hair and sharp Mediterranean features. Felsner has the kind of face that belongs to a courteous gentlemen one moment, then hints at danger, even menace, in the next. He has occasionally affected a monocle because he thinks it dashing.

Brief Personal History:

The heavy forests of Lexington were grown from saplings brought from Terra during the first Exodus. When the planet's precious metal deposits petered out, the planet was virtually abandoned after a century of colonization. This left the trees free to grow where and how they would. Many decades later, settlers returned to Lexington and discovered that the planet was now a major source for hardwoods.

The Felsner family was one of the first to begin logging Lexington's valuable forests. As a boy, Ran spent much of his early life among the woods. As soon as he could walk, his father gave him an ax to swing. According to the father, a Felsner must be able to chop down trees with only an ax in his arms. Years of swinging such axes made Ran strong.

Though Ran would have been excused from service in the AFFS as the only son in his family, he enlisted in 2991. His military career began at the Blackwater Boot Camp near Devil's Hole, Lexington.

Upon graduation in early 2992, he entered the 2nd Lexington Regulars, a motorized infantry regiment then assigned to the 17th Avalon Hussars. The Hussars were stationed on Dobson in the Draconis March, and in September of that year, the Hussars RCT invaded the planet Breed. With minimal support from other Davion units, the regiments of the Hussars seized the planet's spaceports and pushed the Kurita units into the wilds of the planet. During this campaign, Ran Felsner won a battlefield commendation and promotion by the Captain of a 'Mech company. Ran took control of his platoon and held a key defensive position when the platoon's officers were killed in an autocannon blast. Because of his leadership skills and courage, Ran was sent to a Skills Center for cross-training in vehicle repair, missile weapons operation, and communication systems operation.

Sergeant Ran again showed his abilities in 2994 when he led an unofficial but successful attack against an enemy strong point on Klathandu IV. Though some of the officers involved wanted to court-martial Ran for acting without orders, Colonel Denni Frosch, commander of the 2nd Lexington Regulars, disagreed. Impressed with Ran's potential as a tactician, Frosch had him transferred out to the Battle Academy at Robinson. Though an unusual move, it was a stroke of far-sightedness on the Colonel's part.

In 2998, Ran Felsner graduated. It was not with the rank of Sergeant that he left the Academy, but with the rank of Leftenant, which some MechWarriors spend their entire careers trying to attain. He was given an old but serviceable *Griffin* and assigned to the regimental battle group of the 17th Hussars so that he could continue his education under Colonel Frosch.

For the next three years, Ran Felsner learned what it took to command a regiment by acting as the Colonel's aide and bodyguard. On numerous occasions, Ran was given command of the Regimental Command Lance to carry out missions. In 2999, he received the Diamond Sunburst Medal for his actions on Kesai. It also won him a promotion to Major.

In 3001, disaster struck the 17th Avalon Hussars RCT when two *Overlord* DropShips carrying almost all the regimental commanders were lost en route to a combat mission against House Kurita on New Medham. When the regiments dropped onto the planet, they found themselves leaderless. Realizing that something had gone drastically wrong, Major Felsner assumed control of the Hussar 'Mech regiment. From the panicky communications of the other regiments, he soon discovered that everyone else was without command. As leader of the 'Mech regiment, Ran assumed complete control of the RCT. He broke off the planned operation and held the RCT together long enough for an orderly withdrawal to Davion DropShips. For his heroic actions, Ran Felsner received the Medal Excalibur and was promoted to the rank of Leftenant General, though his audacity and rapid rise did not please everyone. He was then sent to the Albion Military Academy on New Avalon to sharpen his administrative skills.

After his stint at the Albion, Ran took command of the Alcyone Capellan March Militia, a collection of regional regiments designed to defend a section of the Capellan March. Defense, however, seemed to be a dirty word to Leftenant General Felsner, and in 3008, he and his March Militia units launched an all-out attack to seize the Capellan world of Daniels. That it was completely against orders did not seem to bother Ran. When later asked why he had risked so much, Leftenant General Felsner replied that he saw a weakness and exploited it, believing that victory was more important to the Federated Suns than obedience to protocol.

Luck was with Ran again. The attack was initially successful, leaving the always ambitious Duke Michael Hasek-Davion aware that he was about to lose a golden opportunity to gain fame and adulation. The Duke eagerly brought in reinforcements to rescue Ran's March Militia, which had started to lose ground. The planet fell to the Federated Suns in 3009. To protect himself, Leftenant General Felsner publicly laid all the credit at the feet of the Duke of New Syrtis, which neatly avoided yet another court martial. The grateful Duke awarded Ran the Syrtis Medal of Honor and recommended him for another promotion.

The promotion passed easily because Ran's meteoric career had by now caught the attention of Prince Hanse Davion. The two met in 3010 when the new Major General arrived in New Avalon to assume control of the Davion Heavy Guards RCT, the Prince's favorite unit. The two men became friends, as Ran Felsner slowly became a military advisor to a Prince who values his unorthodox military strategies. In 3015, Major General Ran Felsner became Marshal Ran Felsner.

In 3022, Marshal Ran participated in the Battle of Rigil Kentarus. He assumed command of the 'Mech regiment and traveled to that Commonwealth world to participate in what was supposed to be a military exercise with the 1st Royal Guards of the LCAF. The purpose of the exercise was to work out procedures for use during future joint LCAF-AFFS missions. No one had counted on the surprise Kurita offensive to seize the world. Defense proved difficult because the Kurita regiments were so numerous and because of poor communications between the defenders. Eventually, Marshal Ran and General Jack Sphire, commander of the Lyran forces, did manage to push off the Kurita forces. This action won Ran the right to carry the McKennsy Hammer, a high Lyran award. It also resulted in his promotion to the rank of Field Marshal as well as his being awarded the Order of Davion by Prince Hanse.

Two years ago, Ran Felsner became the first commander of the Royal Brigade of Guards, which gives him personal control over almost a hundred elite regiments of 'Mechs, armor, and troopers. The posting was a clear and not so subtle hint to Duke Michael Hasek-Davion that he would come up against Ran Felsner if it ever entered his head to challenge Prince Hanse for the throne.

Motivations/Desires/Goals:

Ran Felsner is not an ambitious man, for all his promotions and decorations. These came because he did what was necessary at the time, not because he sought for reward. As a result, Ran is a bit befuddled by all the acclaim.

Last year, he met and fell in love with the beautiful young Countess Cordelia Spenser of Torrence, which has affected many of his actions in recent months. Felsner might be contemplating retirement should he succeed in winning the hand of his Cordelia.

Manner:

Marshal Ran Felsner is a confident man whose manner is self-assured and authoritative no matter what the occasion. He is also quite a wit. Combined with his good looks, these qualities usually win him the admiration and friendship of most people he encounters.

Special Skills/Power:

Ran still fondly remembers his early life in the forests of Lexington. He frequently visits the Royal Game Preserves around Avalon City, aiding the groundkeepers with their forestry duties. He is also an excellent horseman, who will teach the skill to anyone of the Royal Court who might be interested. It was in this context that he first became acquainted with Cordelia Spenser.

Name: Stephen Davion
Title: Duke of Bristol
Position: Commander, 1st Davion Guards
Birthplace: Bristol Manor House, Bristol
Age: 45
Sex: Male

Distinguishing Physical Characteristics:

Marshal Stephen Davion is a stocky man who wears a beard and long black hair. His expression is a perpetual scowl that shows how much he disapproves of what is not up to his standards—which is just about everything. He wears a black patch to cover the loss of an eye in a 'Mech battle, which adds to his already dark and angry appearance.

Brief Personal History:

Stephen Davion is the last member of a family group known as the Ducrimmon-Davions, a line that split from the royal family during the later stages of the Second Succession War. For a time, the Ducrimmon-Davions were prominent and important members of the central government and the AFFS.

When battle damage, bad investments, and the death of several key family members plunged the family into poverty on their homeworld of Benet III, they were far too proud to ask for assistance from the royal family. The Ducrimmon-Davions seemed to vanish from the face of their world, and their disappearance went unremarked by the far-off Davions of New Avalon.

By the time Stephen was born in 2980, his family was destitute and living in a squalid apartment in the Benet capital of Donner. While growing up, Stephen heard many stories of the family's former glories, all the while trying to keep himself and his crippled parents alive through acts of petty theft. Unfortunately, his thievery resulted in several stints in detention homes, with his parents left to fend for themselves. It was during one of these forced holidays that Marie Davion visited the home and happened to notice that the young boy's last name was Davion. Intrigued, Marie met and talked with the child. A subsequent check into Stephen's parentage revealed his noble blood, and before long, the New Avalon Davions rescued the remaining members of the Ducrimmon-Davions.

By the time Stephen was 15, he was alone in the world. The Davions sent him to the Albion Military Academy, thinking a soldier's life would help channel the anger that seemed to fester in his heart. After three years of arguing with his classmates and his instructors, Stephen barely managed to graduate. It was only because of his tactical and strategic skills in preparing battles that his instructors did not fail him.

Now, at last, Stephen could turn his rage against the enemies rather than the friends of the Federated Suns. During his career with the New Ivaarsen Chausseurs, he was a particularly violent commander, who often punished defeated enemies far beyond the limits most military men imposed upon themselves. On several occasions, he was reprimanded for acts unbecoming an officer of the AFFS. One of the incidents involved several Kurita officers who "mysteriously" died just after Stephen Davion had interrogated them.

His promotions have always been heavily debated within the AFFS. Though no one can argue against his skills as a leader and tactician, even the most patriotic Field Marshal pities the enemy that must face the violent Stephen Davion.

His promotion to the First Davion Guards RCT in 3021 was an attempt by Prince Hanse to drag Stephen away from the front lines to the quiet of New Avalon. On New Avalon, the Prince hopes to persuade Stephen to talk with counselors who may be able to help him dissolve the deep cancer of his hate.

Motivations/Desires/Goals:

Stephen has many conflicting desires and goals. On the one hand, he wishes to uphold the Davion name with military honors so that the Ducrimmon-Davions will be remembered. On the other hand, he bears a serious grudge against the royal Davions for having allowed his family to slide so deeply into poverty.

Stephen has kept in touch with Marie Davion, who is the one friend he has made in his life. Some evidence suggests that she has advised him not to speak with the psychologists that Prince Davion has sent to him.

Stephen has never liked Field Marshal Ran Felsner and does not bother to hide it.

Manner:

Stephen treats everyone he meets like a potential enemy. If someone manages to break through his defenses to persuade Stephen that he means no harm, Davion treats them only slightly better.

Special Skills/Power:

Stephen Davion has considerable street savvy because of his early life. He is also an accomplished writer of poetry, though Marie Davion is the only person who knows of or has seen any of his work.

Name: Olivia Fenlon
Title: Duchess of Chesterton
Position: Minister of Foreign Relations
Birthplace: Chesterton Castle
Age: 42
Sex: Female

Distinguishing Physical Characteristics:

Olivia Fenlon is an attractive woman of average height, who has long black hair and curiously glimmering green eyes. The only thing that mars her natural grace and beauty is a long scar left by a laser down the back of her right hand and arm. It is a memento of her last visit to the Lyran Commonwealth.

Brief Personal History:

Olivia Gohalm was born to a middle-class family on Chesterton in 2983. Chesterton, once a major planet in the Capellan Confederation, is today a busy tradeworld with many economic connections to the other realms in the Inner Sphere.

Because of her father's business interests, Olivia grew up surrounded by the diverse people and cultures that mixed with the daily life of Chesterton. So fascinated was she that Olivia filed a Single Offspring Paper when it came time for her to serve in the military. This excused her from service in the AFFS because she was an only child, but she must serve in some other capacity. Olivia entered Chesterton University and began to study the history and languages of the Inner Sphere. She had a gift for both that led to a position as translator in the Chesterton Department of Commerce upon graduation.

It was there she met and wed Duke Charles Fenlon. Soon after, Olivia accompanied her husband to the Royal Court on New Avalon. The Ministry of Foreign Relations put her language ability to good use in translating foreign magazines and books. They also discovered that she had a deep political understanding of the relations between the five Houses.

When Duke Fenlon died in 3011, Duchess Olivia Fenlon decided to remain at the Ministry of Foreign Relations on New Avalon. She therefore passed some of her ducal powers to her husband's brother, Nicholas Fenlon, so that he could handle planetary affairs back on Chesterton.

By 3013, Olivia had become Sub-Minister in charge of relations with the Lyran Commonwealth. When Hanse Davion became Prince that same year, he selected her as the new Minister because he appreciated her ability to suspend common assumptions about the other Houses in order to discover new and exciting possibilities.

When Katrina Steiner made her peace proposal in 3020, it was Minister Olivia Fenlon who persuaded Prince Davion to follow up on the communication. Together, the two began the delicate negotiations that led to the alliance between the Federated Suns and the Lyran Commonwealth.

Since that time, Olivia Fenlon has been in close communications with her counterpart in the Lyran Commonwealth, Grafina Cynthia Eddies, Chancellor of Foreign Affairs.

Motivations/Desires/Goals:

Olivia Fenlon is intent on seeing that the alliance between the Federated Suns and the Lyran Commonwealth is successful. She hopes that it will give both realms such obvious advantages over the three remaining Houses that they will be afraid to harass the Federated Suns. If there is to be competition within the Inner Sphere, Olivia believes it should be in matters of trade rather than through the violence of war.

Despite her peaceful views, she was the target of an assassination attempt in 3024. During one of her visits to Grafina Cynthia Eddies on New Earth, an assassin fired at Olivia as she left the building where the negotiations were taking place. The shot struck her right hand and arm but did little more than scar her.

It did not take the Diplomatic Guards protecting Olivia long to catch the assassin. Upon questioning, it was learned that she was a member of the Brotherhood of Cincinnatus, an anti-government group that violently disapproves of the alliance with the Federated Suns. Though the attack had no affect on the alliance, it has shaken Olivia Fenlon, who now questions many of her former assumptions.

Manner:

Olivia Fenlon is a gracious, elegant women who never overstates her beauty nor understates her intelligence. Her friendship with Prince Hanse made her the logical target of speculation about a possible marriage between the two.

When working, she will not tolerate those who fear to voice their opinions. Those who have failed her repeatedly may find that this unfailingly kind woman has suddenly become cold.

Special Skills/Power:

Aside from her language abilities, Olivia Fenlon is a skilled glass sculptor.

Name: Joan Davion
Title: Duchess of Streator
Position: Minister of Ways and Means
Birthplace: Kremlin, Streator
Age: 56
Sex: Female

Distinguishing Physical Characteristics:

Joan Davion is a plump woman whose years in the federal government have turned her once red hair white with worry. She wears old-style glasses and has perfected the art of pushing them up her nose to convey a sense of stern disapproval.

Brief Personal History:

Joan Davion is a distant cousin to Prince Hanse. From the age of two, she showed an insatiable curiosity for mathematics and sciences, and was programming her father's personal computer by the age of eight. By the time of her sixteenth birthday, her family applied for, and won, an indefinite deferment of service so that Joan could spend the next ten years studying mathematics, earning degrees from several universities.

Upon graduation, she was immediately recruited into the Chancellery of the Exchequer. Her extraordinary abilities won her many promotions, and one day she was assigned the daunting task of improving the economy of the Federated Suns. These projects took a toll on Joan, who began to realize how many people's lives would be affected by what she did or did not do. Soon, she was contemplating leaving the Chancellery for life as a university professor, where her only problems would be abstract mathematical ones.

This was not to be. When Prince Ian died and Prince Hanse came to power, the new Prince pleaded with Duchess Joan to stay on. He even offered her control over the entire Chancellory. This won over Joan Davion, who had always dreamed of what she might do if only she were totally in charge. She has been a particularly successful Minister of Ways and Means ever since.

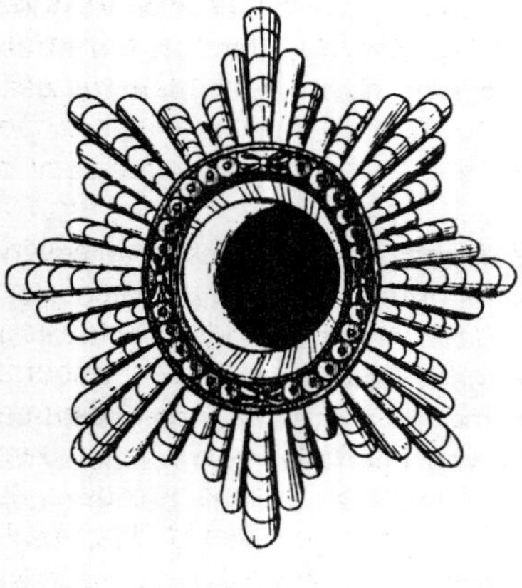

Motivations/Desires/Goals:

Joan Davion wants to see the Federated Suns make use of its enormous economic potential, particularly in developing the potential of many Outback or war-ravaged worlds. During the past twelve years, she has been able to implement her own policies, and their positive impact on the realm have pleased her. The work is so difficult and stressful, however, that Joan has lately begun to contemplate retirement again.

Manner:

Joan Davion is a quiet person who prefers to lead by encouragement rather than by force. Noble society bores her, for she would much rather share a coffee with friends and associates than attend an official function.

Special Skills/Power:

Joan Davion's hobbies include an interest in billiards, whose mathematical aspects intrigue her.

BRIEF ATLAS

INTRODUCTION

The Federated Suns is the largest of the five Successor States in both extent and number of inhabited worlds under its banner. The Davion realm occupies a wedge of space roughly in the lower-right quarter of the Inner Sphere. It shares a long and convoluted border with the Capellan Confederation as well as borders with the Draconis Combine and the Outworlds Alliance. The Federated Suns extends farther than any of the other Successor States into the murkiness of the Periphery, claiming worlds far from the warmth of Terra. Hivrannee, a world in the Crucis March, is over 600 light years from Terra and some 350 light years from New Avalon, capital of the Federated Suns.

This brief atlas provides an historical, political, and economic sketch for 30 worlds in the Federated Suns. These worlds were chosen either because of their importance or because they illustrate a typical feature or particular distinction of the Federated Suns. Lack of space prevents more detailed studies, but these entries will provide at least a glimpse into the vast and fascinating range of Davion worlds.

[NOTE: If there is a working recharge apparatus at one or both of a system's jump points, then the entry will list *Zenith* and/ or *Nadir* to indicate at which jump station the apparatus is located. *Percentage and Level of Native Life* is a rough estimate of how much of the world's planet and animal life is indigenous, as well as a measure of the highest level of development reached by its fauna.

World Name: New Avalon
Star Type: G5V
Position in System: 4
Time to Jump Point: 7 days
Recharging Station: Zenith and Nadir
Noble Ruler: Prince Hanse Davion, Duke of New Avalon
Political Leader: Same as above
ComStar Facility Class: A
ComStar Representative: His Most Exalted Ambassador Huthrin Vandvel, Precentor of New Avalon Station
Percentage and Level of Native Life: 50%, Mammal

New Avalon was first discovered in 2205 by the Terran Alliance explorer ship, the *TAS Camelot*. Even from orbit, the lushness of the planet was plain to see. The report that the *Camelot* filed on the planet was so glowing that New Avalon quickly became one of the prime destinations for shiploads of colonists leaving Terra between the years 2213 and 2221. What the colonists found when they arrived was a world of five continents, four seas, a fairly reasonable weather pattern, abundant fresh water, and a profusion of lifeforms that seemed to accommodate the colonists very well.

It was quickly discovered that the planet was best-suited to agriculture. By crossing Terran plants with varieties indigenous to New Avalon, the colonists had soon bred several important grain crops. The most famous of these is New Avalon wheat, which is extremely hardy and now grows throughout the Inner Sphere. Since then, New Avalon, with its countless square kilometers of open rolling country, has always managed to supply enough food for itself and its five nearest neighbor planets, despite the decline in agricultural technology brought on by the wars. Though its agricultural strength is usually overshadowed by the fact that New Avalon is the capital of the Federated Suns, the planet's fertility is vital to the well-being of the entire Crucis March.

New Scotland is the continent furthest north. Often deep in snow for nine months of the year, the continent is famed for its hardwood trees and animal furs. The most-prized of these furs comes from the legendary Silver Ice Snake-Weasel, a devilish creature nearly two meters long and with a nasty disposition. The continent also has the planet's only large deposit of precious metals. The continental capital of New Scotland is Dundee.

The New Hebrides is a group of islands in the planet's largest sea. Lying near New Avalon's equator, the islands have the hot climate that makes tropical plants and fruits the region's main export. The New Hebrides also have some high mountain ranges. Combined with the islands' location, these mountains attract rather vicious hurricanes that often cause heavy damage to the local communities. The largest of these islands is Carlisle. The continental capital of the New Hebrides is Thomasville.

The continent of Rostock is the third largest on New Avalon. A large and flat land mass, it has the planet's only desert, the Drilands, and the planet's only large canyon, Gaveston's Gorge. Rostock produces a substantial part of New Avalon's food animals. Large sections of the continent are owned by noble families who raise a wide variety of animals that are exported offworld to satisfy hungry people throughout the Federated Suns. The continental capital of Rostock is Flensburg.

Brunswick is the second-largest continent. It was named after the Brunswick family, major landowners during the planet's early history. Today, Brunswick boasts several large centers of oil production, as well as two major spaceports, three major cities, and a majority of the planet's many industries. It also has numerous small farms and ranches. The continental capital of Brunswick is Brunswick City.

Albion is the largest continent on New Avalon. The continent is also the most geographically diverse, with three mountain ranges, many large rivers, thick forests, and large prairies. Albion is also the location of most of New Avalon's industries. Avalon City is triply honored as capital of the continent, capital of the Federated Suns, and ancestral home of the Davion family.

Avalon City began life as a theme park. During the first century of Human life on this world, an eccentric millionaire decided that there should be at least one place on New Avalon that matched the planet's medieval geographical names. For the location of his theme park, he chose a large island where the continent's three major rivers, the Albion, the Thames, and the Cris, came together. This site was baptized the Isle of Avalon. Though the park buildings have long since fallen into ruin, the romantic medievalism has remained. When the Davions took control of the area, they promoted the tradition of festivals extolling the virtues of knighthood in flower.

Lucien's rise to power made it clear that the Davions would rule over the planet for many years, for he decided to incorporate the medievalism of the former theme park into the architecture of a new capital city. Dominating the city from atop Mount Davion is the Royal Palace, which some still refer to as Camelot. It is an enlarged copy of an English medieval fortress that Lucien had seen as a youth during a trip to Terra. The Royal Court is a complex of buildings and halls housing the ministries of the Federated Suns. It was built at the foot of Mount Davion just below the Palace. Protecting the Palace, the Court, and Avalon City is a battalion of Davion's elite Royal Brigade, which is stationed just between the Royal Court and the city.

Since the completion of Avalon City in the late 2390s, expansion has slowly eroded the medieval flavor. Buildings constructed outside the core of the city, which is known as the Old City or the Government District, were not obliged to install medieval facades and so adopted modern styles.

Some 30 kilometers to the south of Avalon City, in what used to be a private game preserve of the Davions, is the New Avalon Institute of Science. The seven colleges each specialize in a particular discipline, though inter-college learning is encouraged. The many university buildings cluster about a large and comfortable campus. Nearby are the training grounds for BatttleMech and AeroSpace Pilots, and the university's private spaceport.

Security is tight around the Institute's many secret projects. Indeed, the NAIS can be reached only via underground subways or by flying into the university's spaceport, which allows the government to keep close track of who enters and leaves the university. As a deterrent and a force of last resort, two BattleMech battalions are stationed within the university grounds to help with security. They also serve as skilled opponents to cadets of the NAIS BattleMech Training School.

Hidden in the mountain range that extends west from Mount Davion is the Davion military headquarters. Known as The Mountain, or as the Fox's Den, it is a complex of huge caves and transportation tunnels deep inside a granite mountain. The magnetically levitated subways lead to and from Avalon City, as well as to the two 'Mech battalions and accompanying fighters hidden in strategically placed caverns throughout the valleys of the mountain range. The need for a heavily fortified military headquarters became evident during the First Succession War, when Draconis Combine forces pushed to within one jump of an under-defended New Avalon.

Despite the collapse of the Star League and the three wars that followed, New Avalon has managed to maintain a fairly high level of technology. The planet's population, though large, lives a comfortable life scattered over the four continents. Public morale is high, with only a few dissidents among the youthful radicals at the Gogh-Bukowski University. Avalon City draws to it many of the realm's best artists, who have embellished the city with their art. New Avalon and Avalon City may be the most agreeable of the five major capitals of the Successor States.

World Name: New Syrtis
Star Type: K5IV
Position in System: 4
Time to Jump Point: 4 days
Recharging Station: Zenith
Noble Ruler: Michael Hasek-Davion, Duke of New Syrtis
Political Leader: Same as above
ComStar Facility Class: A
ComStar Representative: Precentor Roberta Hilo
Percentage and Level of Native Life: 10%, Plant

New Syrtis is a large world in the grip of an ice age. Huge sheets of ice have covered most of the planet, leaving only a narrow band of temperate and tropical weather around the equator. New Syrtis was first discovered during the era of the Terran Alliance, but was not colonized until the late 2260s. Those that did finally settle here were hoping to get rich by discovering deposits of precious metals and minerals said to be hidden beneath all the ice and snow. The gamble paid off when the first settlers found large deposits of metals scattered all over the planet.

New Syrtis became a boom world overnight. Thousands made their way here in hopes of striking it rich. After months of solitary struggle, the miners would trudge their way from the ice and snow with tons of ore and the urge to raise a little hell. Cities like Pink Passion, Lady's Favor, and Hangover Hell began to sprout in the temperate zone. They had only one function—to strip the miners of their wealth in the quickest possible fashion. The deposits of easily mined metals soon played out and the boom was over. Well-organized mining corporations then moved in to mine the deeper deposits, which spelled the end of the independent miner and the raucous sin-cities.

It was at about this time that New Syrtis was entitled to the Hasek family. Always an intelligent, though not particularly scrupulous, family, the Haseks slowly won a stranglehold over the planet's resources. They were able to rein the avarice of the metal companies, who would have stripped away huge chunks of the surface—nearly destroying the planet's fragile ecosystem—in their search for profits. Whether the Haseks did this out of concern for their world or the desire to own all of its wealth is not certain. What is certain is that their policies have resulted in rationing of the vast metal and mineral wealth of the planet over the centuries, with New Syrtis as important to the Federated Suns today as it was in an earlier era.

Because of its importance, New Syrtis is the frequent target of attacks by the nearby Capellan Confederation. When under attack, the planet has as its fearsome ally the almost constant blizzards and bone-freezing cold spells. It has been estimated that one out of every five invading Capellan 'Mechs and tanks is lost because it runs afoul of the weather.

The defensive advantage of the arctic climate is one of the reasons why the Hasek family decided to build their capital city, Saso, on the northern continent instead of the more temperate equator region. The city is in some ways reminiscent of Avalon City, with the Hasek palace set upon a small complex of hills overlooking the city and with the University of Saso nearby. The current Duke of New Syrtis, Michael Hasek-Davion, would like to see Saso University out-distance rival the New Avalon Institute of Science. Near the university and hidden underground beneath the perpetual snow is the Regional Military Headquarters, where Duke Michael coordinates the military activity of the entire Capellan March.

World Name: Robinson
Star Type: G5IV
Position in System: 3
Time to Jump Point: 7 days
Recharging Station: Nadir
Noble Ruler: Aaron Sandoval, Duke of Robinson
Political Leader: Council Leader Hurigu Devlin
ComStar Facility Class: A
ComStar Representative: Precentor Ludmilla Rovta
Percentage and Level of Native Life: 40%, Mammal

Robinson is a large, temperate world. The planet's fertile black soil attracted colonists early in the Exodus, but it did not take long before they realized that Robinson's real treasures were its hidden deposits of oil, precious metals, and rare minerals.

As a result, the farmers, who had already claimed a majority of the available lands, came into conflict with foreign companies eager to get at the hidden wealth. The crisis collapsed the planetary government, and brought David Sandoval, distinguished commander of the planetary militia, to prominence. After assuming the title "Lord Protector" and taking control of the remnants of government, Sandoval then seized the assets and properties of the foreign companies. He justified this with the claim that their negative influence had affected the quality of life on Robinson, and it was his duty to defend his world against them.

Lord Protector Sandoval then worked out a radical land reform bill that would gradually free more land for government exploitation every time land was bought or sold. The government eventually made a fortune mining the lands acquired through these transactions. The land-reform bill also made the Sandovals rich because they owned the only oil and mineral drilling company legally authorized to dig for resources on government land.

More conservative policies eventually balanced the situation toward property owners again, but the Sandoval family had, meanwhile, become the planet's noble rulers. Robinson, because of its political and economic importance, soon became the focus of considerable traffic. When the Federated Suns was split among the Five Princes, Robinson became the capital of the Terran March. The Sandovals' power continued through the chaos of the Five Princes Era. They even survived the Civil Wars with their titles intact, despite the fact that the Sandovals had been staunch Rostov supporters. As a result of the reorganization of the Federated Suns after the Civil War, Robinson became the capital of a greatly enlarged Draconis March.

Robinson has since lost a substantial portion of its surface to mining operations, which has limited its ability to export raw materials and grow its own foods. In exchange, Robinson quickly became a major industrial center. During the good years of the Star League, Robinson was a major source of finished metal products used in the construction of JumpShips and DropShips. It also had many smaller industries producing consumer items such as personal vehicles and appliances.

With the collapse of the Star League, Robinson fell before the invading Draconis Combine forces as they penetrated deep into the Federated Suns. The Kurita forces destroyed most of the planet's heavy industries in an effort to break the population's will to resist. Though many died of starvation or became military casualties, the people of Robinson mounted countless attacks against the Kurita occupiers.

Robinson was eventually liberated, but not before 70 percent of the planet's industrial capacity had been destroyed. Only recently has the damage of the first war been repaired enough to bring the planet's industrial might near to its earlier levels.

World Name: Kesai IV (aka Kasai)
Star Type: A5II
Position in System: 4
Time to Jump Point: 22 days
Recharging Station: None
Noble Ruler: Jamison Trives, Duke of Kasai
Political Leader: Same as above
ComStar Facility Class: B
ComStar Representative: Precentor Reback Tyul
Percentage and Level of Native Life: No native life

Kasai is a hot, dry, desolate world with no native life, few mineral resources, and little to recommend it beside having once been a major Regular Army supply depot.

Amazingly enough, Kesai did have settlers long ago. What attracted them was the discovery of several deposits of radioactive minerals needed for the advanced electronics industry. When the deposits proved to be smaller than expected, only the few settlers who actually liked the hellish surroundings remained on Kesai. For a time, Kesai was used as a prison planet, first by the Draconis Combine and then by the Federated Suns after they took over the planet.

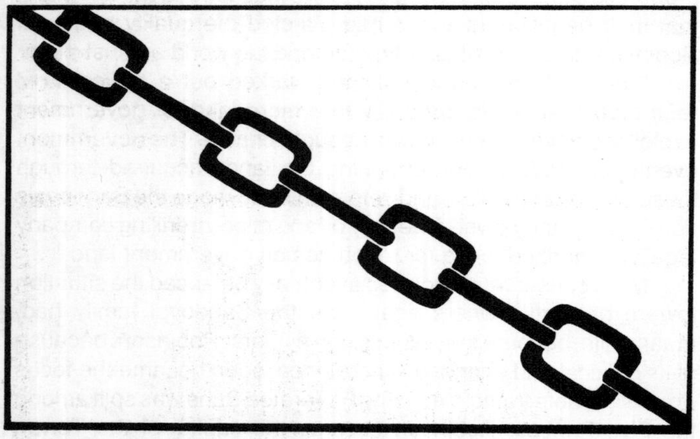

When the Star League came to power, it purchased the rights to Kesai from the Federated Suns. The League built a vast complex of warehouses, factories, maintenance facilities, and DropShip pads around the planet's single sea. The Regular Army used these facilities during most of the Star League era. After the League fell apart and General Kerensky and most of the Regular Army rendezvoused around New Samarkand, the Regulars stripped most of Kesai of its supplies. Soon after Kerensky's exodus, Davion troops moved into the supply depot and assumed control of Kesai. The huge warehouses were not as empty as had been expected, for the Regular Army had been forced to leave much behind in the rush.

When the Federated Suns sought to reinforce Kesai with extra troops, the Draconis Combine guessed why and launched immediate attacks to take the planet. This began a long series of battles between the two realms for control of the planet. The Federated Suns has by now removed most of the supplies left behind by the Star League to more secure warehouses in the Federated Suns. The Davion military still heavily fortifies Kesai to keep the Kuritans from ever being completely certain whether or not anything valuable remains in the warehouses. There have been additional supply caches discovered at the depot over the years, and engineers from the NAIS are studying some of the enormous machinery that Kerensky did not carry off, in hopes of rediscovering its uses.

World Name: Galax
Star Type: F5III
Position in System: 4
Time to Jump Point: 11 days
Recharging Station: Zenith
Noble Ruler: Aleksandr McCorkendale, Duke of Galax
Political Leader: President Simon MacCreedy
ComStar Facility Class: A
ComStar Representative: Precentor Douglas Selv
Percentage and Level of Native Life: 30%, Reptile

In some ways, Galax is the exact opposite of nearby New Avalon. Water- and life-poor, Galax was not much popular as a colony site. The tainted and unsettled atmosphere, constantly churned by the planet's one large moon, kept away all but the staunchest. The perseverance rewarded those who came when huge deposits of almost every valuable mineral, metal, gemstone, jewel, and oil were discovered on Galax. It was not too long before the arrival of a new flood of colonists willing to brave the conditions in search of wealth. It was good that New Avalon, with its abundance of food, was nearby, because the Galax population soon grew far beyond the planet's ability to sustain itself without the help of huge food imports.

During the peaceful years of the Star League, Galax was the focus of the Climate Restructuring and Adjusting Project (CRAP), a project sponsored by the Star League and funded by a Davion government eager to see the planet exploited. In the years 2721–33, CRAP engineers altered the planet's atmosphere with chemical and heat generators that slowly filtered out the most toxic elements while upping the oxygen content. By the end of 2733, the atmosphere was completely tamed, which sparked another wave of colonists. The project was such a success that firms from throughout the Star League began to open factories on Galax. The biggest of these was Boeing Interstellar, which built the huge Galax Megaplex on the planet during the League's last years.

Since the fall of the Star League, Galax has been vital to the Federated Suns war effort. Some of its industries, especially those that depended on the Terran member-state for direction, eventually died out because of the decline of technology. Most of the other industries have survived, though the Star League machinery is now jury-rigged and patched together from years of make-do repairs.

Two things could shut down the many industries of Galax. The first would be disruption in the flow of vital foodstuffs from New Avalon. One estimate indicates that a disruption of a single month would be enough to create panic in a population that has always been politically nervous. The other threat to Galaxian industries is the planet's weather. Over the past decade, a team of meteorologists from the NAIS has proven that the atmosphere of Galax is going toxic again, and at an alarming pace. If it were to become necessary for inhabitants to again wear environmental suits, as was necessary before the CRAP project, there might be a mass exodus offworld. This would, obviously, interrupt the flow of goods from the planet, causing economic and military difficulties throughout the Federated Suns.

World Name: Mallory's World
Star Type: G2I
Position in System: 6
Time to Jump Point: 8 days
Recharging Station: None
Noble Ruler: Rodkel Miller, Duke of Mallory's World
Political Leader: President Katherine DeBusher
ComStar Facility Class: B
ComStar Representative: Precentor Jasper Drexel
Percentage and Level of Native Life: 80%, Mammal

Early in the period known as the Exodus, Mallory's World was discovered and immediately colonized by European and English immigrants. Despite being one of House Davion's most agriculturally rich planets, Mallory's World is usually better known as a waystation for invasion forces.

The first colonists were shocked to find that the host of vicious parasites rampant among the planet's lushness were the carriers of numerous strange diseases. Dr. Beck Mallory was a second-rate parasitologist who had joined the colonists to escape a failed career on Terra. When his fellow colonists consulted him about the parasite problem, Dr. Mallory's first reaction was the desire to flee such responsibility. Gentle and not-so-gentle persuasion eventually convinced him to attempt to discover a solution. Mallory probably amazed no one so much as himself when he discovered a common link among all the native parasites, to which a simple vitamin complex shot would provide immunity. In gratitude, the colonists named their world after Mallory.

They quickly turned the planet into a garden spot, whose surplus food fed many less-fortunate nearby worlds. This might have been a happy world had it not been for the many occupations by repressive Combine or Capellan regimes. Despite the frequent fighting, Mallory's World survived the Age of War and eventually became a member of the Federated Suns. During the Star League era, the people of Mallory's World were able to use the new technologies coming out of the Terran member-state to further boost the yield of their crops and herd animals.

The collapse of the League and outbreak of the Succession Wars would change all that. The First War left the planet, its population, and the majority of the planet's agricultural centers seriously damaged. The second war, though less violent, did even more harm. The strategic location of Mallory's World between the Draconis Combine and the Capellan Confederation made it such a tempting target that those two realms were frequent and brutal in their attacks. The Federated Suns, not willing to part with such a strategically vital world, heavily defended Mallory's World.

By the beginning of the Third Succession War, three of the four continents on Mallory's World have been so trampled that little will ever grow there again. In recent years, the fourth continent has been declared off-limits by the Davion government so that it can eventually begin producing food. Even the Capellans and the Kuritans have decided to respect this liberated zone.

The planet's capital is Mallory City, located on the Northern Oregon continent. Not far from the capital city is Fort Dodger, the headquarters of the planet's BattleMech contingent.

World Name: Gulkana
Star Type: A5V
Position in System: 5
Time to Jump Point: 23 days
Recharging Station: None
Noble Ruler: Kevin Salos, Duke of Gulkana
Political Leader: Captain Thomas Yarborough, President-Elect
ComStar Facility Class: B
ComStar Representative: Precentor Ulivaon Livos
Percentage and Level of Native Life: 30%, Plant

Gulkana is currently the center of much political and religious unrest. A relatively small world with few resources, the planet would probably go unnoticed except for its one industry. The Yeffters Weapons Factory makes DropShip autocannons of all calibers. Most of its products go to Galax, where they are installed on the DropShips of Federated-Boeing.

When the Yeffters factory was shut down last year as the result of a political insurrection, everyone in the Federated Suns, including Prince Davion himself, sat up and took notice. The planet's democractic government had been usurped by a member of the planet's local militia. The nobility, including the weak and apathetic Duke Salos, was mollified with promises that their privileged positions were safe. Captain Yarborough, the new ruler of the civilian government, appeared to be a relatively reasonable man who promised to protect personal liberties and to lower taxes. It looked for a moment as though the change of governments had been for the better.

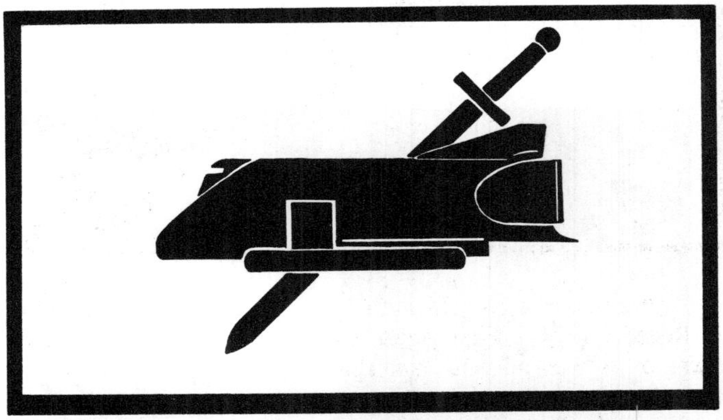

That opinion soon changed when Captain Yarborough proclaimed the Competency Laws, which used a farcical test to classify each inhabitant according to IQ. Those who scored high were given important jobs; those who scored low were given shovels. Though the test was inaccurate and biased against the majority Hindu population, it was acceptable to Yarborough's fanatic and well-armed supporters. The effect of this lunacy was immediate. The entire planet ground to a complete halt, including the vital weapons industry.

Because certain laws prevent the Davion central government from interfering in the affairs of its nobles, the most Hanse Davion has been able to do was attempt to pressure the apathetic Duke Salos to make a stand. The aging Duke has resisted the Prince's pleas for action, however. Some of the planet's lesser nobility would like to use their private forces against Yarborough, but Duke Salos directly controls the major cities, which effectively protects the usurper. At the moment, there is a grass-roots movement dedicated to the overthrow of the crazed Captain and to seek assistance of the Davion government, but it has not been possible to mount a coordinated effort so far.

The capital city of Gulkana is Kanpur.

World Name: Kathil
Star Type: F4V
Position in System: 3
Time to Jump Point: 10 days
Recharging Station: Zenith
Noble Ruler: Koster VanLees, Duke of Kathil
Political Leader: Chairman Dexter Mazon
ComStar Facility Class: B
ComStar Representative: Precentor Indra Seshu
Percentage and Level of Native Life: 30%, Insect

Kathil was once considered a small, unremarkable world near the Capellan border. The few deposits of rare metals were not enough to attract more than a trickle of colonists, most of whom had to turn to farming to survive. With the formation of the Star League and the struggle of the Reunification War, Kathil's importance soared.

The war fought to bring the distant Periphery states into the Star League put a heavy strain on the JumpShips and DropShips of the Star League Navy. The vast distances those ships had to travel for repairs—often all the way to Terra—might keep them out of the fighting for years. For this reason, the Star League decided to build major shipyards outside the Terran member-state in 2596. Among the planets chosen as sites for Star League Naval Repair and Refit Facilities was Kathil. It was chosen over others nearby because its star system is remarkably free of asteroids and dust that might easily damage the delicate JumpShips. It also has a remarkably stable sun with few solar flares or storms. The facilities were officially christened the James McKenna Shipyards in 2602.

The facility is actually composed of two separate shipyards. One was a huge complex of repair docks and machine shops for JumpShips orbiting near the zenith jumpoint. The other was a series of orbiting repair bays for DropShips. On the planet itself was a complex of repair industries and parts warehouses. Throughout the Star League era, the McKenna Shipyards serviced ships of the Star League Navy as well as from the private navies of the Capellan Confederation, the Federated Suns, and civilian trading companies.

Like so much else, that all ended with the collapse of the Star League. Capellan warships made their way to Kathil, and despite heavy resistance from Federated Suns warships, they managed to cripple the JumpShip yards. Several subsequent attacks aimed at both shipyards have further reduced their functioning. Though the JumpShip yards have been repaired enough to do routine ship maintenance and to perform a few major overhauls, so much was damaged and so much has been forgotten that most of the shipyard stands empty and unused. The DropShip yards are in much better condition and remain as the largest and best-kept DropShip repair facility in the Federated Suns.

Several raids by Capellan 'Mechs have destroyed most of the onworld warehouse and parts factories. Almost all the JumpShip parts have been destroyed or carried off. Quite a few parts for DropShips still exist, but the drastic loss of technology has left the use of some of these parts nothing short of a mystery.

The capital of Kathil is District City, near the Kathil General Motors engine factories.

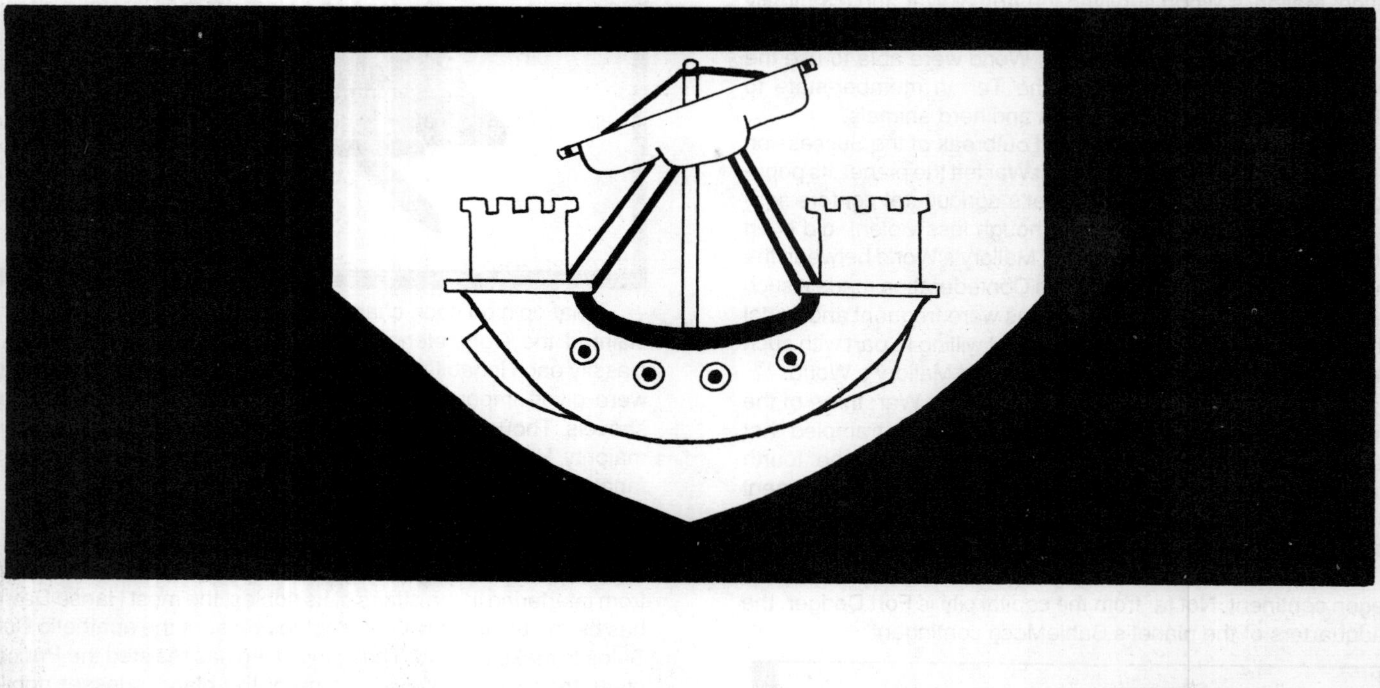

World Name: Redfield
Star Type: K5IV
Position in System: 3
Time to Jump Point: 3 days
Recharging Station: None
Noble Ruler: Jefferson Talbot, Duke of Redfield
Political Leader: Colonel Isolde Denefer (Military Governor)
ComStar Facility Class: B
ComStar Representative: Precentor Michael Opser
Percentage and Level of Native Life: 60%, Mammal

Among the most crucial variables that a strategist must consider when planning to invade a hostile planet is how long it will take DropShips and fighters to reach the target world. The longer it takes, the longer the planetary defenders have to prepare their own reception for the approaching enemy. An ideal world for attack is one whose sun is so small that invading forces could conceivably reach the planet before the planet's defenders realize the enemy had arrived in the system.

Redfield is such a planet. Its small sun makes it possible for DropShips and fighters to make the trip at maximum velocity in a matter of hours. Its terrain, with so many open areas, is also prime for invasion. This explains why Redfield has been the site of so many Capellan invasions, even though star systems like Shoreham or Royalston are closer to them. The suns in those systems are much larger than Redfield's, making their planets far easier to defend.

All the emphasis placed on Redfield's strategic importance is unfortunate because the planet itself has several unique features that the Federated Suns could put to good use if it were not so busy trying to defend it. There are, for example, several large, fertile plains that could easily produce extensive food crops. There are also several indigenous lifeforms that might be important food items for export to other worlds with dim suns. All that means little to the many thousand soldiers on the planet. The Federated Suns is far too concerned with defending Redfield to allow the planet's population to do more than try to grow enough food to keep themselves alive.

World Name: Okefenokee
Star Type: G5V
Position in System: 4
Time to Jump Point: 7 days
Recharging Station: None
Noble Ruler: Grisson Chalex, Duke of Okefenokee
Political Leader: None
ComStar Facility Class: B
ComStar Representative: Precentor Daniel Henderson
Percentage and Level of Native Life: 40%, Mammal

Okefenokee is one of the poorest worlds in the Federated Suns. Located on the edge of the Periphery, it was inhabited until recently only by religious zealots living in isolated villages in the planet's Great Swamp. Most Humans find the world's many native creatures either dangerous or disgusting.

Beneath the many swamps encircling the tropic and temperate zones, there lies a vast pool of oil. For many years, the oil went undiscovered. When Interstellar Star Petroleum, a large petroleum products company, undertook a vast exploration project under government auspices, they found the oil in this unlikely spot. The discovery has touched off an economic boom on Okefenokee, with money, products, and people pouring onworld. The planet's original settlers and inhabitants resent this sudden influx so much that they have taken to shooting at anyone who enters the swamps.

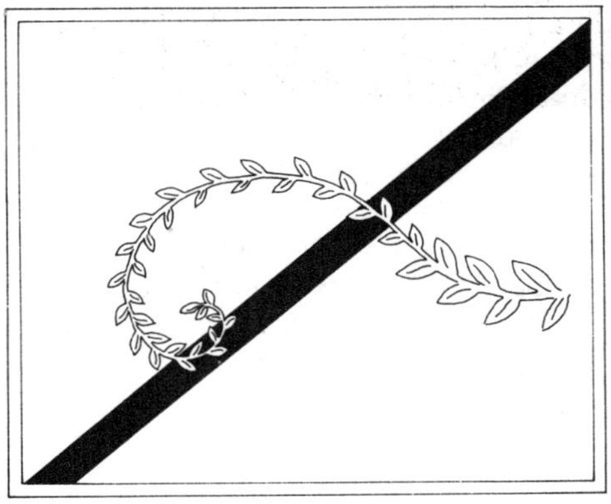

World Name: Parma
Star Type: F2II
Position in System: 5
Time to Jump Point: 10 days
Recharging Station: None
Noble Ruler: Antonio Garibaldi, Duke of Parma
Political Leader: None
ComStar Facility Class: B
ComStar Representative: Precentor Yvonne Euston
Percentage and Level of Native Life: 50%, Reptile

Parma is a large world with a denser atmosphere and higher gravity than Terran-standard conditions. This, combined with the planet's almost perpetual mugginess, makes it a chore to perform even the most simple act of movement. Yet, the natives of Parma somehow find enough energy to make almost constant war on one other.

The ancient ancestors of the present-day Counts and Barons divided the planet into countless little city-states, declaring that no one, not even the Duke himself, could interfere with affairs within the walls of those individual cities. The nobility of Parma have been fighting among themselves ever since. Indeed, violence is such a way of life for the Parmians that they experience a severe shock upon traveling to other worlds where noblemen actually get along well together.

The Duke of Parma has little control over the planet's nobility, nor does he control the planetary militia. Each nobleman has his own private household guard, and so any attempt by the Duke to control the planet through force would lead to yet another war. Instead, Duke Garibaldi has settled for a peacemaker role, using his troops as a buffer between warring city-states until a treaty can be established. The Dukes of Parma are also the Counts of the rich Ace High Continent on the nearby world of Gambler, which accounts for the Garibaldi riches. The Duke's residence on Parma is in the walled city of New Brescia.

World Name: Ozawa
Star Type: K5II
Position in System: 2
Time to Jump Point: 3 days
Recharging Station: None
Noble Ruler: Iona Devries, Duchess of Ozawa
Political Leader: None
ComStar Facility Class: B
ComStar Representative: Precentor Pergil Saundif
Percentage and Level of Native Life: 20%, Mammal

Groups of mostly Japanese colonists settled Ozawa during the first wave of Exodus from Terra. They liked the planet's tough terrain and quickly got used to the extreme weather and tide conditions caused by Ozawa's one large moon. After the discovery of an important deposit of industrial metals, the planet became a major manufacturing center.

Ozawa continued to develop its resources, evolving over time to become a center of high-tech electronics. Its people took great care to preserve the distinctive natural beauty of their world and its fresh air. So great was this compulsion that the planet also became a center of planetary reclamation and reshaping technology.

The collapse of the Star League spelled the end of Ozawa's tranquility. The destructive attacks by both the Draconis Combine and the Capellan Confederation were compounded by a wave of anti-Asian paranoia that spread through Ozawa's non-Japanese population. Violence now characterized a place whose name had become synonymous with a reverence for life. The end came when the Kuritans did not hesitate to make extensive use of nuclear devices during a major push to capture the planet.

Now the landscape of Ozawa is like that of a tortured, twisted corpse. The few inhabitants stay well clear of the Hot Zones, where radiation levels are still lethal.

World Name: Marduk
Star Type: G4II
Position in System: 4
Time to Jump Point: 6 days
Recharging Station: None
Noble Ruler: Dolan Rastkel, Duke of Marduk
Political Leader: None
ComStar Facility Class: B
ComStar Representative: Precentor Jessie Farns
Percentage and Level of Native Life: 70%, Reptile

Located on the Davion-Kurita border, Marduk is the home of the Norse BattleMech Works. It is therefore not surprising that Marduk is the focus of considerable military attention.

The planet was originally exploited for its agricultural resources. Its many regions of warm temperatures, rich soil, and gentle climates have produced a vast array of forests and jungles. Many of Marduk's plant and animal forms were valuable for export, while most of the numerous colonists worked as farmers and ranchers.

Just prior to the creation of the Star League, a deposit of rare metals was discovered in the center of the Tillerbee Jungle. To exploit the discovery, a huge complex of factories was built directly over the metals deposit. After the ore was mined, it was refined and made into metals at a foundry also built nearby, then taken to a BattleMech factory for immediate use in constructing 'Mech frames. This complex of factories was built with such skill that it barely damaged the surrounding jungle. Finished 'Mechs were lifted off Marduk from the spaceport built nearby.

Though the Norse BattleMech Works was one of the first targets of the First Succession War, a factory nestled deep in a huge, swampy jungle proved a formidable target. Since then, the complex has been attacked at least twelve times. Though it has fallen briefly to the Draconis Combine on two separate occasions, the planet has been providing BattleMechs for the Federated Suns almost without interruption.

The capital city of Marduk is New Pontiac, the urban center closest to the Norse BattleMech Works.

World Name: Elidere IV
Star Type: F3V
Position in System: 4
Time to Jump Point: 10 days
Recharging Station: None
Noble Ruler: Michael Hasek-Davion, Duke of New Syrtis
Political Leader: President Jalson Hasek-Davion, Count of Elidere City
ComStar Facility Class: A
ComStar Representative: Precentor Gregory Killos
Percentage and Level of Native Life: 20%, Mammal

Elidere IV was a gift from Hanse Davion to his brother-in-law, Michael Hasek-Davion, in 3022. The previous Dukes of Elidere were members of the Miklos family, and had died to a man in a tragic DropShip accident a year before. This had left the planet, which sits on the border with the Draconis Combine, in a state of uncertainty.

The public reason Prince Davion gave for bestowing Elidere IV on Duke Michael Hasek-Davion was to "keep this vital world in the hand of a strong and trusted ally." There is considerable private speculation that Davion gave Duke Michael the planet in an effort to dilute his growing military strength. There may be some truth to this view, for we have learned that House Davion has asked that Elidere IV be defended by some of Duke Hasek-Davion's elite troops. Whether the Duke will allow a regiment of his Syrtis Fusiliers to go clear across the realm is uncertain.

Geographically, Elidere is a rather small and drab world, with little to make it the focus of so much political maneuvering. Several modest deposits of oil and metals are responsible for the planet's meager metals industry. The largest manufacturer on the planet produces agricultural vehicles for Elidere's farming industry. The capital city of Elidere IV is Jasos City.

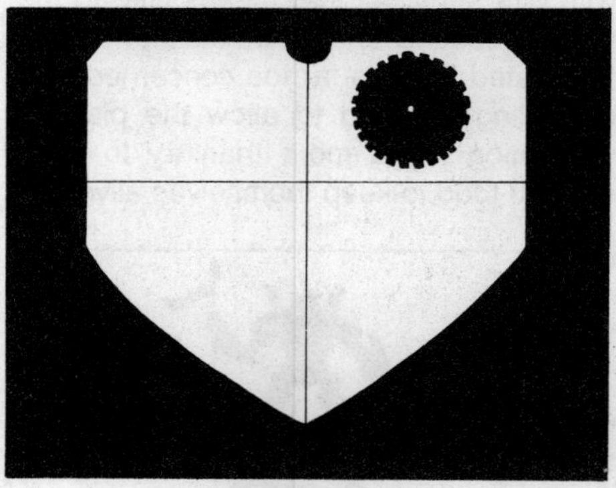

World Name: Brundage
Star Type: G4III
Position in System: 4
Time to Jump Point: 7 days
Recharging Station: None
Noble Ruler: Kenders Kincaid, Duke of Bremond
Political Leader: President Simon Elverson
ComStar Facility Class: B
ComStar Representative: Precentor Ophel Donelson
Percentage and Level of Native Life: 50%, Mammal

Brundage is a water-poor world with a sparse population. It was originally settled by colonists hoping that the planet's many mountain ranges would yield discoveries of metals or jewels. When it did not, most left the planet. Those that stayed behind began to farm the planet's few water-fed valleys.

A recent spate of rumors tells that a huge cache of Star League supplies and equipment lies hidden somewhere in the mountains. There are even unconfirmed reports that government officials have uncovered maps detailing the location of this treasure trove. These rumors have been enough to attract thousands of treasure-hunters from other worlds. It has also brought many difficulties. There was barely enough food grown locally to feed the planet's inhabitants before the influx. Now, with the population nearly doubled, it is impossible to grow enough food for all. There have been appeals to the Ministry of Emergency Services, but even the fastest relief shipment will take several months to arrive.

World Name: Hyalite
Star Type: A4IV
Position in System: 5
Time to Jump Point: 23 days
Recharging Station: None
Noble Ruler: Fredericks Thymus, Duke of Lothair
Political Leader: President Ryer Castleton
ComStar Facility Class: B
ComStar Representative: Precentor Ulior Merson
Percentage and Level of Native Life: 50%, Reptiles

Hyalite sits on the border between the Capellan March and the great unknown of the Periphery. It is an average-sized world orbiting a large sun and has adequate water supplies. What drew the first colonists were Hyalite's many gemstone deposits as well as the unusual chemical compounds to be found there.

The gemstone deposits have played out, which has eliminated most of the fortune hunters among the colonists. The chemical industries are still busy, however. Most of the planet's income is based on the many chemicals exported to other less-endowed worlds. Industries that work with metals and pharmaceutical companies are the two industries most dependent on Hyalite's chemicals.

World Name: Jaipur
Star Type: F1V
Position in System: 4
Time to Jump Point: 8 days
Recharging Station: Zenith
Noble Ruler: Jhenna Feruc, Duke of Jaipur
Political Leader: Council President Dominic Brazzos
ComStar Facility Class: B
ComStar Representative: Precentor Gladis Freshur
Percentage and Level of Native Life: 30%, Mammal

Jaipur is a small planet with modest resources and a reasonably pleasant atmosphere, located in the Capellan March.

The original settlers of Jaipur were from Terran India and from the planet Mars. They settled down to begin farming the dry, but workable land. When the planet became the site of a major battle during the Civil War era of the Federated Suns, the farmers waited quietly until the killing stopped and they could return to their land once more.

For several hundred years, the predominantly agricultural population was contented enough. They did fairly well during the Star League Era, but only barely survived the First and Second Succession Wars. Now that the progress of the Third War has pushed the Capellan border far from Jaipur, the planetary population looks forward to years of peace.

Things began to change recently when the Feruc family decided to become one of the industrial powers of the Federated Suns. They have begun to mine the planet's meager resources and to build huge factories far beyond the capacity of the planet or its people. This has led to growing resentment among the common folk and some of Jaipur's lesser nobles.

The capital of Jaipur is New Amritsar.

World Name: Tancredi IV
Star Type: G5V
Position in System: 4
Time to Jump Point: 7 days
Recharging Station: Nadir
Noble Ruler: Nicholas Brazeau, Duke of Tancredi
Political Leader: None
ComStar Facility Class: A
ComStar Representative: Precentor Jerimiah VanCleff
Percentage and Level of Native Life: 40%, Reptile

Through a fluke of bookkeeping, Tancredi IV was not colonized until the late 2570s. Decades earlier, a Terran Alliance scout ship had made a favorable report after visiting this world, but the report became lost somehow. An historian accidently unearthed it while going through some musty data files in New Avalon one day. How it was possible that no one learned of the planet's existence until then is mystifying.

When the first colonists began to settle Jaipur, they discovered that the original scout ship report had been far too conservative. Not only did the planet have enough pure water to support a large population, but there was plenty of arable land to support food crops. What most surprised the colonists were the riches lying beneath the soil. The discovery of huge metal deposits meant that this planet could quickly become one of the realm's industrial giants. Within 20 years, Tancredi IV had become a major manufacturing force in the region and was daily attracting new companies onworld through an aggressive public relations campaign. Despite being the site of a major battle during the Civil Wars of the Federated Suns, Tancredi IV did not cease its economic expansion.

The Succession Wars stopped all that. When the Star League collapsed, the industrial might of Tancredi was among the first targets of the war, even though the distance to the planet made it difficult for the DCMS forces to make the voyage. Most of Tancredi's industries became casualties of the First War. The military commanders of the Second and Third Wars have avoided damaging the remaining factories, but that has meant using farm and ranchlands for their battles.

Early in the Third Succession War, the Draconis Combine seized Tancredi. For ten years, the world was under the wing of the Dragon, which caused more suffering to the population then to their industries. When the Kuritans began to export most of Tancredi's food crops to feed its own military, it created several major famines. After Davion forces liberated Tancredi in 3023, Hanse Davion approved a massive reconstruction effort to bring the planet back to its former industrial capacity. Because the war has done so much damage to the world's agricultural areas, Tancredi is still in danger of famine.

The capital of Tancredi IV is Dolsonville.

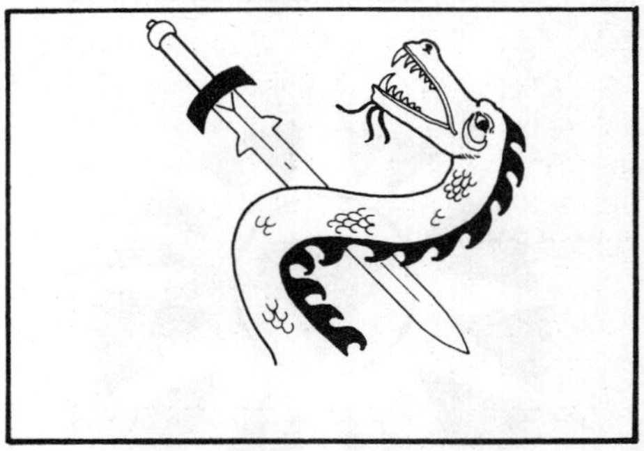

World Name: Rochester
Star Type: G4III
Position in System: 5
Time to Jump Point: 7 days
Recharging Station: None
Noble Ruler: Aaron Sandoval, Duke of Robinson
Political Leader: Prime Minister Tanner Crane
ComStar Facility Class: B
ComStar Representative: Precentor Tulsa Warrens
Percentage and Level of Native Life: None

The first colonists on Rochester were puzzled by the absence of life on a world with such plentiful water and potentially good soil. Then their own crops and animals began to die for unknown reasons. When the experts studied the water and soil to learn the causes, they discovered the presence of a variety of trace elements that, when combined, formed a deadly group of toxins. At that time, there existed no means to filter out or neutralize the poisons.

Though the planet had to be abandoned for years, a purification system developed during the early years of the Star League was capable of removing harmful agents from Rochester's water and soil. Rochester was resettled, and this time, the colony was a success. One of the byproducts of the purification process also became valuable. It was sludge that became known as "Green Jelly Gold" because it contained unusual elements highly sought after by pharmaceutical companies. By the time of the First Succession War, the colony had developed extensively, having spread a carpet of plant and animal life across nearly the entire planet.

The Draconis Combine attacked and occupied Rochester during the First War and quickly destroyed the water and soil purification plants. This led to the rapid loss of plant and animal life, leaving the colonists helpless. The occupying forces of the Draconis Combine refused to feed or evacuate the colonists from their dying world. Fifty-two thousand colonists died in what came to be known as the Rochester Starvation. Only the Kentares Massacre surpassed this event in cruelty or in the anger it aroused in the people of the Federated Suns.

It is rumored that scientists from the NAIS have lately rediscovered the purification process that once gave life to Rochester and that they are about to test it there.

World Name: Lackland
Star Type: G9V
Position in System: 4
Time to Jump Point: 8 days
Recharging Station: None
Noble Ruler: Stanton Defire, Marquess of Lackland
Political Leader: None
ComStar Facility Class: B
ComStar Representative: Precentor Kiletha Trasont
Percentage and Level of Native Life: 70%, Reptile

Lackland is true to its name, for it is almost entirely covered by one immense ocean. The only land that exists is the myriad but scattered island chains.

This unusual planet has caused the downfall of more than a few nobles who could find no use for Lackland. Some have attempted to export the water, but it proved too costly. Others attempted to become successful through aquaculture. When they began to breed marine life as food delicacies, dangerous unknown creatures rose up from the depths of Lackland's ocean and swallowed the pens of fish whole. The people of Lackland like this situation just fine, however, preferring to bask in the sun and to live on the fish they catch themselves.

The current Marquess has taken a new tack in an effort to develop Lackland. He has begun to build underwater habitats against the walls of the undersea cliffs. From these habitats, it may be possible to mine the mountains for chemical compounds and metals. The results have been promising so far, though some of the habitats have been menaced by the huge creatures that the natives know as "The Mothers of the Ocean".

World Name: Hobbs
Star Type: A1III
Position in System: 4
Time to Jump Point: 21 days
Recharging Station: None
Noble Ruler: Ivensen Rethers, Duke of Hobbs
Political Leader: None
ComStar Facility Class: B
ComStar Representative: Precentor Eric Toomy
Percentage and Level of Native Life: 40%, Mammal

Hobbs is a relatively dry world whose native flora and fauna are confined to an extensive system of underground caverns that honeycomb the world. The caverns are host to an amazing ecosystem of animals, some of whose behavior borders on intelligence. The most advanced of the native lifeforms are the Takooma, vaguely ape-like mammals with keen infrared vision and sense of smell. They travel in extended family groups and seem to have a rudimentary language, though scientists from the NAIS and other universities have yet to penetrate its secret.

The Federated Suns military uses the countless interconnected underground caverns of Hobbs as a kind of enormous warehouse, which has put the AFFS in direct conflict with the Takooma. The creatures have waged something akin to a guerrilla war against the alien Humans, ambushing anyone caught alone and stoning them to death. Soldiers who try to hunt down the Takooma often become lost in the maze of tunnels and end up becoming the hunted instead.

The capital city of Hobbs is Luray Caverns, one of many cavern-based cities on the planet.

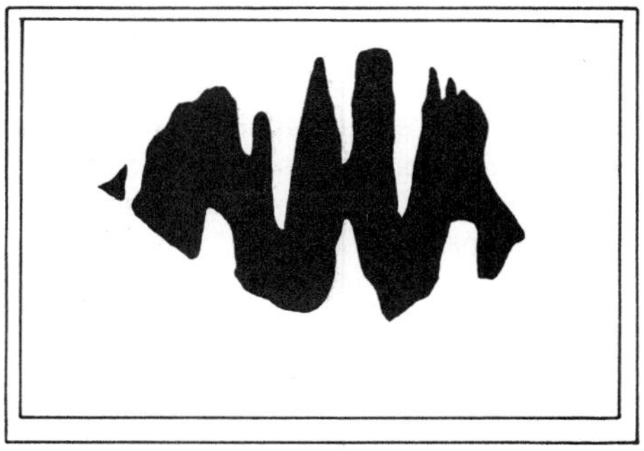

World Name: Kittery
Star Type: G4III
Position in System: 3
Time to Jump Point: 7 days
Recharging Station: None
Noble Ruler: Julian DeBurke, Duke of Gurnet
Political Leader: Colonel Katherine Gral, Military Governor
ComStar Facility Class: B
ComStar Representative: Precentor Hedef Grathos
Percentage and Level of Native Life: 40%, Mammal

The planet Kittery was part of the Capellan Confederation until the Federated Suns seized this small agricultural world in an offensive begun in 3005. Since then, the planet and its people have been a thorn in their side.

Kittery's native carnivores are particularly violent and cunning hunters, which makes travel in the planet's numerous jungle regions extremely dangerous. The planet's population, of mostly Asian ancestry, has been less than cooperative with their new rulers, despite the many freedoms they enjoy as a world within the Federated Suns. In the 20 years since the Davion occupation, Kittery has spawned a particularly vicious guerilla organization known as the Kittery Tong. While the Military Governor would prefer to spend her time preparing against possible attack by the Capellans, she must occupy most of her day worrying over reports of the latest Tong attacks or of jungle predators lately feasting on her troops.

The only positive note in the strained relations between the occupation forces and the planet's population is that the Federated Suns has allowed many Buddhist temples to open in the cities of Kittery. An increasing number of people curious and eager to rediscover the religion of their ancestors have been attending the temple services.

The capital of Kittery is CanFu City.

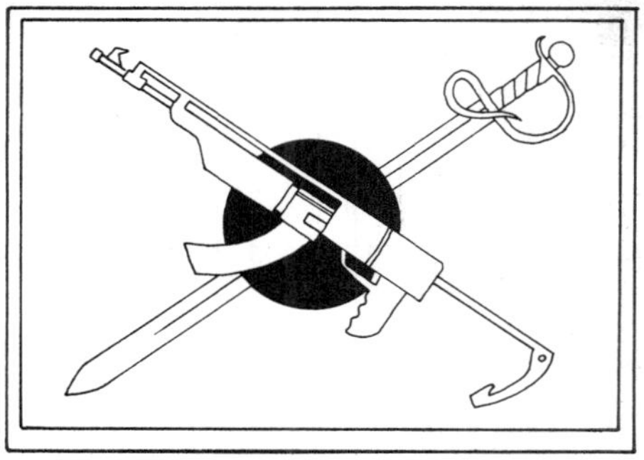

World Name: Great Gorge
Star Type: A3V
Position in System: 5
Time to Jump Point: 21 days
Recharging Station: None
Noble Ruler: Losart Ulitock, Marquess of Great Gorge
Political Leader: None
ComStar Facility Class: B
ComStar Representative: Precentor Paul Thyres
Percentage and Level of Native Life: 40%, Primitive plants

The initial survey of the young world of Great Gorge reported little of interest except for the dimensions of the huge canyon that dominates one of the regions. Most assumed that this planet of marginal atmosphere, little water, and few lifeforms would become a charted but forgotten world.

A religious group from Terra chose Great Gorge as the site of their future Eden. Settling on the planet just before the creation of the Star League, they set about to completely transform the world. Some of their activities, such as scattering desert-grass seeds from airplanes and casting iron-fish eggs into the sea, began to have surprising, even miraculous, effects. The oxygen content of the atmosphere increased because of the plants the colonists cultivated, and fish actually hatched and thrived on the native algae.

By the end of the Star League era, Great Gorge was a comfortable place with plenty of air and plenty of food, thanks entirely to the work of the planet's original colonists. In the meantime, several mining companies had discovered that the walls of the Great Gorge exposed many easily mined metals and chemical compounds. The planet soon became as profitable as it was habitable.

Unfortunately for Great Gorge, its location on the Crucis March border with the Periphery and its abundant resources have made it a prime target of bandit raids. Just recently, the last descendant of the planet's religious colonists died. Without their efforts to keep the planet livable, it is possible that conditions on Green Gorge, particularly its atmosphere, might decline to their previous state. The present Marquess definitely has his work cut out for him.

The capital of Great Gorge is New Stockdon, a city in the Wide Floor section of the Great Gorge.

World Name: Caph
Star Type: G5V
Position in System: 3
Time to Jump Point: 7 days
Recharging Station: None
Noble Ruler: Kindy Gustafuson, Duchess of Caph
Political Leader: Colonel Roberta Gustafson, Military Governor
ComStar Facility Class: B
ComStar Representative: Precentor Lawrence Gelbart
Percentage and Level of Native Life: 20%, Insect

Caph was one of the first worlds that Human space explorers settled after the Kearny-Fuchida drive made interstellar space travel possible. During the Star League era, Caph was a center for advanced engineering research. Indeed, much of the original theory behind the newly invented BattleMechs was developed at the famed Caph Institute of Technology. Caph was also famous for successful medical research and for the development of important new techniques in the field.

With the fall of the Star League, Caph became the center of bitter fighting between the Draconis Combine, the Capellan Confederation, and the Federated Suns. In 2787, the Fifth Battle for Caph became one of the few instances where three realms battled one another on the same planet. The Federated Suns was the eventual victory on Caph, however.

What House Davion inherited was a planet full of ruins and radiation zones. Ninety-nine percent of all the world's buildings had been destroyed, and the little remaining arable land had suffered from the effects of combat. Only now is the planet slowly recovering, with a growth in population and the construction of a few small factories. Even more important is that Caph's location makes it the ideal site for offices and warehouses of many Federated Suns trading companies interested in dealing with the rich Lyran Commonwealth. With the recent alliance between the two realms, this planet's usefulness and importance can only grow greater.

The capital city of Caph is Aswan.

World Name: Hoff
Star Type: G4III
Position in System: 4
Time to Jump Point: 8 days
Recharging Station: Zenith
Noble Ruler: Bradley Cheel, Duke of Hoff
Political Leader: None
ComStar Facility Class: B
ComStar Representative: Precentor Lothec Trive
Percentage and Level of Native Life: 30%, Mammal

Hoff was originally settled for its agricultural possibilities. Its wide plains crisscrossed with small rivers and brooks offer excellent crop and grazing lands. Hoff's other modest resources of metals and coal deposits took a back seat to development of its potential as a farming world.

Early in 3019, a team from the NAIS came to Hoff to reopen the long-abandoned Friden Aerospace Park on the planet's northern continent. The small spaceport had been cleaned up and turned into a test facility where NAIS scientists, including members of the famed Team Banzai, could field-test new military technology. This made Hoff a prime target that the Draconis Combine did not delay in attacking. In 3022, the famous Wolf's Dragoons, then mercenaries in the Draconis Combine, attacked Hoff, where the equally famous Eridani Light Horse was on defense. Though the black 'Mechs of the Dragoons managed to seize the research facility, they were not quick enough to capture any of the scientists or the research projects they were testing.

Since then, the Federated Suns has reinforced Hoff, making it unlikely that the Kuritans will attempt another attack. Reports from our ComStar Precentor state that Davion has reopened the Friden Aerospace Park and that Team Banzai has recently completed tests on a variant of the *Hatchetman* BattleMech.

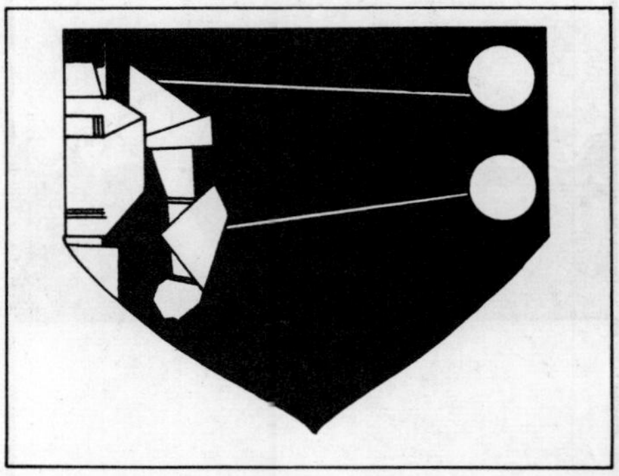

World Name: Baxley
Star Type: K5IV
Position in System: 3
Time to Jump Point: 4 days
Recharging Station: None
Noble Ruler: Jessica DeBrinion, Duchess of Dothan
Political Leader: Ulther DeBrinion, Count of Baxley
ComStar Facility Class: B
ComStar Representative: Precentor Rether Richardson
Percentage and Level of Native Life: 20%, Reptile

Baxley has long been a Skid Row world, whose modest agriculture is barely able to feed its population. It has little mineral or metal resources, and sits on the far end of the Point Barrow-Dothan Desert, a vast volume of space without a single Human colony. These factors contribute to Baxley being one of the poorest and most ignorant of worlds in the Federated Suns.

Because of the poverty and lack of educational resources on Baxley, the inhabitants treat ComStar representatives like demi-gods. Their reverence is so great that the weekly observances and classes at our station enjoy a better turnout than does the local New Avalon Catholic Church. It is not surprising then that this impoverished world provides more aspirants to our Sacred Order than any other in Davion space.

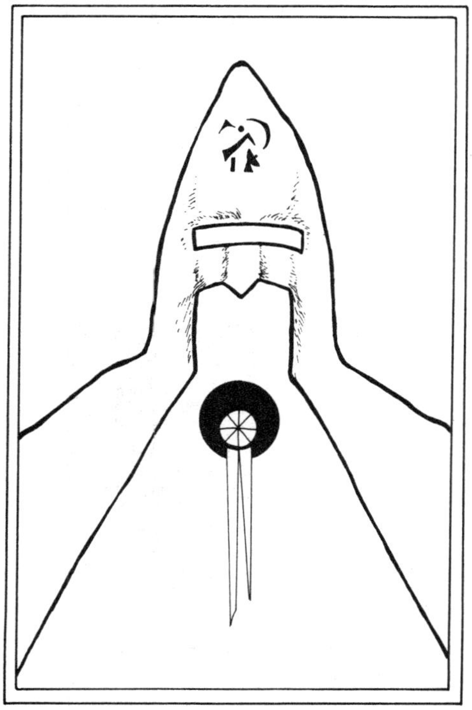

World Name: Zolfo
Star Type: F5V
Position in System: 4
Time to Jump Point: 10 days
Recharging Station: None
Noble Ruler: Vivana Wilzon, Duchess of Zolfo
Political Leader: None
ComStar Facility Class: B
ComStar Representative: Precentor Demarsi Reynas
Percentage and Level of Native Life: 40%, Mammal

Zolfo is one of the success stories of Hanse Davion's recent push for increased economic activity in Outback areas of the Federated Suns. Before the present decade, Zolfo had been primarily an agricultural world unable to live up to its potential because of it lacked skilled workers.

That changed in 3022 when a survey team from BP-Interstellar, a petroleum and steel firm, arrived onworld. In the course of their survey, the team discovered major oil and coal deposits in the planet's various mountain ranges. Since then, more than 30 corporations have bought land for new industries on Zolfo. The Duchess of Zolfo has already begun to use some of the profits from the land sales to build the planet's first schools and colleges. Zolfo is no longer one of the Skid Row worlds, and promises to vindicate Prince Davion's economic development policies by becoming the core of a whole new sphere of economic growth.

World Name: Greeley
Star Type: K5V
Position in System: 2
Time to Jump Point: 4 days
Recharging Station: None
Noble Ruler: Wendell Farh, Duke of Adrain
Political Leader: President Avar Lopper
ComStar Facility Class: B
ComStar Representative: Precentor Hyuku Hiero
Percentage and Level of Native Life: 50%, Mammal

A good part of the planet Greeley is frozen arctic tundra. Its small sun and the exaggerated tilt of the planet's axis produce strong arctic blizzards that actually push their way into the planet's tropical regions to blanket most of the world in snow. Underneath the snow and soil are a few modest deposits of metals and little else.

Recent activity on the planet suggests that Greeley has suddenly become important to the Federated Suns. It appears that they are constructing a military base for BattleMechs, tanks, and infantry units. This new "Fort Greeley" is apparently going to play a role in the training of Davion troops. It is also known that several scientists from the NAIS are at Fort Greeley, which may indicate that the fort will serve as an arctic test center for NAIS-designed devices. This planet definitely bears watching.

World Name: Manteno
Star Type: F5V
Position in System: 4
Time to Jump Point: 10 days
Recharging Station: Nadir
Noble Ruler: Mantor Dasarick,
Duke of Manteno
Political Leader: None
ComStar Facility Class: Closed for repairs
ComStar Representative: Precentor
Olson Fredas
Percentage and Level of Native Life:
20%, Mammals

The planet Manteno is currently experiencing a sudden and continuing outburst of geologic activity. It began about eight years ago with several major earthquakes, and has continued with a series of volcanic eruptions and earthquakes that left nearly everyone on the planet severely shaken. Since the beginning of this period of planetary upheaval, over 20 percent of the population has left Manteno and about 10 percent of the businesses have relocated.

Manteno was not always shaking or threatening to explode. At one time, geologists of the Terran Alliance considered it to be both a stable world and one rich in large and accessible deposits of metals and chemical compounds. The planet was colonized and immediately became an important part of the Federated Suns economy. Manteno's first major setback in the First Succession War came when the Draconis Combine attacked, destroying most of the planet's industries. Manteno was just regaining its industrial might at the time that the current phase of convulsions began. Those who remained on the planet, whether by choice or because they lacked the money to leave, pray that the planetary eruptions are temporary and that one day the ground will stop threatening to buck like a wild animal.

World Name: Cogdell
Star Type: G5V
Position in System: 4
Time to Jump Point: 7 days
Recharging Station: None
Noble Ruler: Edward Sims,
Marquess of Cogdell
Political Leader: People's-Leader Yoloy
Entwhistle
ComStar Facility Class: B
ComStar Representative: Precentor
Timothy Grethel
Percentage and Level of Native Life:
60%, Reptile

With commoners pitted against nobles, Cogdell is a world in deep political turmoil. This conflict has its roots in the fact that the planet's original settlers were a boatload of convicts who had been brought to Cogdell from a Periphery world and dumped there without food or water. The convicts managed to survive and formed their own culture, a semi-anarchy where each citizen had the right to do as he wished and no one had the right to stop him. When the Federated Suns absorbed Cogdell just before the start of the Third Succession War, the Davion government chose a member of the Sims family, a famous native clan, as ruling Marquess. Since then, the Sims family has gradually adopted a more and more pro-Federated and pro-nobility attitude. The public of Cogdell now believes that the Davion government is a threat to their anarchistic way of life. The friction has even erupted recently into open fighting.

Cogdell is rather poor in metals and chemical compounds, but it does offer good fresh-water supplies and plenty of arable soil. The capital city of Cogdell is Hangman's Death on the southern continent.

INDEX

Planet Guide

The Coordinates of Terra are 0,0

World Name	X Coordinate	Y Coordinate	Affiliation
Abbeville	611.32	-40.16	D
Abruzzi	150.74	-306.96	D
Acala	135.36	-125.44	D
Acamar	39.90	-38.86	L
Accrington	-10.17	146.31	S
Achernar	77.98	-32.60	L
Addicks	72.24	-0.26	D
Adelson	477.00	-247.50	D
Adrian	453.27	28.43	D
Agliana	393.29	-292.10	D
Al Na'ir	60.77	31.56	K
Albalii	71.72	79.28	K
Alcyone	130.92	-199.25	D
Aldebaran	25.04	-64.94	L
Aldertaine	45.12	-192.73	L
Algedi	66.76	96.24	K
Algol	65.46	-68.33	L
Algot	86.59	-83.46	D
Allerton	265.76	7.30	D
Almach	101.71	-62.07	D
Alnasi	2.87	117.10	K
Alrakis	-20.08	82.67	K
Alrescha	85.28	-49.81	L
Alsek	318.44	-240.20	D
Alta Vista	511.17	13.04	D
Altair	9.39	14.08	K
Altais	37.55	110.84	K
Altdorf	336.17	120.49	K
Altoona	555.50	-4.17	D
Altorra	-2.87	-265.76	L
Alya	26.60	97.54	K
Ambergrist	84.24	-259.76	L
Amiga	166.65	-85.28	D
An Ting	304.35	110.84	K
Anaheim	302.01	-314.52	D
Ancha	76.15	36.51	K
Andalusia	394.85	-110.06	D
Andarmax	-3.39	-389.90	L
Andro	146.05	-174.48	D
Angol	88.67	-39.12	D
Anguilla	415.72	80.33	D
Anjin Muerto	591.76	-78.76	D
Ankaa	75.37	-9.39	D
Annapolis	166.13	131.96	K
Antietam	333.82	-138.22	D
Apriki	174.74	122.58	K
Arboris	43.55	-51.90	L
Ares	96.24	-175.52	L
Argyle	257.67	-110.06	D
Arlington	214.64	100.67	K
Armaxa	107.45	-207.60	L
Armington	424.58	-306.44	D
Arnaudville	588.89	37.03	D
As Samik	349.99	-289.23	D
Ascella	21.12	75.11	K
Ashio	74.07	68.07	K
Ashkum	112.93	-101.97	D
Ashley	165.87	-311.13	D
Asta	4.17	25.82	K
Asuncion	5.22	-102.49	M
Athenry	17.73	31.30	K
Atlas	158.83	-246.20	D
Atria	-16.17	52.16	K
Aubisson	4.43	147.87	K
Aucara	178.13	-298.88	D
Augusta	284.79	-99.89	D
Avawatz	166.39	-29.73	D
Avigait	154.39	-277.49	D
Axton	128.31	-114.23	D
Azha	54.25	-51.90	L
Bacum	158.57	-291.84	D
Bandora	13.30	-178.91	L
Baranda	453.27	-261.32	D
Barlow's End	307.22	63.64	D
Barlow's Folly	174.74	67.55	K
Barras	21.12	-357.30	L
Barstow	172.13	-41.21	D
Basalt	66.76	-29.21	D
Basantapur	367.73	-286.88	D
Bassfield	594.10	4.96	D
Bastian	299.92	-161.44	D
Bastrop	568.28	61.03	D
Batavia	202.90	-55.81	D
Baxley	436.06	-151.26	D
Beaumont	589.15	40.16	D
Beecher	185.43	-50.60	D
Beenleigh	185.43	-271.23	D
Beid	125.71	-245.93	D
Belaire	337.48	-276.71	D
Bell	105.10	-110.84	D
Belladonna	265.23	-125.71	D
Bellatrix	54.77	-354.69	L
Bellevue	330.96	-58.16	D
Benedict	477.52	-11.48	D
Benet III	265.23	58.68	D
Benjamin	125.18	135.88	K
Bentley	-7.04	-235.76	L
Bergman's Planet	268.36	75.63	D
Berenson	-13.82	-71.72	M
Bernardo	-21.39	-94.93	M
Beta Mensae V	320.00	130.40	K
Betelgeuse	5.74	-309.57	L
Betten Kaitos	167.17	-114.23	D
Bethel	141.88	-185.69	D
Bettendorf	267.32	26.08	D
Bharat	59.46	-41.99	L
Biham	72.50	37.03	K
Birmensdorf	386.51	-273.32	D
Birqash	278.27	-268.62	D
Bithinia	-19.04	-179.95	L
Blandinsville	230.03	-33.12	D
Bluford	451.44	-80.85	D
Boardwalk	-13.82	-199.25	L
Bogard	271.75	-32.60	D
Bonneau	399.81	-167.43	D
Boondock	596.19	21.65	D
Bora	39.12	-159.09	L
Borden	51.90	-373.99	L
Breed	197.16	40.16	D
Bremond	391.72	59.20	D
Brighton	79.80	-202.12	L
Bristol	179.69	-67.55	D
Broaddus	469.18	56.33	D
Brockton	536.73	-120.23	D
Brockway	220.12	-359.12	D
Broken Wheel	539.07	-175.00	D
Bromhead	161.17	-380.77	D
Brookeland	444.14	71.20	D
Brundage	407.11	49.81	D
Brusett	247.76	-330.96	D
Bryant	13.30	-13.30	L
Bryceland	410.24	98.84	D
Buchlau	70.68	-69.37	L
Buckminster	17.73	156.22	K
Buenos Aires	18.26	-345.04	L
Cadiz	123.10	150.48	K
Cahokia	291.31	-73.55	D
Caldwell	334.35	-328.87	D
Calpaca	-20.34	-216.72	L
Calseraigne	-8.35	-370.86	M
Cambiano	354.69	-244.11	D
Cammal	112.40	-166.39	D
Campertown	-5.48	-160.65	L
Capac	312.96	-56.85	D
Capella	38.86	-175.26	L
Capolla	8.08	-41.73	L
Caph	13.04	-0.52	D
Capra	343.47	106.67	K
Capricorn III	88.41	-188.04	L
Carmacks	185.43	-238.89	D
Carmen	28.69	-266.80	L
Carmichael	201.34	-352.08	D
Cartago	141.09	-10.17	D
Carver V	-5.48	-20.86	L
Caselton	117.36	-35.99	D
Cassias	320.52	80.59	D
Castrovla	13.82	-254.80	L
Cavalor	61.55	-391.20	L
Cebalrai	-8.35	120.49	K
Cerulean	416.50	-81.37	D
Chakachamna	302.53	-182.82	D
Chanute	467.88	23.73	D
Charndo	10.43	-153.61	L
Chebanse	227.16	-103.80	D
Chenier	585.76	-41.73	D
Chesterton	115.27	-72.24	D
Chichibu	91.28	75.63	K
Chirikof	338.78	-235.50	D
Cholame	204.21	-31.56	D
Choudrant	339.04	56.59	D
Cimeron	365.90	84.76	D
Claxton	15.65	-266.54	L
Clovis	181.78	3.13	D
Cogdell	526.56	-142.66	D
Cohagen	375.55	-318.96	D
Cohay	615.49	11.48	D
Colchester	308.00	-13.56	D
Colia	360.69	67.55	D
Coloma	291.05	-73.55	D
Colorado	347.65	-200.82	D
Columbine	75.63	-348.95	L
Conquista	-9.65	-309.83	M
Conroe	380.51	92.32	D
Cordiagr	24.52	-192.73	L
Corella	143.44	-253.50	D
Corey	21.65	-137.44	L
Corodiz	136.92	-378.16	L
Corydon	218.55	-39.12	D
Cotocallao	146.05	-302.27	D
Courcellete	142.92	-313.48	D
Covington	342.69	-151.00	D
Crofton	527.60	-78.76	D
Crossing	255.58	47.73	D
Csomad	431.10	-294.70	D
Cumberland	208.90	-267.32	D
Cylene	106.93	25.30	K
Dabih	60.51	84.76	K
Dahar IV	281.92	20.34	D
Damevang	348.69	64.94	D
Daniels	125.71	-198.21	D
Darwendale	314.52	-265.23	D
David	140.57	19.30	D
Davisville	485.61	-27.64	D
De Berry	450.66	63.11	D
Decatur	54.77	-339.04	L
Decus	44.86	-253.76	L
Defiance	428.76	-107.45	D
Delacruz	375.81	114.23	K
Delavan	272.28	-95.97	D
Delcambre	544.55	4.96	D
Delitzsch	329.39	148.13	K
Delos IV	618.88	60.51	D
Delphos	322.61	-52.42	D
Demeter	92.32	-70.16	D
Denbar	69.63	-237.33	L
Deneb Algedi	41.47	26.34	K
Deneb Kaitos	65.20	-13.04	D
Des Arc	501.00	-84.24	D
Deshler	216.99	75.11	D
DeWitt	260.02	-21.91	D
Diboll	479.61	67.81	D
Dicon	20.34	-328.09	L
Die Moot	286.88	-305.40	D
Diefenbaker	200.03	-344.52	D
Dieron	13.04	15.39	K
Dobson	206.03	26.60	D
Donenac	202.12	89.72	K
Doneval II	200.56	8.87	D
Dothan	437.62	-135.36	D
Drienfontein	273.58	-292.88	D
Dromini VI	-21.12	66.76	K
Drozan	31.56	-373.20	L
Dyev	5.74	41.21	K
Ebro	568.80	-130.66	D
Edwards	133.27	-32.60	D
Egress	68.59	-373.47	L
El Dorado	268.88	-146.31	D
Elbar	127.79	-8.08	D
Electra	217.25	-325.22	D
Elgin	1.56	-51.38	L
Eldire IV	282.45	76.68	D
Elnath	18.78	-128.31	L
Eltanin	-1.83	100.67	K
Emerson	171.87	-117.10	D
Emporia	258.97	-6.52	D
Enchi	297.05	-290.27	D
Eom	-18.26	-195.34	L
Epsilon Eridani	18.52	-21.39	L
Epsilon Indi	29.47	-15.65	L
Errai	52.68	4.43	D
Estuan	399.55	-307.74	D
Euclid	352.60	-71.20	D
Eustatius	462.14	-200.29	D
Evansville	296.27	-36.51	D
Exedor	-13.30	-209.68	L
Exeter	190.12	-8.08	D
Fairfax	356.51	31.56	D
Fairfield	289.49	46.16	D
Fallon II	371.12	49.03	D
Falsterbo	172.39	148.92	K
Farnsworth	566.46	31.82	D
Farwell	141.88	-80.33	D
Fellanin II	133.27	33.12	K
Fetsund	450.66	-243.07	D
Filtvet	511.17	-183.08	D
Fincastle	326.00	-164.56	D
Firgrove	252.19	-301.48	D
Flannagan's Nebula	183.34	-393.55	P
Fletcher	47.20	-23.73	L
Flintoft	193.25	-336.17	D
Flushing	197.16	-65.20	D
Fomalhaut	21.12	5.48	D
Foochow	89.72	-98.32	L
Foot Fall	95.45	-110.84	L
Fortymile	195.34	-216.20	D
Franklin	252.98	10.95	D
Frazer	130.40	-336.43	D
Freisland	237.07	-119.71	D
Frondas	-18.00	-249.85	L
Fronde	-6.00	-253.76	L
Fukuroi	111.62	135.36	K
Fuveau	444.40	-234.72	D
Galatia III	91.54	14.87	D
Galax	312.18	-125.18	D
Gallitzin	132.75	-164.83	D
Galtor III	195.86	73.81	D
Gambarare	360.43	-272.28	D
Gambier	385.20	-60.51	D
Gan Singh	41.99	-80.33	L
Gandy's Luck	252.45	93.89	K
Gei-Fu	58.16	-200.82	L
Geifer	37.56	-187.78	L
Genoa	39.38	-60.24	L
Gillingham	505.17	-213.33	D
Glasgow	28.69	-217.51	L
Glenmora	287.92	53.46	D
Glentworth	127.01	-320.52	D
Goderich	145.79	-49.81	D
Goshen	135.88	-95.45	D
Gram	54.77	145.79	K
Grand Base	46.42	-301.48	L
Great Gorge	497.08	-269.15	D
Greeley	453.27	40.68	D
Gronholt	470.22	-263.41	D
Grosvenor	288.44	-226.64	D
Groveld III	390.42	114.49	D
Gulkana	356.77	-179.43	D
Gunthar	39.64	-386.51	L
Gurnet	93.37	-257.93	D
Gurrnazovo	427.19	-242.80	D
Haappajarvi	138.48	-242.80	D
Hachiman	255.06	153.61	K
Hadnall	137.96	-264.97	D
Hagiwawa	124.66	113.19	K
Hahira	554.98	-101.45	D
Hall	-14.08	-44.60	L
Halloran V	73.55	-77.20	D
Halstead Station	78.76	44.34	K
Hamal	65.46	-43.81	L
Handa	277.23	141.88	K
Harloc	42.25	-226.37	L
Harpster	215.16	83.20	D
Harrow's Sun	264.19	71.72	D
Harsefeld	-18.00	-143.18	M
Haynesville	459.27	80.33	D
Hazelhurst	554.98	65.98	D
Hean	75.37	-14.60	D
Hecheng	334.61	-265.49	D
Helen	83.98	10.17	D
Heligoland	86.85	-146.31	L
Helsingfors	145.00	154.13	K
Hephzibah	461.62	-169.00	D
Hexare	31.30	-235.24	L
Hickok	436.06	-4.96	D
Highspire	101.19	-120.23	L
Hivrannee	631.66	-6.00	D
Hobbs	257.67	-246.20	D
Hobson	215.94	-263.15	D
Hoff	266.54	37.03	D
Holloway	65.46	-304.09	L
Homam	147.61	52.42	K
Homestead	58.42	-286.10	L
Hoonaar	464.75	-267.32	D
Horsham	156.74	-358.08	D
Hortense	492.39	-145.79	D
Housekarle	33.64	-279.32	L
Hoyleton	510.65	-52.16	D
Hsien	5.74	-54.77	L
Huan	307.48	99.10	K
Humansville	557.85	-32.60	D
Hun Ho	252.45	137.96	K
Hunan	333.04	-96.76	L
Hurik	78.50	-399.81	L
Huron	317.92	-72.50	D
Hustaing	55.55	-247.76	L
Hyalite	271.23	-357.30	D
Igualada	292.88	115.27	K
Ikast	102.23	-286.62	D
Imalda	30.25	-242.28	L
Imbrial III	296.79	-49.81	D
Imbros III	-2.09	34.43	K
Immenstadt	82.93	-280.36	D
Indicass	73.02	-251.15	L
Ingenstrem	320.00	-227.42	D
Ingersoll	-5.22	-173.69	L
Ingress	50.07	-18.78	L
Inman	560.98	7.30	D
Inner End	592.54	75.89	D
Ipava	261.32	-106.67	D
Irurzun	188.56	98.58	K
Islamabad	324.44	-256.37	D
Ito	7.56	-278.01	L
Jaboatao	485.61	-213.86	D
Jacomarle	13.04	-385.98	L
Jacson	111.62	-349.47	L
Jaipur	177.08	-282.71	D
Jesup	504.91	-122.58	D
Johnsondale	162.22	-23.21	D
Jonathan	101.45	-93.89	L
Jonzac	116.32	-300.18	D
Junction	136.14	79.54	K
June	409.46	-266.54	D
Junior	403.20	1.04	D
Kafr Silim	119.71	-290.53	D
Kaguyak	299.92	-238.11	D
Kaifeng	72.50	-153.61	L
Kaitangata	233.68	-282.19	D
Kaiyuh	365.64	-219.59	D
Kajikazawa	89.19	143.96	K
Kansu	68.33	-60.24	L
Kasdach	21.12	-315.83	L
Kashilla	-1.56	-202.90	L
Kathil	172.91	-149.96	D
Kaus Australis	5.74	84.24	K
Kaus Borealis	11.48	97.28	K
Kaus Media	7.82	80.85	K
Kawabe	242.80	136.92	K
Kawich	54.25	-29.73	D
Kaznejoy	332.52	133.79	K
Keid	3.13	-10.17	L
Kennard	472.83	89.19	D
Kentares IV	150.74	-13.30	D
Kentwood	632.96	-20.34	D
Kervil	34.43	46.42	K
Kesai IV	373.47	106.67	D
Kessel	-13.30	84.24	K
Kestrel	270.71	-47.20	D
Kettering	450.66	-98.32	D
Keuterville	276.71	-315.05	D
Keytesville	378.42	-31.04	D
Kigamboni	208.12	-250.63	D
Kilbourne	576.89	44.08	D
Killarney	405.80	-205.51	D
Kirbyville	511.69	51.64	D
Kirei Na Niwa	300.44	148.66	K
Kirklin	280.88	-68.07	D
Kiserian	288.44	-281.66	D
Kitalpha	113.97	97.80	K
Kittery	82.93	-225.85	D
Klathandu IV	176.82	31.82	D
Kluane	197.95	-201.08	D
Ko	-18.26	44.34	K
Konstance	-20.34	97.28	K
Koping Chian	182.82	117.36	K
Kotzebue	319.48	-207.60	D
Kountze	529.16	27.38	D
Krin	-7.82	-223.24	L
Kurhah	105.10	57.38	K
Kurragin	3.39	-204.99	L
Kurvasa	39.90	-354.69	L
Kuzuu	23.21	67.29	K
Kyrkbacken	2.61	-121.01	M
Lackland	552.90	-175.26	D
Lambrecht	13.56	58.16	K
Larned	297.05	-23.21	D
Lapida II	113.97	51.64	D
Larsha	119.45	-396.68	L
Latexo	396.42	81.11	D
Latice	10.69	-296.27	L
Layover	206.03	-20.08	D
Le Blanc	208.90	18.52	D
Leamington	356.77	-91.80	D
Lee	117.88	-149.44	D
Leipsic	310.61	-104.84	D
Lesalles	9.91	-166.39	L
Lexington	318.44	-22.95	D
Liao	34.43	-68.07	L
Lihue	427.97	-200.29	D
Lima	237.85	54.51	D
Lindsay	208.38	-353.12	D
Linneus	454.31	-8.08	D
Listowel	151.52	-68.07	D
Ljugarn	199.77	121.53	K
Logandale	144.48	-36.25	D
Lothair	268.62	-340.08	D
Lucerne	233.94	12.52	D
Ludwig	163.52	77.72	K
Lyceum	432.41	76.68	D
Lyons	-23.21	39.12	S
Macintosh	361.73	-267.32	D
Macomb	344.52	-20.34	D
Madras	71.20	-328.09	L
Maia	216.46	-326.00	D
Maladar	109.80	-249.85	L
Malagrotta	431.88	-273.58	D
Mallory's World	112.40	6.00	D
Manapire	83.20	-295.49	D
Manassas	318.18	-123.36	D
Mandaree	184.12	-311.66	D
Mandate	89.19	-127.27	L
Mansfield	398.24	-63.11	D
Manteno	190.91	-72.50	D
Mara	122.84	8.87	D
Mararn	531.25	-221.94	D
Marduk	210.20	60.51	D
Marielund	532.29	-207.86	D
Markab	98.58	23.47	K
Markesan	223.51	-75.37	D
Marlette	155.96	-44.86	D
Marlowe's Rift	236.81	88.67	K
Marodzi	294.70	-276.97	D
Masterson	2.61	-187.52	L
Matar	163.52	56.85	K
Matsu	79.02	-137.44	L
Matsuida	280.88	130.14	K
Mauckport	430.84	-53.72	D
Mayetta	408.93	7.04	D
Maynard	307.22	2.35	D
McComb	224.29	58.16	D
McGehee	333.30	75.63	D
McHenry	181.00	-85.28	D
McRae	558.11	-85.28	D
Meinrad	396.16	-85.80	D
Mejicanos	476.48	-183.34	D
Melcher	370.60	38.86	D
Memphis	567.24	-142.92	D
Mendham	149.18	-347.91	D
Menkalinan	0.00	-71.20	M/L
Menke	91.54	-333.82	L
Menker	91.54	-92.84	L
Menkib	76.15	-130.14	L
Mentasta	115.27	-239.41	D
Mermentau	526.03	-39.64	D
Merope	215.16	-326.26	D
Mesartim	103.80	-55.03	D
Metter	538.03	-98.84	D
Midale	191.17	-365.38	D
Milligan	479.35	39.12	D
Millray	616.01	32.08	D
Milos	58.16	-227.16	L
Minakuchi	64.42	117.62	K
Mineite	475.18	-31.30	D
Minnacora	77.98	-172.13	L
Mira	104.32	-52.42	D
Mirach	109.80	-27.12	D
Mirage	479.35	15.65	D
Misery	276.71	85.80	K
Mitchel	109.80	-331.22	L
Mokane	430.84	-23.21	D
Molino	464.48	-135.88	D
Monhegan	117.10	-185.69	D
Monistrol	160.91	124.66	K
Monongahela	166.39	-167.17	D
Monroe	406.85	-45.12	D
Montcoal	250.11	-169.78	D
Montour	351.82	-331.48	D
Moore	5.74	64.68	K
Moravian	110.06	-133.53	D
Mordialloc	221.94	-265.76	D
Morrill	399.55	32.08	D
Morven	549.24	-148.40	D
Moultrie	475.96	-172.13	D
Murchison	75.37	20.08	K
Muridox	107.45	-385.46	L
Muskegon	164.56	-99.89	D
Nadrin	316.09	154.13	K
Nagel	475.18	6.52	D
Nahoni	248.02	-200.29	D
Naka Pabni	381.29	-294.18	D
Nanking	22.95	-52.16	L
Narellan	172.13	-256.11	D
Nashira	48.77	35.99	K
Nashuar	98.84	-184.91	L
Necedah	357.04	-8.08	D
Necromo	103.80	-190.12	L
Neosho	554.72	-25.82	D
Neukirchen	409.98	-249.32	D
New Aberdeen	259.24	79.02	D
New Aragon	70.16	-88.93	D
New Avalon	269.15	-112.14	D
New Canton	8.35	-73.55	L
New Damascus	276.45	-208.38	D
New Earth	-12.52	3.13	S
New Hessen	84.76	-57.12	L
New Home	7.30	-12.78	L
New Ivaarsen	228.98	34.95	D
New Macao	78.24	-123.36	L
New Mendham	180.73	59.98	K
New Procyon	-4.96	-10.43	L
New Rhodes III	103.28	-15.91	D
New Roland	-8.08	-404.76	L
New Sagan	80.59	-179.17	L
New Sytris	233.16	-298.88	D
New Valencia	221.94	-29.47	D
New Vandenberg	157.26	-409.20	P
New Wessex	-12.52	114.49	K
New Westin	38.60	-244.11	L
Newton	320.26	-30.77	D
Nihal	44.34	-313.48	L
Niles	402.15	121.01	D
Ningpo	54.77	-67.29	L
Niomede	-0.26	-352.60	L
Niquinohomo	411.28	-228.72	D
Nirasaki	34.69	19.56	K
Nizina	312.44	-230.29	D
No Return	56.85	-178.91	L
Noatak	333.82	-215.68	D
Nopah	57.64	-33.90	D
Northfield	281.92	-76.15	D
Northwind	34.43	-2.87	D
Notwina	297.57	-195.60	D
Novaya Zemlya	182.30	-184.91	D
Numenor	286.36	-108.49	D
Nunivak	324.96	-189.86	D
O'Fallon	316.35	-83.98	D
Odell	252.72	-98.32	D
Offerman	578.45	-52.68	D
Ogilvie	197.69	-234.20	D
Ohrensen	-16.43	-122.84	M
Okains	243.33	-270.97	D
Okefenokee	612.36	-68.07	D
Olancha	168.74	-6.52	D
Old Kentucky	2.09	-142.92	L
Olindo	467.35	-236.81	D
Oltepesi	254.80	-279.06	D
Orbisonia	151.52	-156.48	D
Ottumwa	402.15	33.64	D
Outreach	-3.13	-34.95	L
Ovan	30.77	-207.60	L
Overton	46.68	-209.68	L
Ozawa	85.54	2.87	D
Pajarito	484.04	77.98	D
Palladaine	-3.13	-290.53	L
Palmyra	454.84	-26.08	D
Palos	51.38	-138.75	L
Panpour	345.56	-294.96	D
Paris	193.51	47.99	K
Parma	363.03	-52.68	D
Pascagoula	433.45	57.90	D
Pattison	593.84	56.33	D
Pattonsbrug	380.51	-6.00	D
Paulding	307.74	-101.97	D
Peabody	333.56	-23.73	D
Perdido	450.66	-114.23	D
Perkasie	131.96	-145.00	D
Petrolia	310.87	-70.94	D
Phact	10.95	-141.09	L
Phalan	-20.08	135.62	S
Piedmont	79.02	78.50	K
Pierce	346.08	-313.22	D
Pike IV	20.34	39.64	K
Pinard	175.52	-392.76	P
Pirates Haven (cluster)		428.76	-
352.86	P		
Pitkin	497.08	77.72	D
Pleiades Cluster	218.29	-322.35	D
Pleione	51.90	-80.33	L
Plymouth	347.12	-69.37	D
Point Barrow	366.16	-130.92	D
Pojos	85.80	-380.51	L
Potwin	310.35	-20.08	D
Poznan	63.11	-81.89	L
Preston	13.30	-215.94	L
Primus	39.90	-322.87	L
Principia	0.00	-360.95	L
Princton	417.02	26.60	D
Prix	38.60	-331.22	L
Propus	-12.52	-185.43	L
Proserpina	161.17	34.16	K
Protection	455.88	5.48	D
Purvo	60.24	-261.06	L
Quemoy	44.86	-152.83	L
Quentin	46.42	12.52	D
Quimberton	83.98	-362.25	L
Quincy	350.52	-34.16	D
Quittacas	159.35	-221.68	D
Raballa	21.91	-159.09	L
Raman	163.00	9.39	D
Ramona	332.00	-38.60	D
Randar	62.33	-177.34	L
Raphael	45.64	-315.31	L
Redfield	141.35	-214.64	D
Redondo	458.75	-216.46	D
Reisling's Planet	177.34	81.37	K
Reisterstown	318.44	-150.48	D
Relevow	70.94	-189.86	L
Remagen	278.27	-157.52	D
Remshield	98.32	-141.09	L
Renown	41.99	-401.37	L
Rentz	575.59	-98.84	D
Repulse	16.43	-400.59	L
Ridgebrook	203.16	-322.35	D
Rigil Kentarus	-2.87	2.61	S
Rio	96.24	-28.95	D
Robinson	234.72	-6.52	D
Robsart	172.39	-343.21	D
Rochester	228.98	3.13	D
Rollis	151.79	-386.77	L
Ronel	96.50	-2.61	L
Rosamond	180.21	-33.90	D
Rosepine	589.41	-22.95	D
Rosiclare	483.26	-47.73	D
Rowe	336.17	48.51	D
Royal	224.55	44.86	D
Royalston	137.96	-226.11	D
Ruchbah	57.90	-22.95	D
Rukbat	42.51	109.80	K
Saaremaa	140.05	138.22	K
Sabanillas	437.62	-177.87	D
Sabik	-12.78	57.64	K
Sadachbia	83.72	37.56	K
Sadalbari	139.01	43.29	K
Safe Port	135.62	-289.49	D
Saffel	25.82	12.00	D
Saginaw	302.27	-96.50	D
Saiph	4.69	-68.59	L
Sakhalin	62.33	-142.92	L
Sakhara V	279.06	34.43	D
Salem	208.64	-162.48	D
Sanilac	148.66	-59.46	D
Sarmaxa	53.72	-164.83	L
Sarna	55.55	-156.22	L
Sauk City	279.06	-3.13	D
Saunemin	248.28	-56.85	D
Savonburg	502.56	-15.65	D
Sax	-1.56	-377.90	L
Scheat	170.04	42.25	K
Schedar	118.92	-21.39	D
Scituate	88.67	-221.94	D
Scudder	197.95	-96.76	D
Second Try	34.69	-116.84	L
Sekulmun	214.12	-222.20	D
Semichi	302.27	-216.72	D
Sendalor	44.08	-269.15	L
Senorbi	348.69	151.26	K
Shaul Khala	221.42	134.57	K
Shaunavon	161.70	-326.26	D
Shawnee	557.85	-65.98	D
Shensi	54.25	-92.84	L
Sheratan	41.21	-21.65	L
Sherwood	568.80	-119.71	D
Shiba	0.26	-334.08	L
Shimonita	94.67	83.20	K
Shimosuwa	46.42	136.66	K
Shiononi	80.59	54.51	K
Shionoha	-5.74	135.62	K
Shipka	80.85	-111.62	L
Shitara	41.99	88.41	K
Shoreham	124.92	-219.59	L
Shubuta	463.96	-124.40	D
Sian	7.30	-231.59	L
Sigma Mare	6.52	-317.65	L
Sirdar	128.31	-312.18	D
Sirius	-2.35	-7.56	L
Skat	94.15	27.12	K
Skepptana	534.64	-192.73	D
Slocum	58.68	-52.68	L
Small World	52.16	-7.04	L
Smolensk	179.95	-130.92	D
Sodertalje	498.39	-237.33	D
Sodus	289.75	-65.72	D
Songgang	330.69	-302.27	D
Sonnia	124.66	-42.25	D
Sparrevohn	346.86	-189.86	D
Spica	93.37	-239.68	D/L
St. Andre	66.24	-101.45	L
St. Ives	112.40	-217.51	L
St. Loris	90.76	-244.11	L
St. Robert	437.36	-31.56	D
Steelesx	294.18	-143.18	D
Stein's Folly	161.17	-202.12	D
Steinhatchee	526.82	-97.80	D
Sterlington	528.90	54.25	D
Sterope	269.15	-407.89	P
Stratford	530.99	39.38	D
Strawn	238.11	-91.28	D
Streator	256.37	-76.41	D
Sturgis	420.67	69.37	D
Styk	42.51	-98.32	L
Styx	26.08	25.82	K
Sullivan	416.76	-31.30	D
Sun Prairie	328.09	29.21	D
Susulatna	307.74	-187.25	D
Sutama	91.54	124.92	K
Suzano	15.91	-114.49	M
Sylvan	426.93	3.13	D
Sylvester	366.68	-115.53	D
Symsonia	491.09	-96.50	D
Taga	89.72	-209.16	L
Talcott	201.60	-124.92	D
Tall Trees	0.78	-67.29	L
Tallin	120.75	-269.41	L
Tallmadge	278.53	42.25	D
Tancredi IV	433.45	94.41	D
Tangipahoa	594.62	-21.65	D
Tannil	123.10	57.90	K
Tantara	102.23	-255.06	L
Tarkio	320.26	0.00	D
Tatlawiksuk	374.77	-191.95	D
Tawas	136.92	-56.33	D
Taygeta	200.03	-298.09	D
Tecumseh	140.57	-105.88	D
Tegaldanas	312.96	-282.19	D
Telos IV	44.34	40.68	K
Teng	128.84	-274.36	L
Tentativa	352.86	-304.88	D
Terra	0.00	0.00	C
Terra Firma	22.69	-33.90	L
Texlos	111.62	-274.88	L
Thestria	282.45	76.68	K
Thibodaux	581.32	-7.56	D
Tigress	75.37	-29.73	L
Tikonov	103.80	-42.51	L
Tishomingo	318.44	46.68	D
Tiskilwa	215.68	-91.02	D
Togura	226.37	154.92	K
Torrence	292.10	-131.96	D
Towne	69.89	8.08	D
Tripoli	201.34	53.46	K
Truth	79.02	-160.65	L
Tsamma	451.97	-45.12	D
Tsinghai	-2.87	-153.61	L
Tsingtao	94.15	-156.48	L
Tsitsang	52.42	-109.54	L
Tsukude	20.34	120.49	K
Turin	21.39	-371.12	L
Tybalt	90.76	-20.60	L
Udibi	318.96	70.42	D
Ulan Batar	126.23	-101.45	D
Ulan Bator	36.25	-140.57	L
Ulysses	417.02	-17.73	D
Umijiri	124.92	107.45	K
Uravan	76.15	-267.84	D
Urich	433.19	19.04	D
Vackisujfalu	470.48	-288.97	D
Valentina	401.37	142.92	K
Valexa	112.67	-88.15	D
Valmiera	190.64	133.27	K
Vandalia	481.96	-66.76	D
Vanern	160.65	142.40	K
Vard	74.85	-358.08	L
Vaucluse	501.26	-163.52	D
Vega	-3.65	107.97	K
Velhas	69.89	-281.14	D
Ventabren	3.39	-124.40	M
Verde	353.12	12.78	D
Verdigreis	376.07	-333.56	D
Verlo	101.45	-305.66	D
Vestallas	67.29	-216.72	L
Vibrunum	536.20	-20.86	D
Vicente	453.01	-70.42	D
Victoria	227.68	-178.39	D
Victoria	93.63	-351.04	L
Waddesdon	97.28	90.50	K
Waimalu	411.80	-179.95	D
Waipahu	430.32	-189.60	D
Waldheim	348.95	130.92	K
Wapakoneta	240.46	69.89	D
Wappingers	176.82	-233.16	D
Ward	72.76	-411.80	L
Warlock	121.27	-260.02	L
Warren	300.70	-331.22	D
Waunakee	320.26	15.13	D
Waxell	281.92	-181.52	D
Wazan	6.00	-141.88	L
Weatogue	99.63	-279.06	L
Wedgefield	374.77	-164.04	D
Weekapaug	108.75	-227.42	D
Wei	64.42	-120.23	L
Weippe	320.26	-309.83	D
Weisau	380.51	135.36	K
Weldon	433.45	-93.63	D
Wernke	250.63	-232.89	D
Westerhand	-5.48	-278.53	L
Wetumpka	532.03	-165.87	D
Willowick	331.22	-75.11	D
Winfield	542.20	23.73	D
Woodbine	508.30	23.99	D
Woodstock	44.60	-37.03	L
Wrentham	172.13	-328.35	D
Wright	26.60	-293.40	L
Wroxeter	167.96	-59.46	D
Xenia	461.62	-89.72	D
Xhosa VII	175.52	14.60	D
Xieng Khouang	86.06	-314.52	L
Yance I	70.42	52.16	K
Yangtze	76.15	-43.81	L
Yorii	-8.08	21.39	K
Yunnah	27.64	-154.13	L
Yuris	135.88	-353.64	L
Zanzibar	120.75	-367.99	L
Zaurak	72.76	-124.40	L
Ziliang	69.11	-266.54	D
Zion	15.39	-83.98	M
Zolfo	532.03	-72.50	D
Zurich	20.34	-63.64	L

Susan Rand
2721 - 2812 (91)

Joseph Davion
2720 - 2743 (23)
Killed in 'Mech Accident

Edwina Winter
2720 - 2786 (66)

M

John Davion
2719 - 2797 (78)
2745 - 2797 (53) XII
Assassinated

Mark Rand-Davion
2744 - 2826 (82)

Rand - Davions

Joshua Davion
2742 - 2797 (55)
Assassinated

M

Mary Halder-Davion
2749 - 2815 (66)

Elise Stuart
2780 - 2861 (81)

M

Paul Davion
2778 - 2842 (64)
2797 - 2842 (45) XIII

Peter Davion
2782 - 2873 (91)

Margaret Duvall
2800 - 2836 (36)

M

Michael Davion
2804 - 2873 (69)
2842 - 2873 (32) XIV

M

Katherine O'Connell
2834 - 2907 (73)

Carl Davion
2828 - 2878 (50)
2873 - 2876 (4) XV
Died in Battle

Melissa Davion
2829 - 2892 (62)
2876 - 2892 (17) XVI

Joseph Davion II
2864 - 2935 (71)
2892 - 2931 retired (40) XVII

M

Olga Zhukoff
2870 - 2968 (98)

Patricia MacPherson
2902 - 2977 (75)

M

Peter Davion
2899 - 2961 (62)
2931 - 2961 (31) XVIII
Assassinated

Simone St. Clair
2935 - 2971 (36)

Andrew Davion
2932 - 2999 (67)
2961 - 2999 (39) XIX

M

Jennifer Campbell
2951 - 3018 (67)

Michael Hasek-Davion
2976 -

M

Marie Davion
2971 -

Ian Davion
2980 - 3013 (33)
2999 - 3013 (15) XX
Killed in Battle
"The Hound"

Hanse Davion
2983 -
3013 - XXI
"The Fox"

Morgan Hasek-Davion

HOUSE DAVION
The Federated Suns

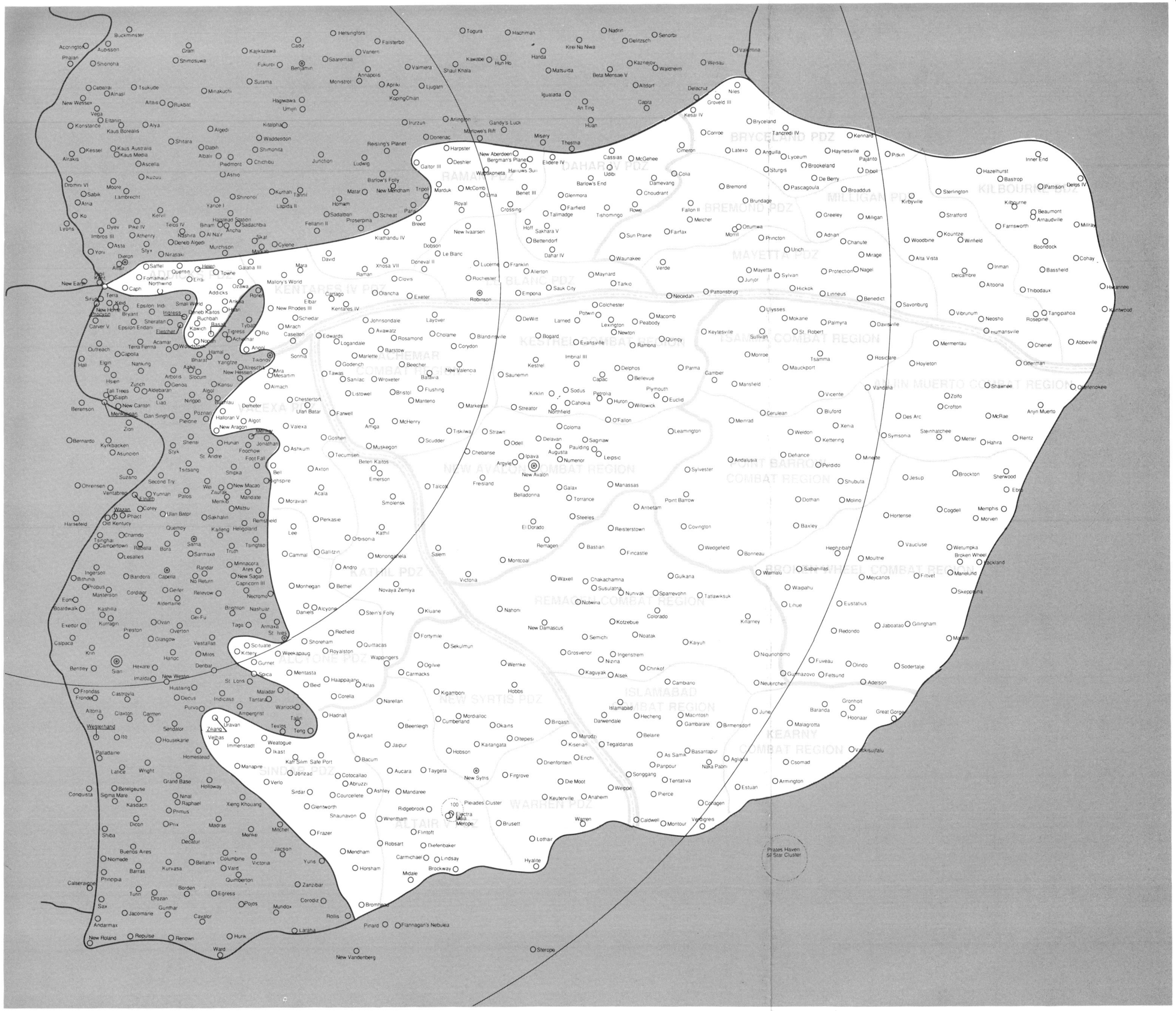

MAP OF THE SUCCESSOR STATES

AND NEARBY KINGDOMS IN THE PERIPHERY

8 PARSECS

40 PARSECS OR 130.4 LIGHT YEARS

MAXIMUM JUMP --- APPROXIMATELY 30 LIGHT YEARS.
FOR NAVIGATIONAL PURPOSES USE 9 PARSECS = 29.34 LY

Map Compiled by COMSTAR.

From information provided by the COMSTAR EXPLORER SERVICE

and the STAR LEAGUE ARCHIVES on Terra.

PDZ BORDERS
THEATER BORDERS
COMBAT REGION BORDERS
MARCH BORDERS

THE FEDERATED SUNS

The Davion Line of Succession and Partial Family Tree

Dates and years in italics indicate time spent in power as primeminister or prince.
Dotted lines represent illegitimate unions.

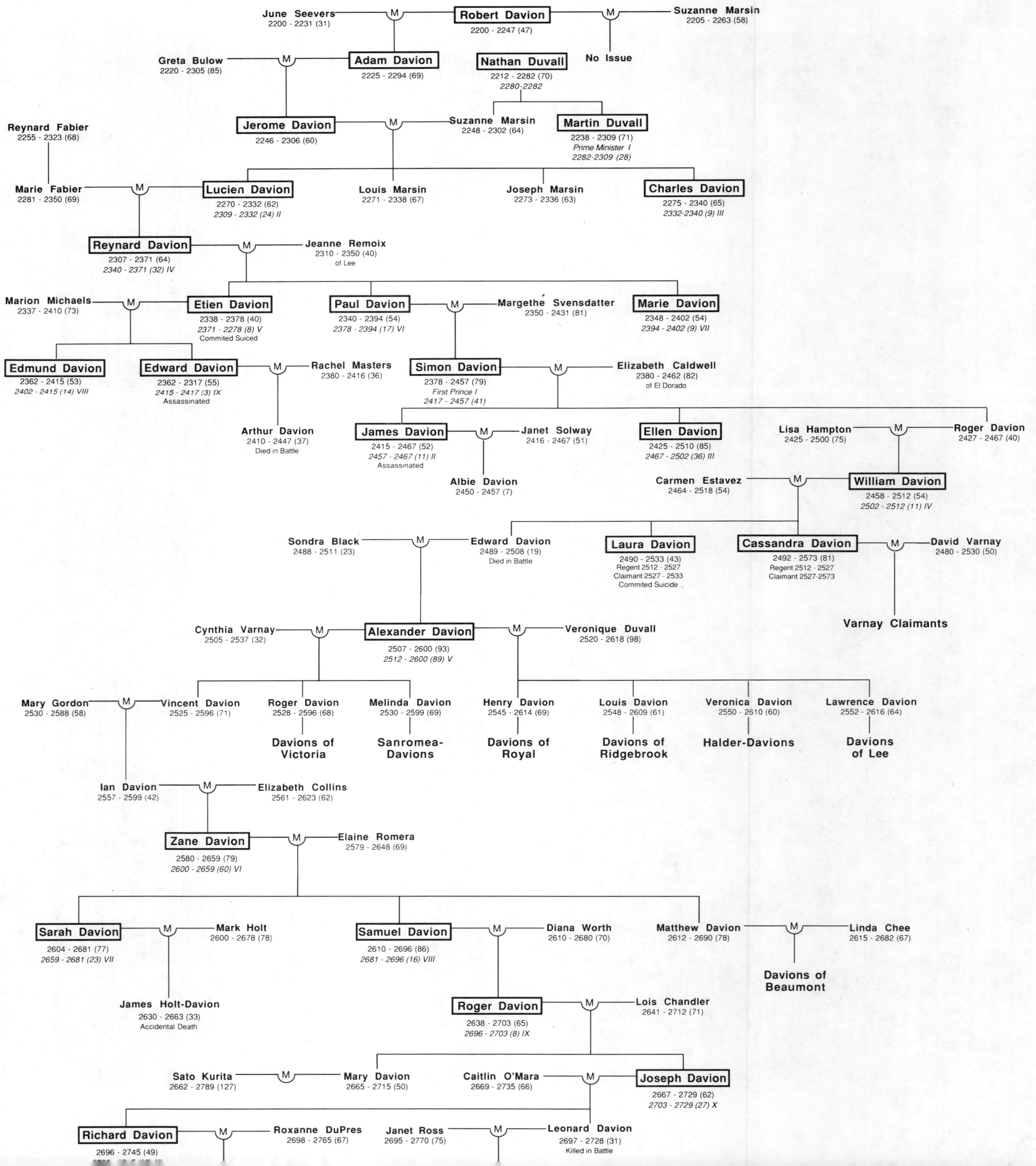

Fill out this card and mail it with a check or money order for $15.95 to:

Mechforce
FASA Corporation
Box 6930
Chicago, IL 60680

NAME ______________________________

ADDRESS ______________________________

CITY ____________ **STATE** ________ **ZIP** ________

HOUSE AFFILIATION (Check One)

☐ **Davion** ☐ **Kurita** ☐ **Steiner** ☐ **Marik** ☐ **Liao**